PROGRESS IN CLINICAL AND BIOLOGICAL RESEARCH

RECENT TITLES

Vol 227: **Advances in Chronobiology**, John E. Pauly, Lawrence E. Scheving, *Editors*. Published in two volumes.

Vol 228: **Environmental Toxicity and the Aging Processes,** Scott R. Baker, Marvin Rogul, *Editors*

Vol 229: **Animal Models: Assessing the Scope of Their Use in Biomedical Research,** Junichi Kawamata, Edward C. Melby, Jr., *Editors*

Vol 230: **Cardiac Electrophysiology and Pharmacology of Adenosine and ATP: Basic and Clinical Aspects,** Amir Pelleg, Eric L. Michelson, Leonard S. Dreifus, *Editors*

Vol 231: **Detection of Bacterial Endotoxins With the Limulus Amebocyte Lysate Test,** Stanley W. Watson, Jack Levin, Thomas J. Novitsky, *Editors*

Vol 232: **Enzymology and Molecular Biology of Carbonyl Metabolism: Aldehyde Dehydrogenase, Aldo-Keto Reductase, and Alcohol Dehydrogenase**, Henry Weiner, T. Geoffrey Flynn, *Editors*

Vol 233: **Developmental and Comparative Immunology,** Edwin L. Cooper, Claude Langlet, Jacques Bierne, *Editors*

Vol 234: **The Hepatitis Delta Virus and Its Infection**, Mario Rizzetto, John L. Gerin, Robert H. Purcell, *Editors*

Vol 235: **Preclinical Safety of Biotechnology Products Intended for Human Use,** Charles E. Graham, *Editor*

Vol 236: **First Vienna Shock Forum,** Günther Schlag, Heinz Redl, *Editors*. Published in two volumes: Part A: *Pathophysiological Role of Mediators and Mediator Inhibitors in Shock.* Part B: *Monitoring and Treatment of Shock.*

Vol 237: **The Use of Transrectal Ultrasound in the Diagnosis and Management of Prostate Cancer,** Fred Lee, Richard McLeary, *Editors*

Vol 238: **Avian Immunology,** W.T. Weber, D.L. Ewert, *Editors*

Vol 239: **Current Concepts and Approaches to the Study of Prostate Cancer,** Donald S. Coffey, Nicholas Bruchovsky, William A. Gardner, Jr., Martin I. Resnick, James P. Karr, *Editors*

Vol 240: **Pathophysiological Aspects of Sickle Cell Vaso-Occlusion,** Ronald L. Nagel, *Editor*

Vol 241: **Genetics and Alcoholism,** H. Werner Goedde, Dharam P. Agarwal, *Editors*

Vol 242: **Prostaglandins in Clinical Research,** Helmut Sinzinger, Karsten Schrör, *Editors*

Vol 243: **Prostate Cancer,** Gerald P. Murphy, Saad Khoury, Réne Küss, Christian Chatelain, Louis Denis, *Editors*. Published in two volumes: Part A: *Research, Endocrine Treatment, and Histopathology.* Part B: *Imaging Techniques, Radiotherapy, Chemotherapy, and Management Issues*

Vol 244: **Cellular Immunotherapy of Cancer,** Robert L. Truitt, Robert P. Gale, Mortimer M. Bortin, *Editors*

Vol 245: **Regulation and Contraction of Smooth Muscle,** Marion J. Siegman, Andrew P. Somlyo, Newman L. Stephens, *Editors*

Vol 246: **Oncology and Immunology of Down Syndrome,** Ernest E. McCoy, Charles J. Epstein, *Editors*

Vol 247: **Degenerative Retinal Disorders: Clinical and Laboratory Investigations,** Joe G. Hollyfield, Robert E. Anderson, Matthew M.LaVail, *Editors*

Vol 248: **Advances in Cancer Control: The War on Cancer—15 Years of Progress,** Paul F. Engstrom, Lee E. Mortenson, Paul N. Anderson, *Editors*

Vol 249: **Mechanisms of Signal Transduction by Hormones and Growth Factors,** Myles C. Cabot, Wallace L. McKeehan, *Editors*

Vol 250: **Kawasaki Disease,** Stanford T. Shulman, *Editor*

Vol 251: **Developmental Control of Globin Gene Expression,** George Stamatoyannopoulos, Arthur W. Nienhuis, *Editors*

Vol 252: **Cellular Calcium and Phosphate Transport in Health and Disease,** Felix Bronner, Meinrad Peterlik, *Editors*

Vol 253: **Model Systems in Neurotoxicology: Alternative Approaches to Animal Testing,** Abraham Shahar, Alan M. Goldberg, *Editors*

Vol 254: **Genetics and Epithelial Cell Dysfunction in Cystic Fibrosis,** John R. Riordan, Manuel Buchwald, *Editors*

Vol 255: **Recent Aspects of Diagnosis and Treatment of Lipoprotein Disorders: Impact on Prevention of Atherosclerotic Diseases,** Kurt Widhalm, Herbert K. Naito, *Editors*

Vol 256: **Advances in Pigment Cell Research,** Joseph T. Bagnara, *Editor*

Vol 257: **Electromagnetic Fields and Neurobehavioral Function,** Mary Ellen O'Connor, Richard H. Lovely, *Editors*

Vol 258: **Membrane Biophysics III: Biological Transport,** Mumtaz A. Dinno, William McD. Armstrong, *Editors*

Vol 259: **Nutrition, Growth, and Cancer,** George P. Tryfiates, Kedar N. Prasad, *Editors*

Vol 260: **EORTC Genitourinary Group Monograph 4: Management of Advanced Cancer of Prostate and Bladder,** Philip H. Smith, Michele Pavone-Macaluso, *Editors*

Vol 261: **Nicotine Replacement: A Critical Evaluation,** Ovide F. Pomerleau, Cynthia S. Pomerleau, *Editors*

Vol 262: **Hormones, Cell Biology, and Cancer: Perspectives and Potentials,** W. David Hankins, David Puett, *Editors*

Vol 263: **Mechanisms in Asthma: Pharmacology, Physiology, and Management,** Carol L. Armour, Judith L. Black, *Editors*

Vol 264: **Perspectives in Shock Research,** Robert F. Bond, *Editor*

Vol 265: **Pathogenesis and New Approaches to the Study of Noninsulin-Dependent Diabetes Mellitus,** Albert Y. Chang, Arthur R. Diani, *Editors*

Vol 266: **Growth Factors and Other Aspects of Wound Healing: Biological and Clinical Implications,** Adrian Barbul, Eli Pines, Michael Caldwell, Thomas K. Hunt, *Editors*

Vol 267: **Meiotic Inhibition: Molecular Control of Meiosis,** Florence P. Haseltine, *Editor*

Vol 268: **The Na, K-Pump,** J.C. Skou, J.G. Nørby, A.B. Maunsbach, M. Esmann, *Editors*. Published in two volumes: Part A: *Molecular Aspects*. Part B: *Cellular Aspects*.

Vol 269: **EORTC Genitourinary Group Monograph 5: Progress and Controversies in Oncological Urology II,** Fritz H. Schröder, Jan G.M. Klijn, Karl H. Kurth, Herbert M. Pinedo, Ted A.W. Splinter, Herman J. de Voogt, *Editors*

Vol 270: **Cell-Free Analysis of Membrane Traffic,** D. James Morré, Kathryn E. Howell, Geoffrey M.W. Cook, W. Howard Evans, *Editors*

Vol 271: **Advances in Neuroblastoma Research 2,** Audrey E. Evans, Giulio J. D'Angio, Alfred G. Knudson, Robert C. Seeger, *Editors*

Vol 272: **Bacterial Endotoxins: Pathophysiological Effects, Clinical Significance, and Pharmacological Control,** Jack Levin, Harry R. Büller, Jan W. ten Cate, Sander J.H. van Deventer, Augueste Sturk, *Editors*

Vol 273: **The Ion Pumps: Structure, Function, and Regulation,** Wilfred D. Stein, *Editor*

Vol 274: **Oxidases and Related Redox Systems,** Tsoo E. King, Howard S. Mason, Martin Morrison, *Editors*

Please contact the publisher for information about previous titles in this series.

THE ION PUMPS
Structure, Function, and Regulation

THE ION PUMPS
Structure, Function, and Regulation

Proceedings of the First Shoresh Workshop on Ion Pumps, Held at Shoresh, Israel, August 30 Through September 2, 1987

Editor

Wilfred D. Stein
Department of Biochemistry
Institute of Life Sciences
Hebrew University
Jerusalem, Israel

ALAN R. LISS, INC. • NEW YORK

Address all Inquiries to the Publisher
Alan R. Liss, Inc., 41 East 11th Street, New York, NY 10003

Printed in the United States of America

Library of Congress Cataloging-in-Publication Data

Shoresh Workshop on Ion Pumps (1st : 1987)
The ion pumps: structure, function, and regulation: proceedings of the First Shoresh Workshop on Ion Pumps, held at Shoresh, near Jerusalem, August 30th through September 2, 1987 / editor, Wilfred D. Stein.

p. cm. — (Progress in clinical and biological research; v. 273)
Includes bibliographies and index.
ISBN 0-8451-5123-1
1. Ion pumps—Congresses. 2. Biological transport, Active—Congresses. I. Stein, Wilfred D. II. Title. III. Series.
QH601.S45 1987
574.87′5— dc19 88-1721
CIP

Contents

Contributors

Claude Aflalo, Department of Chemistry, University of California at San Diego, La Jolla, CA 92093 **[337]**

Tapani Alatossava, Department of Genetics, University of Oulu, Linnanma, 90570 Oulu 57, Finland **[373]**

Jens P. Andersen, Institute of Physiology, University of Aarhus, 8000 Aarhus C, Denmark **[33, 147]**

Anita C. Aperia, Department of Pediatrics, Karolinska Institute, St. Göran's Children's Hospital, 112 81 Stockholm, Sweden **[427]**

Yoram Avi-Dor, Department of Biology, Technion-Israel Institute of Technology, Haifa 32000, Israel **[369]**

Giovanni Felice Azzone, C.N.R. Unit for the Study of the Physiology of Mitochondria and Institute of General Pathology, University of Padua, Padua, 35131 Italy **[395]**

Ernst Bamberg, Max-Plank-Institut für Biophysik, 6000 Frankfurt am Main, Federal Republic of Germany **[225]**

Timothy G. Beeker, Laboratory of Membrane Biology, Cedars-Sinai Medical Center, Los Angeles, CA 90048 **[349]**

J. A. Berden, Laboratory of Biochemistry, University of Amsterdam, B.C.P. Jansen Institute, 1018TV Amsterdam, The Netherlands **[81]**

Mervyn C. Berman, Medical Research Council Biomembrane Research Unit, Department of Chemical Pathology, University of Cape Town Medical School, Cape Town, Republic of South Africa **[189, 195]**

Egbert Boekema, Fritz-Haber-Institut der Max-Planck Gesellschaft, 1000 Berlin 33, Federal Republic of Germany **[67, 75]**

D.C. Buster, Department of Chemistry and Biochemistry, University of Arkansas, Fayetteville, AR 72701 **[331]**

W.F. Buurmeijer, Laboratory of Plant Physiological Research, Agricultural University, 6703 BW Wageningen, The Netherlands **[293]**

S. Roy Caplan, Department of Membrane Research, Weizmann Institute of Science, Rehovot 76100, Israel **[377]**

C. Carmeli, Department of Biochemistry, Tel Aviv University, Tel Aviv 69978, Israel **[307]**

The numbers in brackets are the opening page numbers of the contributors' articles.

Rita Casadio, Department of Biology, University of Bologna, I-40126 Bologna, Italy **[201]**

Philippe Champeil, Service de Biophysique, Département de Biologie, Centre d'Etudes Nucleaires de Saclay, 91191 Gif Sur Yvette, Cedex, France **[183]**

D. Charlemagne, INSERM U 127, Hôpital Lariboisière, 75010 Paris, France **[45]**

David M. Chipman, Department of Biology, Ben Gurion University of the Negev, Beer Sheva 84105, Israel **[155, 171, 177]**

Eugenia Chirieri-Kovács, Department of Biophysics, Medical Faculty, Bucharest 76241, Romania **[327]**

George A. Davidson, Medical Research Council Biomembrane Research Unit, Department of Chemical Pathology, University of Cape Town Medical School, Cape Town, Republic of South Africa **[189, 195]**

Marlene DeLuca, Department of Chemistry, University of California at San Diego, La Jolla, CA 92093 **[337]**

F. de Wolf, Biological Laboratory, Vrije Universiteit, 1081 HV Amsterdam, The Netherlands **[289]**

Alexandru Dinu, Department of Biophysics, Medical Faculty, Bucharest 76241, Romania **[327]**

Laszlo Dux, Institute of Biochemistry, Szent-Gyorgyi University Medical School, H-6701 Szeged, Hungary **[7]**

P. Easton, Department of Chemistry and Biochemistry, University of Arkansas, Fayetteville, AR 72701 **[331]**

Konrad S. Famulski, Department of Cellular Biochemistry, Nencki Institute of Experimental Biology, 02-093 Warsaw, Poland **[355]**

Klaus Fendler, Max-Planck-Institut für Biophysik, 6000 Frankfurt am Main, Federal Republic of Germany **[225]**

Claus J. Fimmel, Laboratory of Membrane Biology, Cedars-Sinai Medical Center, and University of California, Los Angeles, CA 90048 **[349]**

Sidney Fleischer, Department of Molecular Biology, Vanderbilt University, Nashville, TN 37235 **[435]**

Petra Fromme, Max-Volmer-Institut, Technische Universität Berlin, 1000 Berlin 12, Federal Republic of Germany **[67]**

J.M. Galmiche, DB/SBPH, Centre d'Etudes Nucleaires de Saclay, 911 Gif Sur Yvette Cedex, France **[289]**

Preeti Gangola, Department of Biochemistry, Wayne State University School of Medicine, Detroit, MI 48201 **[105]**

Klaus Gerwert, Max-Planck-Institut für Ernährungsphysiologie, 4600 Dortmund, Federal Republic of Germany **[321]**

G. Girault, DB/SBPH, Centre d'Etudes Nucleaires de Saclay, 91191 Gif Sur Yvette Cedex, France **[289]**

Erik Goormaghtigh, CP206/2, Service de Chimie Physique des Macromolécules aux Interfaces, Université Libre de Bruxelles, B1050 Brussels, Belgium **[51]**

Peter Gräber, Max-Volmer-Institut, Technische Universität Berlin, 1000 Berlin 12, Federal Republic of Germany **[67, 75]**

Wolfgang Grampp, Department of Physiology and Biophysics, University of Lund, S-223 62 Lund, Sweden **[233]**

N. Michael Green, Laboratory of Protein Structure, National Institute for Medical Research, Mill Hill, London NW7 1AA, United Kingdom **[15, 165]**

Zippora Gromet-Elhanan, Department of Biochemistry, The Weizmann Institute of Science, Rehovot 76100, Israel **[299]**

Florent Guillain, Service de Biophysique, Département de Biologie, Centre d'Etudes Nucleaires de Saclay, 91191 Gif Sur Yvette Cedex, France **[183]**

Tadeusz Gulik-Krzywicki, Centre de Genètique Moléculaire, C.N.R.S., 91190 Gif Sur Yvette, France **[99]**

Menachem Gutman, Department of Biochemistry, Tel Aviv University, Ramat Aviv 69978, Israel **[269]**

A. F. Hartog, Laboratory of Biochemistry, University of Amsterdam, B.C.P. Jansen Institute, 1018TV Amsterdam, The Netherlands **[81]**

Steve J. Hersey, Department of Physiology, Emory University, Atlanta, GA 30322 **[247]**

Benno Hess, Max-Planck-Institut für Ernährungsphysiologie, 4600 Dortmund, Federal Republic of Germany **[321]**

R. Hiller, Department of Biochemistry, Tel Aviv University, Tel Aviv 69978, Israel **[307]**

J. F. Hinton, Department of Chemistry and Biochemistry, University of Arkansas, Fayetteville, AR 72701 **[331]**

Joanne Hullihen, Department of Biological Chemistry, Johns Hopkins University School of Medicine, Baltimore, MD 21205 **[87]**

Makoto Inui, Department of Molecular Biology, Vanderbilt University, Nashville, TN 37235 **[435]**

William P. Jencks, Graduate Department of Biochemistry, Brandeis University, Waltham, MA 02254 **[177]**

Peter Leth Jørgensen, Institute of Physiology, Aarhus University, 8000 Aarhus C, Denmark **[25, 33, 417]**

Helga Jütte, Department of Microbiology, Biozentrum, University of Basel, CH-4056 Basel, Switzerland **[373]**

Cyrus Karkaria, Department of Biochemistry, Wayne State University School of Medicine, Detroit, MI 48201 **[105]**

Steven J.D. Karlish, Biochemistry Department, Weizmann Institute of Science, Rehovot 76100, Israel **[207]**

D.J. Keeling, Smith Kline & French (Research) Ltd., Welwyn, Herts AL6 9AR, United Kingdom **[255]**

Eduard Kellenberger, Department of Microbiology, Biozentrum, University of Basel, CH-4056 Basel, Switzerland **[373]**

Shifra Ken-Dror, Department of Biology, Technion-Israel Institute of Technology, Haifa 32000, Israel **[369]**

T.L. Kirley, University of Cincinnati College of Medicine, Cincinnati, OH 45267 **[39]**

Dan A. Klaerke, Institute of Physiology, University of Aarhus, DK-8000 Aarhus C, Denmark **[417]**

Bente S. Kopeliovitch, Hebrew University, Faculty of Agriculture, Department of Agricultural Botany, Rehovot 71600, Israel **[409]**

R. Kraayenhof, Biological Laboratory, Vrije Universiteit, 1081 HV Amsterdam, The Netherlands **[289]**

S.M. Laing, Smith Kline & French (Research) Ltd., Welwyn, Herts AL6 9AR, United Kingdom **[255]**

Joseph F. Lamb, Department of Biology and Preclinical Medicine, University of St. Andrews, St. Andrews, KY16 9T6 Scotland **[361]**

P. Läuger, Department of Biology, University of Konstanz, D-7750 Konstanz, Federal Republic of Germany **[217]**

Jade Li, Department of Structural Studies, Medical Research Council Laboratory of Molecular Biology, Cambridge CB2 2QH, United Kingdom **[93]**

Pia Lorentzon, Hassle A.B., S-43183 Molndahl, Sweden **[247]**

M. Lübben, Institute of Biochemistry, Medical University of Lübeck, 2400 Lübeck 1, Federal Republic of Germany **[57]**

Siro Luvisetto, C.N.R. Unit for the Study of the Physiology of Mitochondria and Institute of General Pathology, University of Padua, Padua 35131, Italy **[395]**

D.H. MacLennan, Banting & Best Institute for Medical Research, Toronto, Ontario, Canada M5G 1L6 **[15]**

Stephania Markus, Departments of Biology and Chemistry, Ben Gurion University, Beer Sheva 84105, Israel **[171]**

Anthony N. Martonosi, Department of Biochemistry and Molecular Biology, State University of New York, Health Science Center, Syracuse, NY 13210 **[7]**

E. Mayoux, INSERM U 127, Hôpital Lariboisière, 75010 Paris, France **[45]**

David McIntosh, Department of Chemical Pathology, University of Cape Town, Cape Town, Republic of South Africa **[183]**

Yoshinori Moriyama, Department of Biochemistry, Roche Institute of Molecular Biology, Roche Research Center, Nutley, NJ 07110 **[387]**

Shimon Moshiach, Department of Biochemistry, Tel Aviv University, Ramat Aviv 69978, Israel **[269]**

Shmuel Muallem, Laboratory of Membrane Biology, Cedars-Sinai Medical Center, Los Angeles, CA 90048 **[349]**

Esther Nachliel, Department of Biochemistry, Tel Aviv University, Ramat Aviv 69978, Israel **[269, 283]**

Georg Nagel, Max-Planck-Institut für Biophysik, 6000 Frankfurt am Main, Federal Republic of Germany **[225]**

Nathan Nelson, Department of Biochemistry, Roche Institute of Molecular Biology, Roche Research Center, Nutley, NJ 07110 **[387]**

P. Nieboer, Laboratory of Biochemistry, University of Amsterdam, B.C.P. Jansen Institute, 1018TV Amsterdam, The Netherlands **[81]**

Taka-aki Ohkawa, Department of Biology, College of General Education, Osaka University, Osaka, Japan **[423]**

P. Oliviero, INSERM U 127, Hôpital Lariboisière, 75010 Paris, France **[45]**

J.J.J. Ooms, Department of Plant Physiological Research, Agricultural University, 6703 BW Wageningen, The Netherlands **[293]**

Stepen Pandol, University of California, Los Angeles, School of Medicine, Los Angeles, CA 90048 **[349]**

Peter L. Pedersen, Department of Biological Chemistry, Johns Hopkins School of Medicine, Baltimore, MD 21205 **[87]**

Harvey S. Penefsky, Department of Biochemistry, The Public Health Research Institute, New York, NY 10016 **[261]**

Janne Petersen, Institute of Physiology, University of Aarhus, 8000 Aarhus C, Denmark **[33]**

Valeria Petronilli, C.N.R. Unit for the Study of the Physiology of Mitochondria and Institute of General Pathology, University of Padua, Padua 35131, Italy **[315]**

Daniela Pietrobon, C.N.R. Unit for the Study of the Physiology of Mitochondria and Institute of General Pathology, University of Padua, Padua 35131, Italy **[395]**

Slawomir Pikula, Nencki Institute of Experimental Biology, Polish Academy of Sciences, Warsaw, Poland **[7]**

M. Poyard, INSERM U 127, Hôpital Lariboisière, 75010 Paris, France **[45]**

Zvi Priel, Department of Chemistry, Ben Gurion University, Beer-Sheva 84105, Israel **[171]**

T. Privett, Department of Chemistry and Biochemistry, University of Arkansas, Fayetteville, AR 72701 **[331]**

Alberte Pullman, Laboratoire de Biochimie Théorique associé au CNRS, Institut de Biologie Physico-Chimique, 75005 Paris, France **[113]**

Edd Rabon, Wadsworth Veterans Administration Hospital, Los Angeles, CA **[247]**

Ronit Regev, Department of Biology, Technion-Israel Institute of Technology, Haifa 32000, Israel **[369]**

Kurt R. H. Repke, Energy Conversion Unit in Central Institute of Molecular Biology, Academy of Sciences, DDR 1115 Berlin, German Democratic Republic **[401]**

Jean-Louis Rigaud, Service de Biophysique, Département de Biologie, Centre d' Etudes Nucleaires de Saclay, 91191 Gif Sur Yvette Cedex, France **[99]**

Barry P. Rosen, Department of Biochemistry, Wayne State University School of Medicine, Detroit, MI 48201 **[105]**

Jürg P. Rosenbusch, Biozentrum, University of Basel, CH-4056 Basel, Switzerland **[121]**

Jean-Marie Ruysschaert, Service de Chimie Physique des Macromolécules aux Interfaces, Université Libre de Bruxelles, B1050 Brussels, Belgium **[51]**

George Sachs, Wadsworth Veterans Administration Hospital, Los Angeles, CA 90073 **[247]**

Gene A. Scarborough, Department of Pharmacology, School of Medicine, University of North Carolina, Chapel Hill, NC 27514 **[51]**

G. Schäfer, Institute of Biochemistry, Medical University of Lübeck, 2400 Lübeck 1, Federal Republic of Germany **[57]**

Hedva Schickler, Hebrew University, Faculty of Agriculture, Department of Agricultural Botany, Rehovot 71600, Israel **[409]**

Günter Schmidt, Max-Volmer-Institut, Technische Universität Berlin, 1000 Berlin 12, Federal Republic of Germany **[67]**

Mordechay Schonfeld, Hebrew University, Faculty of Agriculture, Department of Agricultural Botany, Rehovot 71600, Israel **[409]**

David Scott, CURE, Wadsworth Veterans Administration Hospital, Los Angeles, CA 90073 **[247]**

Michel Seigneuret, Service de Biophysique, Département de Biologie, Centre d' Etudes Nucleaires de Saclay, 91191 Gif Sur Yvette Cedex, France **[99]**

Hans Seiler, Department of Inorganic Chemistry, University of Basel, CH-4056 Basel, Switzerland **[373]**

J. Senn-Bilfinger, Byk Gulden Lomberg Chemische Fabrik GmbH, D-7750 Konstanz, Federal Republic of Germany **[255]**

Varda Shoshan-Barmatz, Department of Biology, Ben Gurion University of the Negev, Beer Sheva 84105, Israel **[155]**

Staffan Skogvall, Department of Physiology and Biophysics, University of Lund, S-223 62 Lund, Sweden **[233]**

Esther Stauffer, Department of Microbiology, Biozentrum, University of Basel, CH-4056 Basel, Switzerland **[373]**

Wilfred D. Stein, Department of Biochemistry, Institute of Life Sciences, Hebrew University, Jerusalem 91904, Israel **[xxi, 3, 141, 239, 345]**

M. Steinberg, Department of Pharmacology, State University of New York Health Science Center, Syracuse, NY 13210 **[39]**

Kenneth A. Taylor, Department of Anatomy, Duke University Medical Center, Durham, NC 27710 **[7]**

W.R. Taylor, National Institute for Medical Research, Mill Hill, London NW7 1AA, United Kingdom **[15]**

H. Tiedge, Center for Neurobiology and Behavior, Columbia University, New York, NY 10032 **[57]**

Phillip Tooth, Department of Structural Studies, Medical Research Council Laboratory of Molecular Biology, Cambridge CB2 2QH, United Kingdom **[93]**

Katalin Török, Laboratory of Protein Structure, National Institute for Medical Research, Mill Hill, London NW7 1AA, United Kingdom **[165]**

Izuo Tsutsui, Department of Biology, College of General Education, Osaka University, Osaka, Japan **[423]**

P.A. Tyson, Department of Pharmacology, State University of New York Health Science Center, Syracuse, NY 13210 **[39]**

Marin van Heel, Fritz-Haber-Institut der Max-Planck Gesellschaft, 1000 Berlin 33, Federal Republic of Germany **[75]**

Vasile Vasilescu, Department of Biophysics, Medical Faculty, Bucharest 76241, Romania **[327]**

Bente Vilsen, Institute of Physiology, University of Aarhus, 8000 Aarhus C, Denmark **[33, 147]**

W.J. Vredenberg, Laboratory of Plant Physiological Research, Agricultural University, 6703 BW Wageningen, The Netherlands **[293]**

Bonnie A. Wallace, Department of Chemistry and Center for Biophysics, Rennselaer Polytechnic Institute, Troy, NY 12180 **[133]**

E. T. Wallick, University of Cincinnati College of Medicine, Cincinnati, OH 45267 **[39]**

Bjorn Wallmark, Hassle A.B., S-43183, Molndahl, Sweden **[247]**

J. Weber, Institute of Biochemistry, Medical University of Lübeck, 2400 Lübeck 1, Federal Republic of Germany **[57]**

Ulrich Weigel, Department of Biochemistry, Wayne State University School of Medicine, Detroit, MI 48201 **[105]**

Noreen Williams, Department of Biological Chemistry, Johns Hopkins University School of Medicine, Baltimore, MD 21205; present address: Uniformed Services University of the Health Sciences, Department of Biochemistry, Bethesda, MD 20814 **[87]**

Antoni Wrzosek, Department of Muscle Biochemistry, Nencki Institute of Experimental Biology, 02-093 Warsaw, Poland **[355]**

Ruth Yam, Department of Biochemistry, Tel Aviv University, Ramat Aviv 69978, Israel **[279]**

Mario Zoratti, C.N.R. Unit for the Study of the Physiology of Mitochondria and Institute of General Pathology, University of Padua, Padua 35131, Italy **[315]**

Preface

The structure, function, and regulation of the ion pumps are subjects that have been intensively studied for a long time. In the last few years, with the isolation and reconstitution of the pump proteins and, later, the cloning of many of the relevant genes, molecular solutions to the central questions of ion pumping appear within grasp. This was a good time, therefore, for many of the leading workers in the field to gather at the First Shoresh Workshop on Ion Pumps, Shoresh, Israel, August 30 through September 2, 1987 to discuss and compare current findings in several systems that are the focus of attention.

We were fortunate in being able to bring together experts working on the sodium-potassium pump, the calcium pumps, the cell-membrane proton-potassium pump, and proton pumps (all these comprising the so-called E_1E_2 class of pumps), and those working on the F_1F_0 proton pumps of bacterial cells, mitochondria, and chloroplasts. Other speakers discussed the light-driven bacteriorhodopsin pump and an anion pump, as well as some general aspects of the design and function of membrane transport proteins.

This book represents the cumulative fruits of the lectures, discussions, and posters presented during the three days of the workshop. The first day was devoted to reviewing current concepts on the structure of transport proteins, and the functional models based on the structural data. On the second day, we concentrated on function itself, discussing the biochemical processes and kinetic behaviour of the pump-enzymes in their dual roles as ATP hydrolases and ion translocators. On the third day the theme was regulation in its diverse aspects, including both the short-term kinetic and the long-term synthetic modes of pump regulation. The daytime hours were devoted to lectures and subsequent discussions, and in the evening sessions moderators introduced relevant posters for general discussion.

When we planned the workshop, we had thought that it might be possible to discover mechanisms common to the E_1E_2 and F_1F_0 classes of ion pumps. The diversity of structure and function between these two classes of pump meant that we had to content ourselves with the finding of common themes, approaches, and techniques. All of us, however, agreed that the comparison itself was a valuable exercise, as readers of this book will discover.

Many of the participants of the workshop were concerned with the basic scientific problems of ion pumping, but all of us are aware of the great significance of ion pumping when applied to physiological questions in health and disease, particularly with regard to the heart and the kidneys.

This meeting was made possible by funding from the Weizmann Institute Renal Research Fund, the Goldshlegger Fund, and the International Union of Pure and Applied Biophysics. We are particularly indebted to Mr. Raymond Sagov and Mrs. Jeanette Pomeraniec-Sagov for their generous support of the Renal Research Fund.

Ruth Bar-Adon of the Biochemistry Department of the Weizmann Institute provided invaluable secretarial help. Chana Stein was a devoted and ingenious organizer at the workshop itself and gave much help with the editorial work.

THE ORGANIZERS:

S.J.D. Karlish
Rehovot, Israel

W.D. Stein
Jerusalem, Israel

STRUCTURE

The Ion Pumps: Structure, Function,
and Regulation, pages 3–6

EDITOR'S INTRODUCTION: STRUCTURE OF THE ION PUMPS

The papers in this first section of the book are largely addressed to the question of ion pump structure. Now, we do not know the structure of any ion pump in full detail. All researchers are convinced, however, that with this information we could advance much faster towards reaching a full understanding of the mechanism of action of ion pumping. Much effort is, therefore, currently being put into the elucidation of pump structure, the work progressing along two main fronts. In the first, exemplified here by the papers of Michael Green and of Peter Jorgensen, the data on the DNA sequences of the cloned genes coding for ion pumps are analysed and attempts made to build models for the possible structures of the pumps. The work proceeds on the basis of analogies between these sequences and the sequences of proteins for which three-dimensional structures are known with certainty from X-ray diffraction analysis. Very valuable is the demonstration that many ion pumps have substantial sequences which show much homology between one pump and another, suggesting that these repeating motifs are crucial to pump function. These probably define domains which share particular functions, such as ATP-binding or the transduction of the information contained in the fact of ATP-binding or of phosphorylation. (As these chemical events control the conformation changes concerned in ion movements, the occurrence of the chemical event has to be transduced by the pump into the conformation change that translocates the ions.) Homologies are found, too, between domains of the ATP-driven ion pumps and domains in soluble enyzmes which bind and hydrolyse ATP, again suggesting particular roles for these domains in the action of the ion pumps. But we are still far from having the knowledge which would enable us to build with confidence a structure from the sequence data, even with the ingenious use of comparative anatomy data for the proteins. What is crucially lacking is the identification of those regions of the ion pump proteins which are involved in the movement of the ions themselves. (Karlish's paper in the next (function) section takes up this point among others.) We have, as yet, no tool or reagent with which we can confidently label such ion-binding and ion-translocation regions, in the way that we can for the ATP itself. Indeed, Rosenbusch's stimulating, even provocative, paper shows us that the ion movements might possibly occur through a quite novel region of the protein - a sodium wire - which, by analogy with the current-carrying movement of silver through crystals of $RbAg_4I_5$, might occur via a lattice of carboxyls and guanidyl (or amino) residues contributed by the protein's amino-acid side-chains. How the polypeptide chains of the membrane proteins may pack together is considered from a theoretical point of view by Pullman. Her calculations of the energetics of the close-packing of membrane-embedded α-helices predict that these will generally pack in anti-parallel fashion, this being the more stable conformation, although parallel packing can certainly not be ruled out. The minimum length of polypeptide that will give stable helix-helix associations can be computed (and the amount of α-helix present in a pump measured, using infra-red measurements, in Goormaghtigh's paper). Finally, it is shown that the routes of cation movement through a membrane-embedded protein need not be via a lattice of carboxyl groups, since the carbonyl groups of the peptide backbone offer an environment for cations which is sufficiently favorable to enable them to leave the aqueous phase.

The second approach uses the electron microscopic analysis (by image processing methods) of crystals of the pumps. The current status of the work on the F_1F_0 ATPases is presented in important papers by Schafer and his colleagues and by Graber, Boekema and their collaborators, the structural data being used to assess models for pump function in these systems. The F_1F_0 systems, in total contrast to the E_1E_2 pumps of the plasma membranes, are built on a three-fold symmetry axis, an intriguing finding which must have significance for pump function. Fascinating is the demonstration that cells as primitive as the archaebacteria have pump structures that bear great similarity to those of more highly evolved forms, demonstrating the persistence over aeons of time of succesful solutions to the biological problem of ion pumping. (Rosen shows that an anion pump induced in bacteria in response to the need for detoxicative extrusion of arsenate and antimonate is probably also built out of an F_1-like ATPase and an F_0-like membrane-associated polypeptide. It will be interesting to see if the detailed structure of this anion transporter turns out to resemble that of the classical F_1F_0 ATPases.) Berden's finding that there are two catalytic subunits in the F_1 portion of the ATPase and Williams' finding of five magnesium-binding sites both need to be reconciled with the three-fold symmetry that the electron microscope reveals.

Martonosi's paper shows us how far we have gone in the image analysis of the E_1E_2 ATP-driven ion pumps, while Li does this for the lactose permease. Martonosi's group finds that the calcium pump actually crystallises in different forms, depending on the concentrations of substrates and substrate congeners, two of the different forms being correlated with the two major different structures (E_1 and E_2) revealed by biochemical studies, and clearly crucial for pump function. The hope here is that, in the fulness of time, detailed structural information will be available for each of these different forms of the pump proteins, information which will lead to the understanding of those changes of conformation which are the transport events. (Here Steinberg's paper is of interest. It pinpoints the amino-acid in the sodium pump's sequence which binds to a fluorescent label that reports the protein's change of conformation.) Rigaud and his colleagues use a different electron microscopic approach to investigate the association between monomers of the light-driven proton pump bacteriorhodopsin to form oligomers. The pump, embedded in membranes of different lipid-to-protein compositions, associates in a concentration-dependent oligomerisation, which can be seen by direct visualisation of the membranes. The sodium pump, too, can form oligomers in a lipid membrane but Vilsen and her compatriots show that the monomer of composition $\alpha\beta$ suffices to carry out all the conformation changes that an ion pump must do to carry out ion pumping. That different sodium pumps contain different forms of the α-chain is becoming clear and Charlemagne's study shows this to be the case for the pump from heart muscle.

Science, we are taught, is the Art of the Soluble. This means that all of us work with the methods which are at the current cutting edge of our several disciplines, always struggling with the difficult new problems which these methods themselves have enabled us to confront. The ability to clone the transport genes has given an immense stimulus to our studies, but we now grapple with the problem of building valid structures on the basis of these sequences. Crystallisation of some membrane proteins is now in some cases possible, and we look forward to the structural analysis of crystals of the ion pumps. But we will then be faced with the problems that will arise in identifying those parts of the protein that carry out particular functions. With static structures solved, we will need to identify those conformation changes involved in the transport events themselves. Finally, we will be in a position to ask what are the structural concomitants of the ATP-driven phosphorylations (for the E_1E_2 pumps) or the debinding/ binding of ATP for the F_1F_0 pumps, that is, to ask what are the structural bases of the transduction of chemical and osmotic energies. The papers of this section direct us to these future goals.

W.D. Stein
Editor

The Ion Pumps: Structure, Function,
and Regulation, pages 7–14

THE Ca^{2+} PUMP OF SKELETAL MUSCLE SARCOPLASMIC RETICULUM

Anthony N. Martonosi, Laszlo Dux, Kenneth A. Taylor and Slawomir Pikula

Department of Biochemistry and Molecular Biology, State University of New York, Health Science Center, Syracuse, New York 13210 USA (A.N.M); Institute of Biochemistry, Szent-Gyorgyi University Medical School, H-6701 Szeged, Hungary (L.D); Department of Anatomy, Duke University Medical Center, Durham, North Carolina 27710 USA (K.A.T); Nencki Institute of Experimental Biology, Polish Academy of Sciences, Warsaw, Poland (S.P).

INTRODUCTION

The regulation of muscle contraction in skeletal muscle is accomplished by ATP-dependent transport of calcium across the sarcoplasmic reticulum membrane (Ebashi, 1960, 1961; Hasselbach and Makinose, 1961; Ebashi and Lipmann, 1962). The process is of universal significance, not only in the control of motility, but also in the metabolic regulation of most living cells (for review, see Martonosi, 1983).

The complete molecular machinery required for the ATP-dependent translocation of Ca^{2+} from the cytoplasm into the lumen of sarcoplasmic reticulum is contained in the Ca^{2+} transport ATPase (Martonosi and Beeler, 1983). The Ca^{2+}-ATPase is present in different iso-forms in the sarcoplasmic reticulum of slow or fast skeletal, cardiac and smooth muscles, with an average molecular weight of close to 110,000 (MacLennan et al., 1985; Brandl et al., 1986, 1987).

The active Ca^{2+} transport is stoichiometrically linked to the hydrolysis of ATP. The process involves the binding

---

Supported by research grants from the National Institutes of Health, USPHS (AM 26545, GM 30598), the U. S. National Science Foundation (PCM 84-03679, Int. 86-17848) and the Muscular Dystrophy Association. K.A.T. is an Established Investigator of the American Heart Association.

of $2Ca^{2+}$ to a high affinity site on the enzyme in the E_1 conformation, followed by transient phosphorylation of a β-carboxyl group of an aspartyl residue in the active site and the entrapment of Ca^{2+} in a $Ca_2 - E_1 \sim P$ intermediate. A change in enzyme conformation from the E_1 to the E_2 state ($Ca_2 - E_2 \sim P$) leads to the translocation and release of Ca^{2+} in the interior of the sarcoplasmic reticulum (Fig. 1).

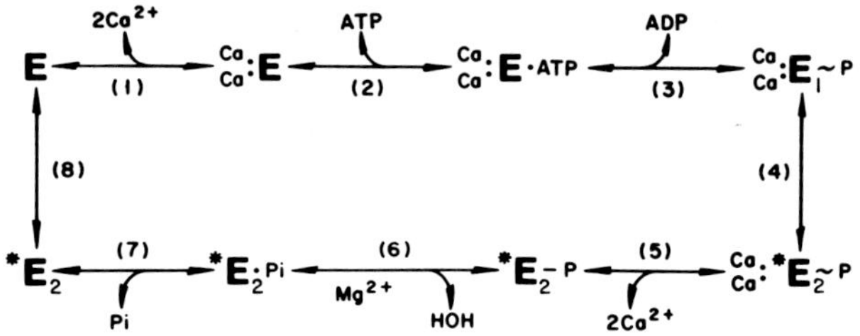

Fig. 1. Elementary steps of Ca^{2+} translocation. The intermediates on the top line represent substructures in the E_1 state, while those on the bottom line are in the E_2 state (from de Meis and Vianna, 1979).

During this process the affinity constant of the binding site for Ca^{2+} decreases from $\simeq 10^7$ M^{-1} to about 10^3 M^{-1}, satisfying the thermodynamic requirements for the accumulation of Ca^{2+} in the interior of sarcoplasmic reticulum. The process is completed by the Mg^{2+}-dependent hydrolysis of the $E_2 - P$ intermediate, with the release of inorganic phosphate and the isomerization of the enzyme from the E_2 into the E_1 state (Inesi and de Meis, 1985). The Ca^{2+} transport is reversible, yielding the synthesis of 1 mole of ATP for 2 calcium ions released from the sarcoplasmic reticulum (for review, see Martonosi, 1984).

The kinetic and thermodynamic parameters of the elementary steps of the process have been determined in considerable detail, both in the forward and in the reverse reaction (Inesi, 1985), and much information is available about the regulation of the concentration and transport activity of the system under physiological conditions (Martonosi et al., 1985, 1987a,b).

The structure of the Ca^{2+}-ATPase. The amino acid sequence of Ca^{2+}-ATPase isoenzymes was recently deduced from the complementary DNA sequences (MacLennan et al., 1985; Brandl et al., 1986, 1987). The structures predicted for the fast and slow isoenzymes are 84% identical. A hypothetical model of the tertiary structure of Ca^{2+}-ATPase was constructed (Brandl et al., 1986) that consists of 3 major structural regions:

1. The cytoplasmic headpiece that contains the domains for ATP binding, phosphorylation and the transduction mechanism.

2. A pentahelical stalk that is assumed to contain the high affinity binding site for Ca^{2+}.

3. The intramembranous region that may contain as many as 10 transmembrane helices; these anchor the Ca^{2+}-ATPase in the lipid bilayer and form the transmembrane channel for the passage of Ca^{2+}.

The purpose of our studies was to define the structure of Ca^{2+}-ATPase by electron microscope reconstruction from images of Ca^{2+}-ATPase crystals.

TWO-DIMENSIONAL CRYSTALS OF Ca^{2+}-ATPase IN THE NATIVE SARCOPLASMIC RETICULUM IN THE E_1 AND E_2 STATES.

a. The E_2-type crystals induced by vanadate. The vanadate (V) ions, in the absence of calcium, stabilize the Ca^{2+}-ATPase in the E_2 conformation, and promote the formation of Ca^{2+}-ATPase crystals (Fig. 2) both in native sarcoplasmic reticulum and in reconstituted vesicles containing the purified Ca^{2+}-ATPase (Dux and Martonosi, 1983a-e; 1984). The formation of the vanadate-induced crystals is aided by inside-positive membrane potential, suggesting an influence of transmembrane potential on the

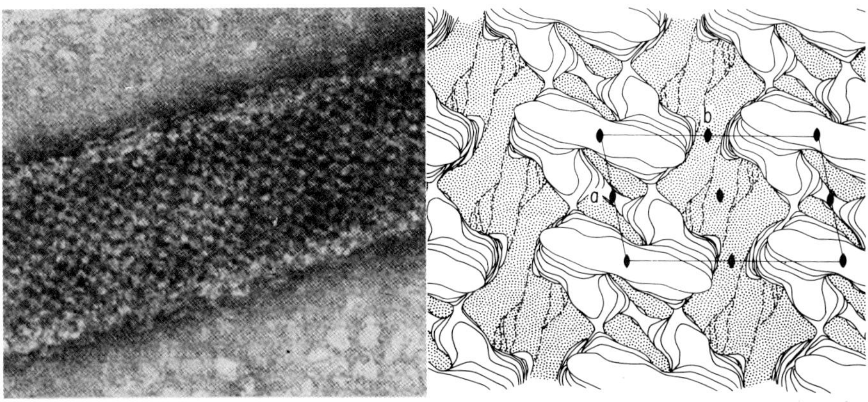

Fig. 2. Views of the Ca^{2+}-ATPase from reconstructions of E_2-type crystals. Negative staining with uranyl acetate. (Taylor et al., 1986a).

---

conformation of the Ca^{2+}-ATPase (Dux and Martonosi, 1983d; Beeler et al., 1984). The crystals are disrupted by Ca^{2+} at concentrations sufficient to saturate the high affinity

Ca^{2+} binding sites of the Ca^{2+}-ATPase (Dux and Martonosi, 1983c; Varga et al., 1987). The E_2-type crystals contain dimer chains of ATPase molecules forming an oblique surface lattice with a spacegroup of P2 (Taylor et al., 1984, 1986a,b; Peracchia et al., 1984); the space group and the unit cell dimensions (a = 65.9 Å, b = 114.4 Å and γ = 77.9°) are consistent with Ca^{2+}-ATPase dimers as structural units. The Ca^{2+}-ATPase molecules are linked together to form a dimer by a stain-excluding bridge (Fig. 2) above the cytoplasmic surface of the bilayer (Taylor et al., 1986a,b). The gap under the bridge is likely to be accessible in the native membrane to calcium, ATP and other solutes. The bridge is oriented parallel with the long axis of the tubules. Three-dimensional reconstruction of Ca^{2+}-ATPase molecules in unstained frozen-hydrated tubules reveals both the cytoplasmic and the intramembranous regions of the protein (Taylor et al., 1986b). The most striking feature of the cytoplasmic surface is the helical groove between the dimer chains, while on the luminal surface the groove is located between the two ATPase molecules constituting a dimer. As a result, the intrabilayer domains of the Ca^{2+}-ATPase molecules within the dimer are farther apart on the luminal than on the cytoplasmic side.

Decavanadate binds to the Ca^{2+}-ATPase with higher affinity and stoichiometry than monovanadate, and it is more potent in inducing crystallization of Ca^{2+}-ATPase (Varga et al., 1985; Csermely et al., 1985a,b). The Ca^{2+}-ATPase crystals induced by mono- and decavanadate appear similar, but the decavanadate-induced crystals are more stable in the presence of 10^{-5} M calcium (Varga et al., 1987).

b. The E_1-type crystals of Ca^{2+}-ATPase induced by Ca^{2+} or lanthanides. A different crystal form of Ca^{2+}-ATPase develops in sarcoplasmic reticulum vesicles exposed at pH 8.0 to 10^{-5} - 10^{-4} M Ca^{2+} or 10^{-6} - 10^{-5} M lanthanides (Dux et al., 1985a). Ca^{2+} and lanthanides are presumed to bind to the Ca^{2+}-ATPase in the E_1 form. The space group of the Ca^{2+}- or lanthanide-induced crystals is P1, and the unit cell dimensions are consistent with ATPase monomers as structural units (Fig. 3). The formation of the E_1 crystals is promoted by inside-negative potential. A similar crystal form is induced by 20 mM Ca^{2+} at pH 6.0 (Dux et al., 1987).

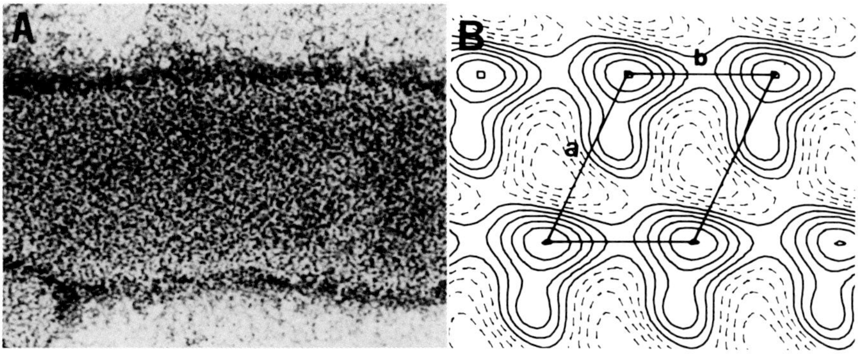

Fig. 3. Micrograph and projection map of praseodymium-induced Ca^{2+}-ATPase crystals. A. The image is a complex superposition of the lattices from the top and bottom of the flattened crystalline tubule; 276,000X magnification. B. Projection map shows unit cell; a = 66.8 Å, b = 50.4 Å and γ = 114.6° (Dux et al., 1985).

The differences between the E_1 and E_2-type crystals indicate different types of interactions between ATPase molecules in the two conformations.

MICROCRYSTALS OF Ca^{2+}-ATPase IN DETERGENT-SOLUBILIZED SARCOPLASMIC RETICULUM.

Three-dimensional microcrystalline arrays of Ca^{2+}-ATPase molecules develop in detergent-solubilized sarcoplasmic reticulum upon exposure to 20 mM Ca^{2+} for several weeks at 2°C, in a storage medium of 0.1 M KCl, 10 mM K-MOPS, pH 6.0, 3 mM $MgCl_2$, 3 mM NaN_3, 5 mM DTT, 25 IU/ml Trasylol, 2 μg/ml 1,6-di-tert-butyl-p-cresol, 20% glycerol and 2-4 mg detergent ($C_{12}E_8$ or Brij 36T) per mg protein (Dux et al., 1987). Preparations stored under these conditions retain significant Ca^{2+}-stimulated ATPase activity when assayed in the presence of asolectin. The stability of the enzyme is a necessary but not sufficient condition for crystallization, since the activity of solubilized Ca^{2+}-ATPase was preserved under a variety of conditions that did not lead to crystallization (Pikula et al., 1987).

The crystals consist of stacked lamellar arrays of Ca^{2+}-ATPase molecules (Fig. 4). Analysis of electron micrographs obtained from edge-on views of these stacked sheets yield a sheet thickness of ≃ 180 Å and show lamellae with ATPase molecules extending out both sides. The core of the lamellae probably contains a lipid-detergent phase into which the hydrophobic tail portions of ATPase molecules are symmetrically inserted (Mullner et al., 1987).

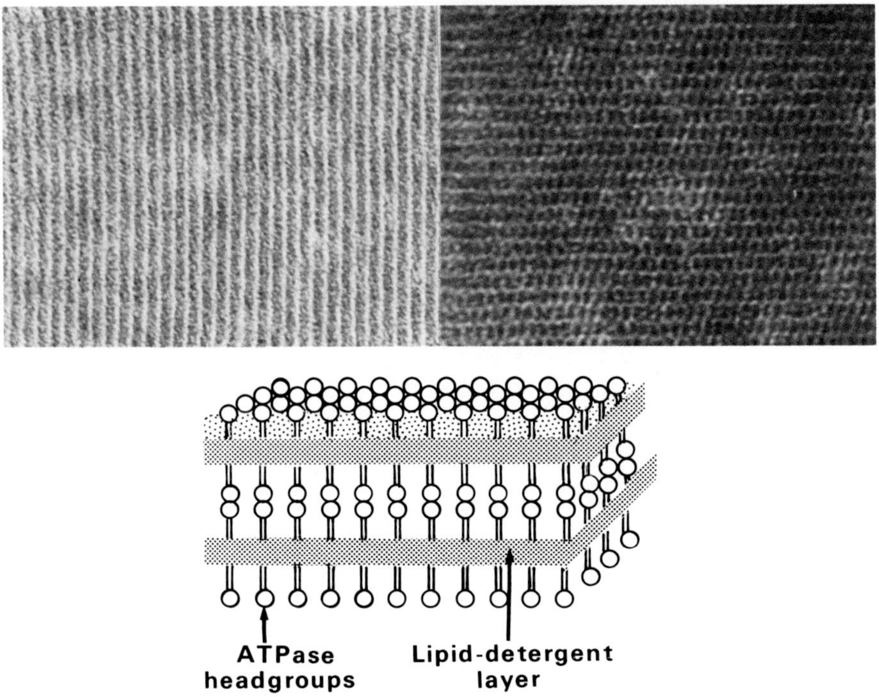

Fig. 4. Three-dimensional Ca^{2+}-ATPase crystals in detergent-solubilized sarcoplasmic reticulum. Top left: thin sections; magnification: 207,000x. Top right: negatively stained lamellae; magnification: 308,000x. Bottom: schematic representation of a crystalline lamella.

Diffraction patterns, when obtained, have indicated that the unit cell is rectangular, with sides of 164.2 ± 2.2 and 55.5 ± 1.5 Å. Interaction between the hydrophilic head portions of the ATPase molecules are expected to play a role in the stacking of the lamellae. The high concentration of Ca^{2+} (20 mM) and low pH (pH 6.0) may promote these interactions by minimizing charge repulsions. Experiments are in progress to generate three-dimensional crystal forms from fully delipidated Ca^{2+}-ATPase that are suitable for x-ray diffraction analysis using detergent-small amphiphile combinations (Michel, 1983; Deisenhofer et al., 1985).

CONCLUSIONS AND PROSPECTS.

The structure of Ca^{2+}-ATPase was determined at low resolution in three distinct crystal forms. At this stage the aesthetic pleasure of seeing the shape of the molecule emerge from what used to be blurred lollypops, overshadows the tantalizing paucity of information that these structures yield about the mechanism of Ca^{2+} transport. The next phase, the production of stable, well-ordered, three-dimensional crystals for x-ray crystallography is likely to be slow and difficult, considering the large size and lability of the protein. In the meantime, kinetic and spectroscopic studies will continue to remain the primary sources of information on structure-function relationships.

"Heaven gives its glimpses only to those
Not in position to look too close." (R. Frost)

REFERENCES

Beeler TJ, Dux L, Martonosi A (1984). J Membr Biol 78:73-79.
Brandl CJ, Green NM, Korczak B, MacLennan DH (1986). Cell 44:597-607.
Brandl CJ, de Leon S, Martin DR, MacLennan DH (1987). J Biol Chem 262:3768-3774.
Csermely P, Varga S, Martonosi A (1985a). Eur J Biochem 150:455-460.
Csermely P, Martonosi A, Levy GC, Ejchart AJ (1985b). Biochem J:230, 807-815.
Deisenhofer J, Epp O, Miki K, Huber R, Michel H (1985). Nature 318:618-624.
De Meis L, Vianna AL (1979). Annu Rev Biochem 48:275-292.
Dux L, Martonosi A (1983a). J Biol Chem 258:2599-2603.
Dux L, Martonosi A (1983b). J Biol Chem 258:10111-10115.
Dux L, Martonosi A (1983c). J Biol Chem 258:11896-11902.
Dux L, Martonosi A (1983d). J Biol Chem 258:11903-11907.
Dux L, Martonosi A (1983e). Muscle & Nerve 6:566-573.
Dux L, Martonosi A (1984). Eur J Biochem 141:43-49.
Dux L, Taylor KA, Ting-Beall HP, Martonosi A (1985a). J Biol Chem 260:11730-11743.
Dux L, Pikula S, Mullner N, Martonosi A (1987). J Biol Chem 262:6439-6442.
Ebashi S (1960). J Biochem (Tokyo) 48:150-151.

Ebashi S (1961). J Biochem (Tokyo) 50:236-244.
Ebashi S, Lipmann F (1962). J Cell Biol 14:389-400.
Hasselbach W, Makinose M (1961). Biochem Z 333:518-528.
Inesi G (1985). Annu Rev Physiol 47:573-601.
Inesi G, de Meis L (1985). In Martonosi A (ed): "The Enzymes of Biological Membranes", Vol. 3, New York: Plenum, pp 157-191.
MacLennan DH, Brandl CJ, Korczak B, Green NM (1985). Nature 316:696-700.
Martonosi AN (1983). In Stracher A (ed): "Muscle and Nonmuscle Motility", Vol. 1, New York: Academic Press, pp 233-258.
Martonosi A (1984). Physiol Rev 64:1246-1320.
Martonosi AN, Beeler TJ (1983). In Peachey LD, Adrian RH (eds): "Handbook of Physiology. Skeletal Muscle", Bethesda, Maryland: American Physiological Society, pp 417-485.
Martonosi A, Kracke G, Taylor KA, Dux L, Peracchia C (1985). In Graves JS (ed): "Regulation and Development of Membrane Transport Processes", New York: Wiley, pp 57-85.
Martonosi A, Dux L, Taylor KA, Ting-Beall HP, Varga S, Csermely P, Mullner N, Papp S, Jona I, (1987a). In Fambrough D, Hille B (eds): "Proteins and Excitable Membranes", New York: Wiley, pp 257-286.
Martonosi A, Taylor KA, Varga S, Ting-Beall HP (1987b). In Harris JR, Horne RW (eds): "The Electron Microscopy of Proteins", Vol. 6, Membranous Structures, London: Academic Press, pp 255-376.
Michel H (1983). Trends in Biochem Sci 8:56-59.
Mullner N, Pikula S, Dux L, Peracchia C, Taylor KA, Martonosi A (1987). Fed Proc 46:2131.
Peracchia C, Dux L, Martonosi A (1984). J Musc Res Cell Motil 5:431-442.
Pikula S, Mullner N, Dux L, Martonosi A (1987). Fed Proc 46:2131.
Taylor KA, Dux L, Martonosi A (1984). J Mol Biol 174:193-204.
Taylor KA, Dux L, Martonosi A (1986a). J Mol Biol 187:417-427.
Taylor KA, Ho MH, Martonosi A (1986b). Ann NY Acad Sci 483:31-43.
Varga S, Csermely P, Martonosi A (1985). Eur J Biochem 148:119-126.
Varga S, Csermely P, Mullner N, Dux L, Martonosi A (1987). Biochim Biophys Acta 896:187-195.

The Ion Pumps: Structure, Function, and Regulation, pages 15–24

A CONSENSUS STRUCTURE FOR CATION PUMPS

N.M. Green*, W.R. Taylor* and D.H. MacLennan+

*National Institute for Medical Research, Mill Hill, London NW7 1AA, UK; +Banting & Best Institute for Medical Research, Toronto, Ontario, Canada M5G 1L6

INTRODUCTION

The intensive cloning activities of the last four years have led to the amino acid sequences of at least sixteen ion pumps. These include seven sodium pumps (e.g.Kawakami et al., 1985; Shull et al., 1986; Ovchinnikov et al., 1987), five proton pumps (e.g. Serrano et al., 1986), two calcium pumps (Brandl et al., 1986) and two bacterial potassium pumps (e.g. Hesse et al., 1984; Solioz et al., 1987), all related by sequence homology and by the coupling of ion transport to formation of an aspartyl phosphate intermediate. The listed cations except K^+ are required for phosphorylation and are pumped out of the cytoplasm. In the bacterial pumps potassium is required for dephosphorylation and is pumped in, but although no cation requirement for phosphorylation was demonstrated protons may be involved. In this respect, K^+ is playing the same counterion role as it does in the Na^+/K^+ and H^+/K^+ pumps. Although K^+ is not a counterion for the Ca^{++} and fungal H^+ pumps it does activate the dephosphorylation step, which may represent a vestige of its primitive role as a counterion. The number of ions transported per cycle varies from 3 Na^+ to 2 Ca^{++} to 1 H^+. These varied pumping activities are performed by proteins with many common features.

Pumps of the same ion specificity have highly conserved sequences (>80%), except for the two prokaryotic potassium pumps. The sequences of pumps of differing specificity show only 20%-30% identity concentrated in about twelve homologous regions scattered through the molecule, with up to 60%

of identities in pairwise comparisons. The acid secreting pump of gastric mucosa (H^+/K^+) is an interesting exception, showing a greater overall identity to the sodium pump (62%) than to the other proton pumps (22%).

SEQUENCE IDENTITY AND STRUCTURAL RELATIONSHIP

Several families of proteins have been analysed by pairwise superimposition of their known backbone structures (Chothia and Lesk, 1986). This defined a conserved core, which included between 50 and 100% of the sequence. A smooth curve was obtained which correlated with RMS distance between alpha carbons with the percentage identity, in pairwise comparisons of the core sequences. This relation persisted even when the identity was below 20%. There is an apparent conflict between this observation and other studies (e.g. Argos, 1987) which have found that pairs of identical pentapeptides derived from unrelated proteins often fold differently. However, this is but one more example of the often remarked influence of distant interactions on folding, which frustrates the accurate prediction of structure from sequence. It does not invalidate the conclusion that, when considering the alignment of proteins with common functions, a moderate fraction of identities over a domain-sized segment is sufficient to justify the assumption of a common core structure, particularly when supported by predicted secondary structure.

STRATEGY OF STRUCTURE PREDICTION

Simple linear progression from sequence via secondary and tertiary structure to a final model is seldom rewarding because of the low precision of the first step (<60%). There can be no improvement without some way of allowing for the effects of distant interactions on folding. Examination of some 200 known structures has shown that over 80% of domains can be assigned to one of five types (α, β, $\alpha\beta$, $\alpha+\beta$ and cysteine rich, Richardson, 1981). The domain type can often be assigned from the predicted secondary structure and then be used, together with the template fitting procedure of Taylor and Thornton (1984), to clarify ambiguities in the original prediction. The correct assignment procedure can be aided by ancillary information from which domain boundaries and functions can be deduced (Fig. 1).

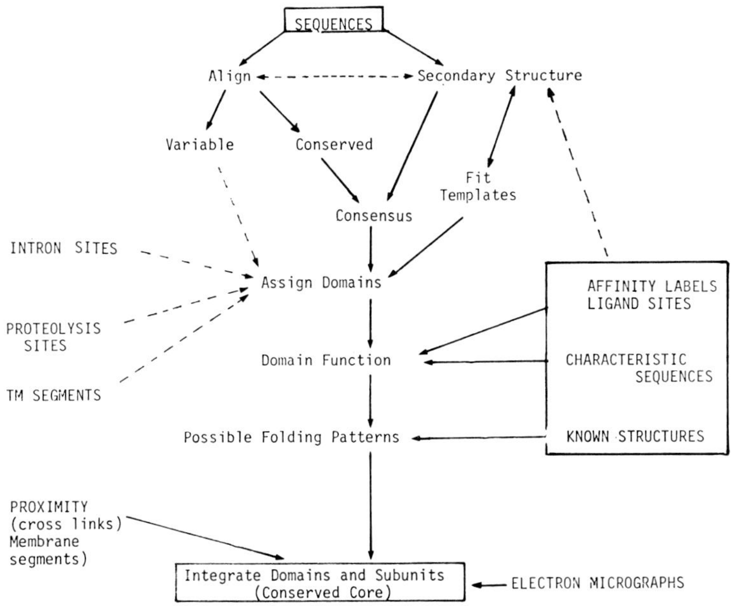

Figure 1. A strategy for structural analysis. Experimental input is shown peripherally (upper case). The arrows indicate the paths by which information is combined to build up an overall molecular picture.

A MODEL FOR THE Ca^{++} PUMP (MacLennan et al., 1985)

Definitive domain boundaries are provided by two clear hydrophobic hairpins, which divide the sequence into (1) a poorly conserved N-terminal domain (60 residues), (2) a well conserved predominantly β-stranded domain (140 residues), (3) the major cytoplasmic region (435 residues), which probably consists of two or three domains. This is followed by (4) a hydrophobic C-terminus (240 residues).

The major cytoplasmic region contains all the well defined functional sites in the ATPase. These are the highly conserved phosphorylation region near the N-terminus (40 residues), the somewhat less conserved site of labelling by fluorescein isothiocyanate which marks the nucleotide binding

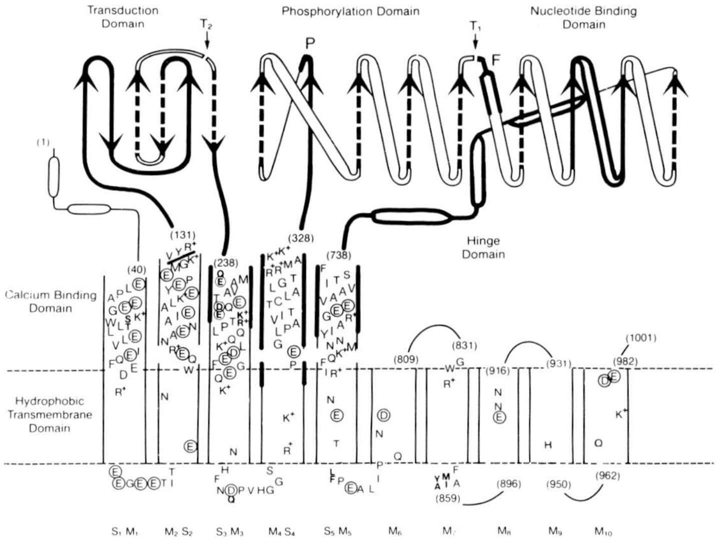

Figure 2. Provisional structure of the calcium pump. Taken from Green et al. (1986). This two-dimensional diagram illustrates the major domains and their predominant secondary structure. The topology of the β-strand connections is arbitrary. A more exact analysis of the ATP domain has been made elsewhere (Taylor and Green, 1988). Heavy continuous lines indicate regions which are well conserved between different cation pumps. Negatively charged residues are ringed. In the transmembrane region only the charged residues plus asparagine (N) and glutamine (Q) are shown in their appropriate locations on the helical nets. In this region the ringed negatively charged residues have homologous residues in the corresponding positions in the transmembrane helices of the Na pump.(Reprinted by permission of John Wiley & Sons.)

site, and the most highly conserved C-terminal region (65 residues) which in the Na^+/K^+ ATPase is labelled by FSBA (Ohta et al., 1986). This double domain has a functional resemblance to kinases, which suggested the basis for a gating mechanism. Many kinases (Bennet and Huber, 1984) consist of hinged domains which move during the reaction cycle. They are always doubly linked, with the second link from the C-terminus of domain 2 interacting non covalently with domain 1. In the Ca^{++} pump the nucleotide domain is formed mainly from alternating α-helices and β-strands. The resulting parallel β-sheet is preceded by the phosphorylation region, demarcated by a tryptic cleavage site and a deletable region. It is followed by the most highly conserved region of all, which, by virtue of its C-terminal location must be near to the membrane and to the conserved N-terminal phosphorylation site. Because of this proximity, and by analogy to the doubly linked hinge seen in kinases (Blake and Rice, 1981), we proposed that it interacts with both nucleotide binding and phosphorylation domains, permitting movement between them and transmitting the effects of such movements to the calcium binding sites.

The location of the calcium sites and their relation to the transmembrane channel has yet to be defined. All that is known is that they are probably in the N-terminal tryptic fragment (A_2) (Ludi and Hasselbach, 1984), that they are close to the membrane (Teruel and Gomez-Fernandez, 1986) and distant (>40Å) from the FITC which labels the nucleotide site (Scott, 1985). They are not directly connected to the kinase region. Energy transduction must therefore be mediated either through interactions in the stalk region or through the directly connected transduction domain, or through both. The operation of the pump must be dependent on the segments which connect the kinase and transduction domains to the membrane and form the narrow stalk, seen in electron micrographs of both Ca^{++} and Na^{+} pumps. In the Ca^{++} pump these segments are all predicted to be long (15-20 residues) α-helices and we proposed that they formed the stalk, together with other helical segments from the N-terminus (MacLennan et al., 1985). The helical character is only partially conserved in the other ATPases and in the three dimensional reconstructions (Taylor et al., 1984) the stalk is shorter and wider than could be accounted for simply by 5 parallel α-helices. A more complex structure is probable.

COMPARISON WITH OTHER PUMPS

In this model we have defined a number of separate domains, each of which is essential to the function of the Ca^{++} pump. How well are these domains conserved in pumps of different ion specificity? We have aligned all the known sequences, supplementing sequence identity with clues from secondary structure, and restricting gaps to unstructured regions. An overall map of four representatives (Fig. 3) shows the percentage of identities between the various domains in the neighbouring sequences.

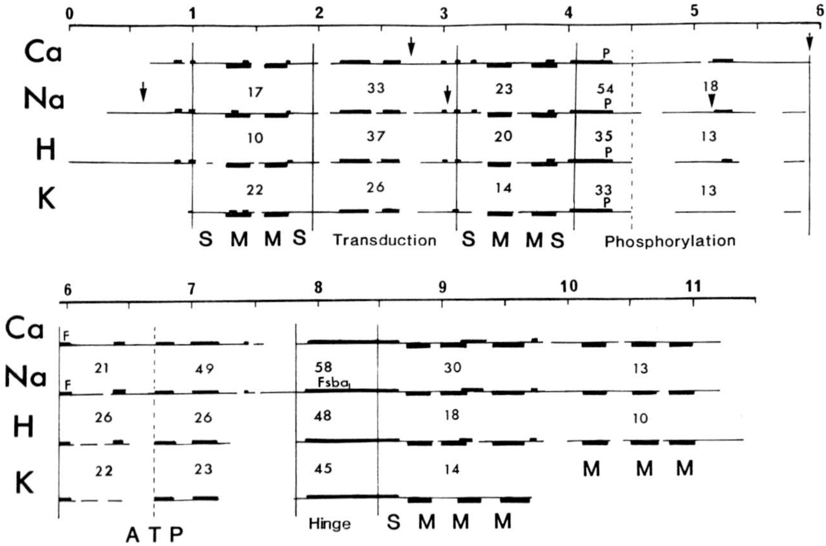

Figure 3. Alignment of sequences of four cation pumps. Ca^{++} pump, fast skeletal (Brandl et al., 1986); Na pump, sheep kidney (Shull et al., 1985); proton pump, yeast (Serrano et al., 1986); K pump, E. coli (Hesse et al., 1984). Conserved segments are shown as bars above the line, hydrophobic segments are bars below the line. [conserved segment symbol] conserved segments; [hydrophobic segment symbol] hydrophobic segments (M); S, stalk segments; ↓ sites of proteolytic cleavage (Jørgensen and Collins, 1986); P, phosphorylation; F, FITC; | domain boundary; ⁞ subdivision of phosphorylation and ATP binding domains. The numbers indicate the percentage identity between the individual domains above and below.

It can be seen that all the major regions of conservation are present in all the pumps, as are the first seven hydrophobic segments. The relative conservation of the different domains is the same for all pairs of pumps. It decreases from a mean of 47% for the hinge regions to 20% for the N-terminal halves of the nucleotide binding domains. The stalk segments are not well conserved (22%) and the membrane segments are so diverse that, except for M_1 and M_4, there is no significant conservation (<15%). In contrast, where ion specificity is preserved, the stalk and membrane are among the most highly conserved segments (92%) (Brandl et al., 1986). We conclude that the cores of the ATPase domains are likely to match within 1.5-2Å and that we can propose a consensus structure for the motor based on the Ca^{++} pump.

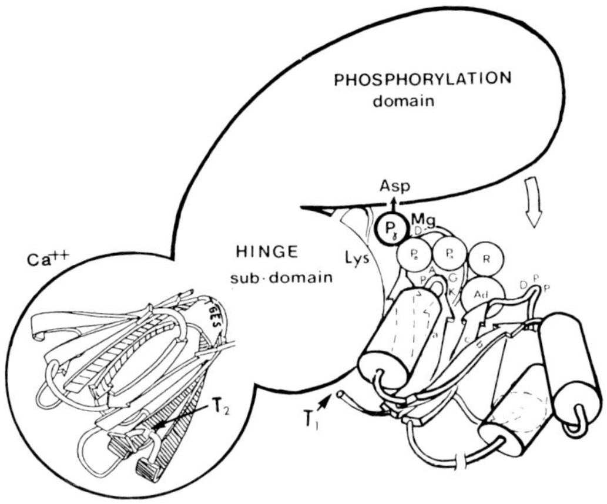

Figure 4. The relation between the extramembranous domains of cation pumps. The stalk and transmembrane segments are invisible. The topology of the nucleotide domain comes from the analysis of Taylor and Green (in press). It is based on the structure of adenylate kinase, with the ATP bound near three conserved bends at the C-terminus of the β-sheet. The γ-phosphate interacts with Mg^{++} and points towards the phosphorylation site and the hinge.

A schematic view of the relation between domains has been shown above (Fig. 2). A more realistic model (Fig. 4), viewed from above, satisfies an important general principle, namely that the main interacting regions should be derived from conserved segments of the sequence, the remainder being accommodated on exposed surfaces. The highly conserved hinge provides a focus for interaction with the three domains.

At present, hypotheses about the relations between the gate and channel are purely speculative. The pattern of conservation suggests that they are restricted to stalk and transmembrane regions. The two prokaryotic ATPases lack the last three transmembrane segments, but otherwise all the pumps contain the same set of segments. We favour a model in which the ion binding sites and the gate are located in the stalk region and a separate polar channel provides a route of escape for the occluded ions. A model with the gate as an integral part of the channel (e.g. Tanford, 1982) is also possible. The density of polar and charged residues in the hydrophobic segments is about the same as in bacterial rhodopsin (about 2 per helix). Their nature and locations vary with the ion specificity. There are six to eight negative charges in all the eukaryotic pumps but only two or three positive charges, except for the Ca^{++} pump of sarcoplasmic reticulum, which has six. A channel with a network of ion pairs may be possible for the Ca^{++} ATPase, but is unlikely for any of the others, unless extra charged groups from more polar segments are introduced into the transmembrane region. Critical roles for charged residues can be tested by mutagenesis.

A tentative interpretation of the reaction cycle in terms of this model has been proposed (Green et al., 1986). A related model for the sodium pump has also been discussed (Karlish et al., 1985).

REFERENCES

Argos P (1987). Analysis of sequence-similar pentapeptides in unrelated protein tertiary structures. J Mol Biol 197:331-348

Bennett WS, Huber R (1984). Structural and functional aspects of domain motions in proteins. CRC Crit Rev Biochem 15:291-384

Blake CCF, Rice DW (1987). Phosphoglycerate kinase. Phil Trans Roy Soc B 293:93-104

Brandl CJ, Green NM, MacLennan DH (1986). Two Ca^{2+} ATPase genes: homologies and mechanistic implications of deduced amino acid sequences. Cell 44:597-607

Chothia C, Lesk AM (1986). The relation between the divergence of sequence and structure in proteins. EMBO J 5:823-826

Green NM, Taylor WR, Brandl CJ, Korczak B, MacLennan DH (1986). Structural and mechanistic implications of the amino acid sequence of calcium-transporting ATPases. CIBA Foundation Symp 122, pp 93-114. Wiley, Chichester

Hesse JE, Wieczorek L, Altendorf K, Reicin AS, Dorus E, Epstein W (1984). Sequence homology between two membrane transport ATPases, the kdp-ATPase of *Escherichia coli* and the Ca^{2+} ATPase of sarcoplasmic reticulum. Proc Natl Acad Sci USA 81:4746-4750

Jørgensen PL, Collins JH (1986). Tryptic and chymotryptic cleavage sites in sequence of α-subunit of (Na^+ + K^+)-ATPase from outer medulla of mammalian kidney. Biochim Biophys Acta 860:570-576

Karlish SJD, Rephaeli A, Stein WD (1985). Transmembrane modulation of cation transport by the Na,K-pump. In: The Sodium Pump (Glynn IM, Ellory CD, eds) pp 487-499, The Company of Biologists Ltd, Cambridge

Kawakami K, Noguchi S, Noda M, Takahashi H, Ohta T, Kawamura M, Nojima H, Nagano K, Hirose T, Inayama S, Hayashida H, Miyata T, Numa S (1985). Primary structure of the α-subunit of Torpedo californica (Na^+ + K^+)ATPase deduced from cDNA sequence. Nature 316:733-736

Ludi H, Hasselbach W (1984). Separation of the tryptic fragments of sarcoplasmic reticulum ATPase with high performance liquid chromatography. FEBS Letts 167:33-36

MacLennan DH, Brandl CJ, Korczak B, Green NM (1985). Amino acid sequence of a Ca^{2+} + Mg^{2+}-dependent ATPase from rabbit muscle sarcoplasmic reticulum, deduced from its complementary DNA sequence. Nature 316:696-700

Ohta T, Nagano K, Yoshida M (1986). The active site structure of Na^+/K^+-transporting ATPase: location of the 5'-(p-fluorosulfonyl)benzoyladenosine binding site and soluble peptides released by trypsin. Proc Natl Acad Sci USA 83:2071-2075

Ovchinnikov Yu.A, Arzamazova NM, Arystarkhova EA, Gevondyan NM, Aldanova NA, Modyanov NN (1987). Detailed structural analysis of exposed domains of membrane-bound Na^+, K^+-ATPase. FEBS Letts 217:269-274

Richardson JS (1981). The anatomy and taxonomy of protein structure. Adv Protein Struct 34:167-339

Scott TL (1985). Distances between the functional sites of the Ca^{2+} + Mg^{2+} ATPase of sarcoplasmic reticulum. J Biol Chem 260:14421-14423

Serrano R, Kielland-Brandt MC, Fink GR (1986). The yeast plasma-membrane ATPase is essential for growth and has homology with Na^+, K^+ and Ca^{2+} ATPases. Nature 319:689-692

Solioz M, Mathews S, Furst P (1987). Cloning of the K^+-ATPase of Streptococcus faecalis. J Biol Chem 262:7358-7362

Shull GE, Greeb J, Lingrel JB (1986). Molecular cloning of three distinct forms of the Na^+, K^+-ATPase α-subunit from rat brain. Biochemistry 25:8125-8131

Shull GE, Schwartz A, Lingrel JB (1985). Amino acid sequence of the catalytic subunit of the $(Na^+ + K^+)$ATPase deduced from a complementary DNA. Nature 316:691-695

Tanford C (1982). Simple model for the chemical potential change of a transported ion in active transport. Proc Natl Acad Sci USA 79:2882-2884

Taylor KA, Dux L, Martonosi A (1984). Structure of the vanadate-induced crystals of sarcoplasmic reticulum Ca^{2+}-ATPase. J Mol Biol 174:193-204

Taylor WR, Thornton JM (1984). Recognition of super-secondary structure in proteins. J Mol Biol 173:487-514

Teruel JA, Gomez-Fernandez JC (1986). Distances between the functional sites of sarcoplasmic reticulum $(Ca^{2+} + Mg^{2+})$-ATPase and the lipid/water interface. Biochim Biophys Acta 863:178-184

The Ion Pumps: Structure, Function,
and Regulation, pages 25–32

STRUCTURE OF NUCLEOTIDE SITE AND RELATIONSHIP BETWEEN E_1–E_2 TRANSITION AND CATION OCCLUSION IN Na,K–ATPase

Peter Leth Jørgensen
Institute of Physiology
Aarhus University, 8000 Aarhus C, Denmark

INTRODUCTION

A wealth of structural information about the Na,K–pump proteins has appeared during the last 2–3 years from sequencing cloned cDNA and identification of genes. Several isoforms of the genes of α–subunit of Na,K–pump were identified in the human and rat genomes (Sverdlov et al., 1987; Shull and Lingrel, 1987) and the sequences of α–subunit (Kawakami et al., 1985, 1986a; Ovchinnikov et al. 1986; Shull et al., 1985, 1986a) and β–subunit (Brown et al., 1987; Kawakami et al., 1986b; Ovchinnikov et al., 1986; Shull et al., 1986b) were deduced from cDNA of mRNA of tissues in piscine and mammalian species. Crystals of the proteins of membrane bound Na,K–pump (Hebert et al., 1982) has been prepared by incubation in vanadate solution, but only low resolution (>20–25 Å) three dimensional models can be constructed on this basis. The details of the organization of the $\alpha\beta$–unit in the membrane therefore remain unknown and important problems related to the mechanism of active Na,K–transport are unsolved. This article will focus on the organization of the nucleotide and cation binding sites in the $\alpha\beta$–unit of Na,K–ATPase and the relationship between cation occlusion and E_1–E_2 conformational transitions in the protein.

EVOLUTION OF CATION PUMP PROTEINS

A comparison of Na,K–, H,K–, Ca–, H– and K–pumps shows that overall sequence homology is moderate, 17–24% except for 63% between H,K–ATPase and Na,K–ATPase, while tertiary structure is remarkably similar with respect to localization of transmembrane segments relative to domains involved in ATP binding and phosphorylation. The homologies are most pronounced in segments connecting transmembrane helices (M4 and M5) with the phosphorylation site and the nucleotide binding area, respectively, while there is little or no homology between the aminoacid sequences of N-termini, transmembrane helices, and

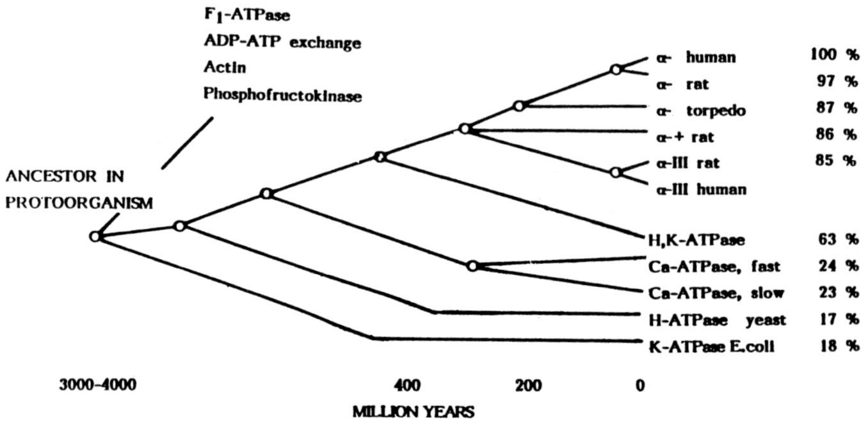

Figure 1. Evolutionary tree for cation pump proteins with overall homologies relative to α–subunit of Na,K–ATPase shown to the right. Sequence homologies and references are shown in Table I. Time scales are based on information about divergence of filetic lines (Dayhoff et al., 1978).

presumptive cation binding areas in Na,K–ATPase, Ca–ATPase, H–ATPase, and K–ATPase. Homologies between cation pumps and other ATP binding proteins suggest that they have a common origin as shown in Fig. 1 (Cf. Tonomura, 1986). The molecular basis for evolution of the family of cation pumps from a common ancestor may be gene duplication and drift or fusion and splicing of the gene sequences (Doolittle, 1985).

CYTOPLASMIC DOMAINS AND NUCLEOTIDE BINDING AREA

A characteristic structural feature of the cation pump proteins is a cytoplasmic protrusion with approximate dimensions 45 x 65 Å in the plane of the membrane and a length of 50–60 Å in the plane perpendicular to the membrane. The bulk of the protrusion is formed by the large central domain (Res. 340–780 in α–subunit) that forms sites for ATP binding and phosphorylation. Studies of

ATP binding sites in proteins of known molecular structure show that functional residues are contributed by regions of different protein sequences and secondary structure. In dehydrogenases, phosphofructokinase (Walker et al. 1982) and in particular, adenylate kinase (Fry et al. 1985) an important feature is a hydrophobic pocket for accommodation of the adenine and ribose moieties which is formed by Ile, Val, His and Leu residues. The triphosphate moiety is flanked by a hydrophobic strand of parallel β-pleated sheet terminated by Asp (Fry et al. 1985). Such a segment in adenylate kinase and β-subunit of F1-ATPase shows some homology with respect to charges and hydrophobic residues to segments in α-subunit of Na,K-ATPase (543-561). In addition to the segment around Asp 369, labeling with ATP analogues provides evidence for contribution from three additional segments in α-subunit of Na,K-ATPase. The first indication came from ATP sensitive covalent insertion of FITC into Lys-501 in α-subunit (Karlish, 1980). The C-terminal part of the central domain contributes two peptide segments (655-664 and 704-722) that are labeled by FSBA (Ohta et al. 1986) 5-(p-fluorosulfonyl)-benzoyl-adenosine, an ATP analogue that labels the Na,K-ATPase with concomitant inhibition of ATP hydrolysis, while K-phosphatase and ouabain binding remain intact. The peptides are located at residues 655-664 and 704-722 of the α-subunit.

CATION SITES AND CONFORMATION OF OCCLUDED FORMS

The problem of assigning Na^+, K^+ or Ca^{2+} binding sites to residues in the aminoacid sequences has not been solved. It will also be important to learn if cation sites are located in the cytoplasmic protrusion or if they are part of the intramembrane portion of the pump protein. Studies of cation binding at equilibrium and competition between Na^+ and K^+ for binding and occlusion (Forbush, 1987; Jørgensen, 1982) are in agreement with the notion that the Na,K-pump possesses a single set of sites for binding of either Na^+ or K^+. A cavity for occlusion of cations within the structure of the pump may be part of the transport pathway, because occlusion and release of cations from the cavity is governed by physiological ligands (Karlish et al., 1982). The $\alpha\beta$-unit is the minimum functional unit for occlusion of ^{22}Na or ^{86}Rb in Na,K-ATPase (Vilsen et al. 1987). It is generally accepted that Na^+ ions are occluded in $E_1P[Na]$ phosphoforms and that Rb^+(or K^+) are stabilized in an $E_2[K]$ form, but below evidence is presented for occlusion of Na^+ in E_2P-forms and for occlusion of K^+ in E_1-forms of the protein that may have a strong impact on our understanding of the transport mechanism.

Using CrATP as phosphorylating substrate values of ^{22}Na occlusion as high as 11 nmol/mg protein, were reached in selected purified preparations of Na,K–ATPase with 6 nmol EP/mg protein (Vilsen et al., 1987). Examination of tryptic cleavage patterns of the CrADP–E_1P[Na] form, confirms that the α–subunit is cleaved according to the

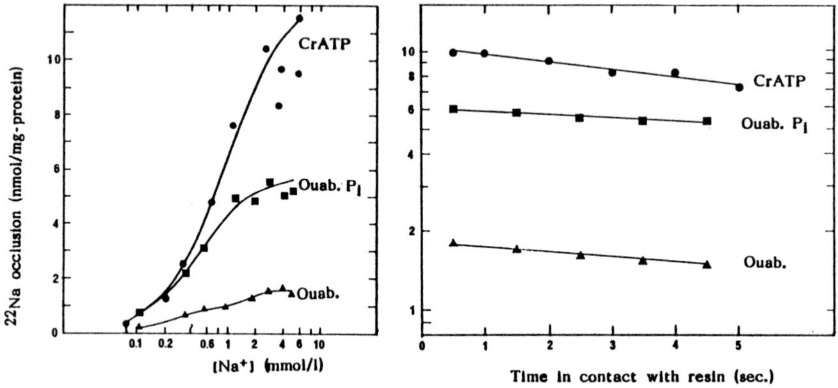

Figure 2. ^{22}Na Occlusion and deocclusion of membrane–bound Na,K–ATPase. Procedures as described by Vilsen et al. (204). CrATP: incubation with increasing concentrations of Na for 3 h at 20° C with 0.5 mM CrATP, 1 mM $MgCl_2$, in 10 mM Tris, 0.1 mM EDTA, pH 7.5. Ouabain: incubation in the same buffer for 15 min at 20° C with 1 mM ouabain, 1 mM $MgCl_2$ with or without 1 mM P_i–Tris. For determination of the rate of deocclusion the time spent in the Dowex was varied using a electronically operated automated syringe.

pattern that is characteristic for the E_1–form (unpublished). In Fig. 2, occlusion of Na^+ in CrADP–E_1P[Na] is compared with occlusion of Na^+ in complexes with ouabain. It is seen that the capacity for occlusion of Na^+ is low in presence of Mg^{2+} and ouabain, but that it is increased to 5–6 nmol/mg protein in presence of 1 mmol/l phosphate without changing the apparent affinity for Na^+. Phosphorylation of the aspartyl residue may thus be necessary for occlusion or directly involved in coordinating the cation in the E_2–form. The capacity for Na^+ occlusion by the Ouab–E_2P[Na] complex is lower than that of the CrADP–E_1P[Na] complex. This may suggest that the transition from E_1P to E_2P involves the release of one of the occluded Na^+ ions as shown in Fig. 3. Rb^+(or K^+) ions are occluded in the E_2[Rb] form but there is also evidence for occlusion of Rb^+ in E_1–forms. Formation of a ternary complex of Na,K-ATPase, K^+ and ATP is clearly demonstrated after cleavage by chymotrypsin or after blocking sulfhydryl groups (Jørgensen and Petersen, 1985).

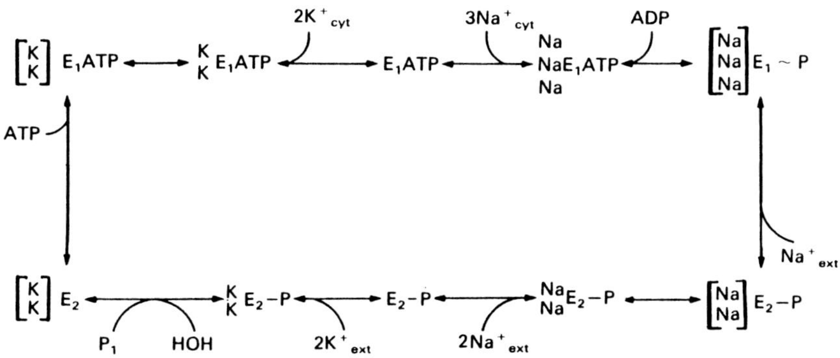

Figure 3. E_1–E_2 reaction cycle of the Na,K-pump with 4 major occluded conformations and ping-pong sequential cation translocation. The phosphoforms occlude Na^+ and dephosphoforms occlude K^+ or Rb^+. Na^+ and K^+ without brackets are cations bound to an open form such that they can exchange with medium cation. [Na] or [K] within brackets are occluded and prevented from exchanging with medium cations. It is proposed that release of Na_{ext} accompanies transition from E_1P[Na3] to E_2P[Na2] since capacity for occlusion of Na^+ in the ouabain stabilized form is lower than after incubation with CrATP (Fig. 2).

E_1–E_2 TRANSITION AND CATION TRANSLOCATION

In the reaction scheme for Na,K-pumping in Fig. 3, the E_1–E_2 transitions constitute the cation translocation steps, Na-extrusion involving the E_1P–E_2P transitions and K-uptake the E_2K–E_1ATP transition. Na^+ and K^+ are transported in sequence, Na^+ first and then K^+ in a ping-pong, sequential reaction. Recent developments has allowed studies of the correlation between E_1–E_2 transitions and cation translocation. New techniques have been introduced for studying effects of membrane potential in reconstituted vesicles (Raphaeli et al. 1986), pre–steady state cation fluxes in vesicles (Forbush, 1984), and measurement of transient Na–transport currents in planar lipid bilayers reconstituted with purified Na,K–ATPase (Fendler et al. 1985). Data from this work provide convincing evidence for correlation of ping pong transport of Na^+ and K^+ with E_1–E_2 transitions in the protein.

The scheme in Fig. 3 shows that Na^+ ions are occluded in both the E_1P[Na3] and E_2P[Na2] forms and the E_1P[Na3]–E_2P[Na2] transition is the charge translocating step. Dehydration of the cation and coordination in an occluded conformation are therefore considered to be principal conditions for translocation. Occlusion of Na^+ in the inward facing E_1–configuration of the cation sites is assumed to occur at the level of transition between cytoplasmic and intramembrane portions of the protein. As part of the transposition of mass accompanying E_1–E_2 transition, the cation binding sites move, but only over a fraction of the transmembrane length of the protein. The outward facing E_2–configuration of the cation sites is connected to the extracellular aqueous phases through an access mouth or channel. The cation binding sites are assumed to consist of electrophilic carbonyl groups which are located in a cavity to accommodate the cations inside the protein. Transition from open to occluded forms of the inward (E_1) or outward (E_2) facing configurations involves only limited changes in conformation of carbonyl residues and a substitution of solvent molecules from the inner coordination sphere of the cation (Jørgensen and Petersen, 1985). Dehydration of Na^+ takes place on the cytoplasmic surface and after coordination of the cation, the carbonyl groups rearrange from open to occluded form.

The essential part of the E_1–E_2 transition is the shift between inwardly exposed, occluded E_1–forms and outwardly exposed, occluded E_2–forms. At the extracellular mouth of the cation pathway the site shifts from closed to open configuration and Na^+ is rehydrated and released. The same set of coordinating groups, except phosphate, bind K from the outside. This considerations illustrate why the clarification of the structure and conformation of cation occluded forms of Na,K–ATPase may be a key to understanding the cation transport mechanism.

ACKNOWLEDGEMENTS

Work in the authors laboratory is part of the Biotechnology

Research Center for Membrane Proteins and supported by the Danish Medical Research Council and Novo's Foundation.

REFERENCES

Brown TA, Horowitz B, Miller RP, McDonough AA, Farley RA (1987) Molecular cloning of Na,K-ATPase β-subunit from dog kidney. Biochim. Biophys. Acta, in press

Dayhoff MO, Schwarts RM, Orcutt BC (1978) Atlas of Protein Sequence and Structure, Natl.Biomed.Res.Found.Washington DC,pp.345-363

Doolittle RF (1985) The genealogy of some recently evolved vertebrate proteins. TIBS 10: 233-237

Fendler K, Grell E, Haubs M, Bamberg E (1985) Pump currents generated by the purified Na,K-ATPase from kidney on black lipid membranes. EMBO J. 4: 3079-3085

Forbush B III (1987) Formation of a Na,K-ATPase·ouabain·^{22}Na complex. Biophys. J. 51: 179a.

Fry DC, Kuby SA, Mildvan AS (1985) ATP-binding site of adenylate kinase. Proc. Natl. Acad. Sci. USA 83: 907-911

Glynn IM (1985) The Na,K-transporting adenosine triphosphatase. In:The Enzymes of Biological Membranes, Ed.: A.N. Martonosi, Plenum Press, New York, Vol. 3, pp. 35-114

Hebert H, Jørgensen PL, Skriver E, Maunsbach,AB (1982) Crystallization patterns of membrane bound Na,K-ATPase. Biochim. Biophys. Acta 689:571-574

Hebert H, Skriver E, Maunsbach AB (1985) Three-dimensional structure of renal Na,K-ATPase from membrane crystals. FEBS Lett. 187: 182-187

Jørgensen PL (1982) Mechanism of the Na,K-pump: protein structure and conformations of the pure Na,K-ATPase. Biochim. Biophys. Acta 694:27-68

Jørgensen PL, Andersen JP, (1988) Structural basis for E_1-E_2 conformational in Na,K-pump and Ca-pump proteins. J. Membr. Biol., in press

Jørgensen PL, Collins JH (1986) Tryptic and chymotryptic cleavage sites in sequence of α-subunit of Na,K-ATPase from outer medulla of mammalian kidney. Biochim. Biophys. Acta 860: 570-576

Jørgensen PL, Petersen J (1985) Chymotryptic cleavage of α-subunit of renal Na,K-ATPase: effects on enzymatic properties, ligand binding and cation exchange. Biochim. Biophys. Acta 821: 319-333

Karlish SJD (1980) Characterization of conformational changes in Na,K-ATPase labeled with fluorescein at the active site. J. Bioenerg. Biomembr. 12:111-136

Karlish SJD, Lieb WR, Stein WD (1982) Combined effects of ATP and phosphate on rubidium exchange mediated by Na,K-ATPase reconstituted into phospholipid vesicles. J. Physiol. 328:333–350

Kawakami K, Noguchi S, Noda M, Takahashi H, Ohta T, Kawamura M, Nojima H, Nagano K, Hitose T, Inayama S, Hayashida H, Miyata T, Numa S (1985) Primary strucrture of α-subunit of Torpedo californica Na,K-ATPase. Nature 316:733–736

Kawakami K, Ohta T, Nojima H, Nagano K (1986a) Primary structure of α-subunit of human Na,K-ATPase. J. Biochem. 100:389–397

Kawakami K, Nojima H, Ohta T, Nagano K (1986b) Molecular cloning and sequence analysis of human Na,K-ATPase β-subunit. Nucleic Acids Research 14:2833–2844

Ohta T, Nagano K, Yoshida M (1986a) Active site structure of Na,K-ATPase: location of 5-p-fluorosulfonyl-benzoyl-adenosine binding site. Proc. Natl. Acad. Sci. USA 83: 2071–2075

Ovchinnikov YA, Modyanov NN, Broude NE, Petrukhin KE, Grishin AV, Arzamazova NM, Aldanova NA, Monastyrskaya GS, Sverdlov ED (1986) Pig kidney Na,K-ATPase, primary structure and spatial organization. FEBS Lett. 201:237–245

Rephaeli A, Richards D, Karlish SJD (1986) Conformational transitions in fluorescein-labeled Na,K-ATPase reconstituted into phospholipid vesicles. J. Biol. Chem. 261: 6248–6254

Shull GE, Greeb J, Lingrel JB (1986) Molecular cloning of three distinct forms of Na,K-ATPase α-subunit from rat brain. Biochemistry 25:8125–8132

Shull GE, Lane LK, Lingrel JB (1986) Aminoacid sequence of the β-subunit of Na,K-ATPase. Nature 321:429–431

Shull GE, Schwartz A, Lingrel JB (1985) Aminoacid sequence of the catalytic subunit of Na,K-ATPase. Nature 316:691–698

Shull MM, Lingrel JB (1987) Multiple genes encode the human Na,K-ATPase α-subunit. Proc. Natl. Acad. Sci. US, in press

Sverdlov ED, Monarstyrskaya GS, Broude NE, Yshkaryov YA, Allikmets RL, Melkov AM, Smirnov YV, Malyshev IV, Dulobova IE, Petrukhin KE, Grishin AV, Kijatkin NI, Kostina MB, Sverdlov VE, Modyanov NN, Ovchinikov YA (1987) The family of human Na,K-ATPase genes. Febs Lett. 217:275–278

Tonomura Y, (1986) Energy transducing ATPases-Structure and Kinetics, Cambridge Univ. Press. London

Walker JD Saraste M, Ronswick MJ, Gay, NJ (1982) Sequences of ATP-requiring enzymes and a common nucleotide binding fold EMBO J. 1:945–951

Vilsen B, Andersen JP, Petersen J, Jørgensen PL (1987) Occlusion of ^{22}Na and ^{86}Rb in membrane-bound and soluble protomeric αβ-units of Na,K-ATPase. J. Biol. Chem. 262: 10511–10517

The Ion Pumps: Structure, Function,
and Regulation, pages 33–38

OCCLUSION OF $^{22}Na^{+}$ AND $^{86}Rb^{+}$ IN MEMBRANE-BOUND AND SOLUBLE PROTOMERIC αβ-UNITS OF Na,K-ATPase

Bente Vilsen, Jens P. Andersen, Janne Petersen
and Peter L. Jørgensen

Institute of Physiology, University of Aarhus,
8000 Aarhus C, Denmark

INTRODUCTION

In the process of active Na,K-transport, the cations pass states, in which they are trapped in the pump protein, exchanging with medium cations only at a very low rate. Post et al. (1972) originally suggested the term "occlusion" for the state of Rb^{+} in Na,K-ATPase after dephosphorylation. Beauge and Glynn (1979) were able to measure occlusion of K^{+} or Rb^{+} using Dowex columns for removal of freely exchangeable ions. Kinetic studies show that the Rb^{+}-occluded form of the protein is an energy well from which exit in either direction is slow (Karlish and Stein, 1982). The direction of movement of the Rb^{+} ions out of the occluded state is controlled by physiological ligands (Karlish et al., 1982).

Occlusion of Na^{+} has been demonstrated after chymotryptic cleavage (Glynn et al., 1984) and after treatment with oligomycin (Esmann and Skou, 1985). Both procedures stabilize the Na,K-pump protein in the E_1P conformation in presence of ATP.

If a cavity for occlusion of cation is part of the pathway for transport of Na^{+} and K^{+} across the membrane, the minimum protein unit required for occlusion defines the unit that contains this cation pathway. Previously, it was shown that the soluble monomeric αβ-unit possesses Na,K-ATPase activity (Brotherus et al.,1983), and that it can assume three (E_1Na, E_2K and E_2P) of the four major conformational states in the E_1-E_2 reaction cycle of the Na,K-pump (Jørgensen and Andersen, 1986). It was, however, not possible to stabilize the E_1P conformation in the soluble state, and thus determine if soluble αβ-units can assume the E_1P form or if this involves association between αβ-units.

The purpose of this investigation has been to examine whether soluble monomeric $\alpha\beta$-units can occlude Rb^+ and Na^+ or if this requires formation of oligomeric $(\alpha\beta)_2$-units. For occlusion of $^{86}Rb^+$, we used established procedures (Glynn and Richards, 1982). The Dowex technique was also used for occlusion of $^{22}Na^+$ and the phosphorylated complex was stabilized by use of CrATP as substrate (Pauls et al., 1980). The optimum conditions for occlusion of $^{22}Na^+$ in membrane-bound and soluble Na,K-ATPase were determined, and the sedimentation properties of the Rb^+-or Na^+-occluded phosphorylated complexes were studied in the analytical ultracentrifuge.

RESULTS AND DISCUSSION

Na^+ Occlusion

Reaction of Na,K-ATPase with CrATP leads to occlusion of Na^+ ions. Fig. 1A shows the amount of occluded Na^+ as a function of Na^+ concentration for membrane-bound and soluble Na,K-ATPase. The data obtained with membranous Na,K-ATPase have been fitted to Hill's equation, resulting in a $K_{\frac{1}{2}}$ of 0.95 mM, a maximum site concentration of 12.9 nmol of occluded Na^+/mg of protein, and a Hill number of 1.5. Optimum conditions for occlusion of Na^+ were found to be 3 hours incubation with 0.5 mM CrATP in presence of 1 mM Mg^{2+} and absence of K^+. Thus, Na^+-occlusion is a slow process in accordance with previous data demonstrating a slow rate of phosphorylation of Na,K-ATPase by CrATP (Pauls et al., 1980).

In Fig.1B, we have determined the rate of release of occluded Na^+ from membrane-bound and soluble Na,K-ATPase by varying the rate of passage through the Dowex column. The occluded ions are released from the soluble enzyme at a slow rate, that is similar to the rate observed in membranous Na,K-ATPase and comparable to the rate measured by Glynn et al. (1984) after modification with chymotrypsin. In Fig.1B we have used preparations with particularly high specific activities of Na,K-ATPase (46.3 µmol/mg/min) in order to determine the maximum level of Na^+ occlusion. Extrapolation of the lines to zero time permits correction for loss of Na^+ during the passage down the Dowex column. This results in maximum values of 10.8 and 8 nmol Na^+ occluded/mg protein for membrane-bound and soluble Na,K-ATPase, respectively. Previously, evidence of occlusion of Na^+ has been incomplete due to difficulties involved in isolation of the ADP-sensitive phosphoenzyme. Maximum values of only 2-3 nmol of occluded Na^+/mg of protein have been measured on enzyme pretreated with either α-chymotrypsin or with N-ethylmaleimide (Glynn et al., 1984). Using CrATP as phosphorylating substrate, thus enabled us to measure 3-4 times higher values of Na^+ occlusion. With soluble enzyme we obtained 10-25 % lower maximum values of Na^+ occlusion than with membrane-bound Na,K-ATPase. Studies of Na,K-ATPase activity have shown that this is due to irreversible inactivation occuring upon solubilization at low cation concentrations (Vilsen et al., 1987).

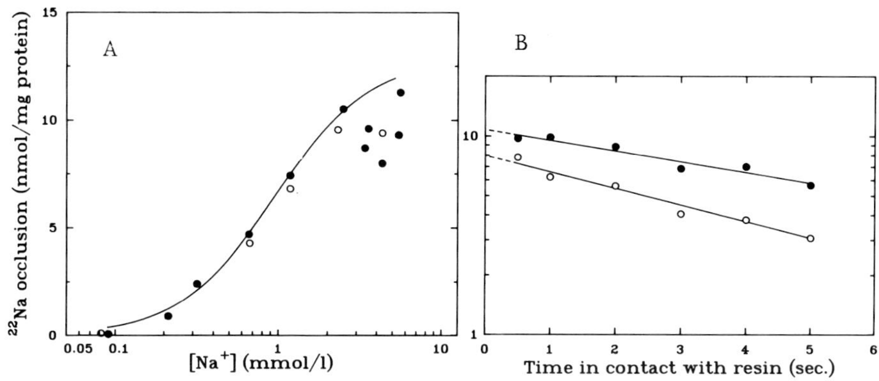

Figure 1. CrATP induced Na^+ occlusion in membrane-bound (●) and soluble Na,K-ATPase (○). (A) Na^+ concentration dependence. The Na^+ occluded complex was formed by incubation of purified Na,K-ATPase in membrane-bound form with $^{22}Na^+$ and CrATP for 3 hours. Solubilization with $C_{12}E_8$ (3 mg/mg protein) was performed 5 min. prior to termination of the incubation. (B) Release of occluded Na^+.

Determination of Stoichiometry

This requires at least three corrections of the measured values of Na^+ occlusion. The first correction is for the loss of Na^+ that occurs during the passage of the enzyme down the ion exchange column (see above). The second correction is for lack of saturation of Na^+-binding sites (Fig. 1A). The third correction is for variation of enzyme purity as indicated by the specific Na,K-ATPase activity (for details see Vilsen et al., 1987). Our data yield stoichiometries ranging from 1.8 to 2.7 mol of Na^+/mol of EP. The latter value is obtained after the three corrections of the measured values mentioned above. The value of 1.8 is a lower limit calculated directly from the measured numbers. There was no significant difference between the stoichiometries obtained with soluble enzyme and membrane-bound Na,K-ATPase. On the basis of measured values for Na^+ occlusion and EP level which were 5-10 fold lower than those determined in the present study, Glynn et al. (1984) calculated a stoichiometry of 2.8 mol of occluded Na^+/mol of EP in chymotrypsin-cleaved enzyme. This result, as well as ours, agrees reasonably well with the accepted transport stoichiometry of 3 Na^+ ions transported per molecule of ATP hydrolyzed.

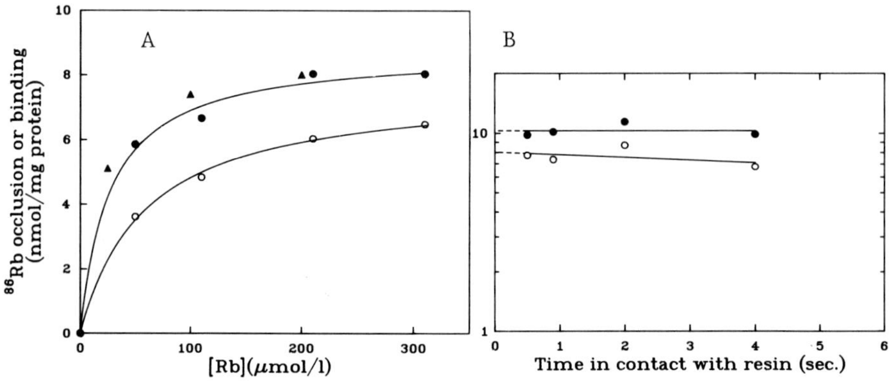

Figure 2. Rb^+ occlusion in membrane-bound (●,▲) and soluble (○) Na,K-ATPase. (A) Rb^+ concentration dependence. The Rb^+ occluded complex was formed by incubation purified Na,K-ATPase in membrane-bound form with $^{86}Rb^+$ for 15 min. Solubilization with $C_{12}E_8$ was performed as described in the legend to Fig.1A. (B) Release of occluded Rb^+.

Occlusion of Rb^+

The relation between the amount of Rb^+ occluded after passage of the enzyme down the ion exchange column and the Rb^+ concentration is shown in Fig. 2A for membrane-bound as well as for soluble Na,K-ATPase. Also shown in Fig. 2A are data representing total Rb^+ binding (▲) at equilibrium i.e. including freely exchangeable Rb^+. It is obvious that there is no significant difference between the levels of binding and occlusion, indicating that the sites involved in Rb^+ occlusion are not distinct from those of Rb^+ binding. Fig. 2B shows that the rate of deocclusion of Rb^+ is extremely slow ($k<0.030\ s^{-1}$) for both membrane-bound and soluble enzyme. The extrapolated initial values are 10.3 and 8.0 nmol of occluded Rb^+/mg of protein for membrane-bound and soluble enzyme, respectively (specific activity 46.3 µmol/mg/min; as for the enzyme used in Fig.1B). After appropriate corrections we obtain a stoichiometry of 2.0 and 2.4 mol of Rb^+/mol of enzyme capable of forming phosphoprotein for membranous Na,K-ATPase and soluble enzyme, respectively. Our results on Rb^+ occlusion indicate that membrane-bound as well as soluble Na,K-ATPase in the dephosphoform is capable of occluding 2 mol of Rb^+/mol of active site. This agrees with the transport stoichiometry of 2 Rb^+ or K^+ions transferred across the membrane in each transport cycle (Glynn and Karlish, 1975; Sen and Post, 1964). In contrast Glynn and Richards (1982) have found a stoichiometry of 3 Rb^+ ions occluded per phosphorylation site for membrane-bound enzyme.

Self-association state of soluble Na,K-ATPase

In the analytical ultracentrifuge, the soluble Na,K-ATPase in the phosphorylated Na^+ occluded form as well as in the Rb^+ occluded form was found to have a sedimentation coefficient ranging between 6.8 and 7.4 S. It was found to be independent of the duration of incubation with CrATP. As previously documented, this range of sedimentation coefficients is consistent with the protein existing predominantly as αβ-units with molecular weights in the range 143,000-170,000 (Jørgensen and Andersen, 1986).

Concluding Remarks

The αβ-unit can assume the dephosphoforms E_1Na and E_2K (Jørgensen and Andersen, 1986), and this paper shows that it occludes 2 Rb^+ ions. The soluble monomeric αβ-unit also assumes the phosphorylated E_2P form (Jørgensen and Andersen, 1986), and the present data demonstrate that both in membranous and soluble monomeric Na,K-ATPase, phosphorylation from CrATP stabilizes a Na^+ occluded conformation reminiscent of the E_1P form. Thus, it is possible to conclude that the soluble αβ-unit can assume all of the four major conformation that Na,K-ATPase must pass to carry out ion translocation in the E_1-E_2 reaction cycle. This together with the demonstration that soluble αβ-units of the Na,K-ATPase occlude Na^+ and Rb^+ with the same stoichiometry and kinetics as membrane-bound enzyme indicate that occlusion is not due to segregation of the ions in a channel-like compartment formed between subunits in an $(\alpha\beta)_2$-dimer, neither is it due to movement of ion-binding sites into the bilayer structure, since the latter has been broken down by formation of mixed micelles of lipid and detergent in the solubilization process. The present data suggest that the ions are occluded in a presumptive "cavity" that is formed within the structure of the αβ-unit.

REFERENCES

Beauge LA, Glynn IM (1979). Occlusion of K ions in the unphosphorylated sodium pump. Nature 280:510-512.

Brotherus JR, Jacobsen I, Jørgensen PL (1983). Soluble and enzymatically stable $(Na^{+}+K^{+})$-ATPase from mammalian kidney consisting predominantly of protomer αβ-units. Biochim Biophys Acta 731:290-303.

Esmann M, Skou JC (1985). Occlusion of Na^{+} by the Na,K-ATPase in the presence of oligomycin. Biochem Biophys Res Commun 127:857-863.

Glynn IM, Hara Y, Richards DE (1984). The occlusion of sodium ions within the mammalian sodium-potassium pump:its role in sodium transport. J Physiol 351:531-547.

Glynn IM, Richards DE (1982). Occlusion of rubidium ions by the sodium-potassium pump:its implications for the mechanism of potassium transport. J Physiol 330:17-43.

Jørgensen PL, Andersen JP (1986). Thermoinactivation and aggregation of αβ Units in soluble and Membrane-bound (Na,K)-ATPase. Biochemistry 25:2889-2897.

Karlish SJD, Lieb WR, Stein WD (1982). Combined effects of ATP and phosphate on rubidium exchange mediated by Na-K-ATPase reconstituted into phospholipid vesicles. J Physiol 328:333-350.

Karlish SJD, Stein WD (1982). Effects of ATP or phosphate on passive rubidium fluxes mediated by Na-K-ATPase reconstituted into phospholipid vesicles. J Physiol 328:317-332.

Pauls H, Bredenbrocker B, Schoner W (1980). Inactivation of $(Na^{+}+K^{+})$-ATPase by Chromium (III) complexes of nucleotide triphosphates. Eur J Biochem 109:523-533.

Post RL, Hegyvary C, Kume S (1972). Activation by adenosine triphosphate in the phosphorylation kinetics of sodium and potassium ion transport adenosine triphosphatase. J Biol Chem 247:6530-6540.

Sen AK, Post RL (1964). Stoichiometry and localization of adenosine triphosphate-dependent sodium and potassium transport in the erythrocyte. J Biol Chem 239:345-352.

Vilsen B, Andersen JP, Petersen J, Jørgensen PL (1987) Occlusion of ^{22}Na or ^{86}Rb in membrane-bound and soluble protomeric αβ-units of Na,K-ATPase. J Biol Chem 262: 10511-10517.

The Ion Pumps: Structure, Function,
and Regulation, pages 39–44

IDENTIFICATION AND LOCALIZATION OF THE 5-IODOACETAMIDOFLUORESCEIN REPORTER SITE ON DOG KIDNEY Na,K-ATPase

M. Steinberg[1,2], P.A. Tyson[2], E.T. Wallick[3], and T.L. Kirley[3]
[2]SUNY Health Science Center at Syracuse, Syracuse NY, 13210 and [3]U. of Cincinnati College of Medicine, Cincinnati, OH 45267

INTRODUCTION

A fluorescent derivative of iodoacetamide, 5-iodoacetamidofluorescein (5-IAF) has been shown to label the catalytic subunit of the Na,K-ATPase without loss of enzymatic activity (Kapakos and Steinberg, 1982). The fluorescence signal of the enzyme-bound probe decreases with K and increases with Na or ATP. These changes are conformationally dependent, in that K stabilizes E_2, and Na or ATP, E_1. Due to the unique properties of this probe, it was of interest to determine if the labeling is specific, i.e., whether a single labeled residue functions as a reporter site. If this is the case, then knowledge of the location of the labeled amino acid in the sequence would be a valuable tool in structural studies. In the course of our work with 5-IAF, it became clear that the stoichiometry of labeling was 13-20 nmoles mg^{-1}, indicating that perhaps 2-3 residues per catalytic subunit were modified. In this report, we define conditions used to increase the specificity of 5-IAF labeling; in addition, the cysteine residue modified by 5-IAF to act as a reporter site is identified. A preliminary report of this work has appeared elsewhere (Tyson and Steinberg, 1987).

METHODS

Na,K-ATPase was purified from frozen dog kidneys

[1]Publishing before 1981 as M.S. Flashner

essentially as described by Jorgensen (1974). Specific activity was 8-12 µmol P_i released min^{-1} mg $protein^{-1}$. Treatment with iodoacetate was performed according to Castro and Farley (1979): 3-4 mg/ml protein was incubated with 25 mM iodoacetate in 100 mM Na phosphate buffer, pH 7.0, in the dark for 30 min at room temperature. Excess iodoacetate was removed before fluorescent labeling by passage over a Sephadex G-50 centrifuge column equilibrated with 25 mM imidazole-HCl buffer at pH 7.5. Labeling with 5-IAF was done essentially as described by Kapakos and Steinberg (1982), except that the reaction was for 4 hours at room temperature. Fluorescence was measured with a Perkin-Elmer MPF-66 fluorescence spectrophotometer interfaced to a model 7500 computer. Excitation was at 495 nm and emission was monitored at 520 nm, with 5 nm slits.

IAF-labeled Na,K-ATPase (1.7 mg/ml) was digested with TPCK-trypsin at a ratio of 1:100 trypsin:ATPase (w/w) for 2 hrs at 37°C; additional trypsin was added after the first hour, and the reaction was stopped with 0.5 mM PMSF. The soluble peptide fragments released by this treatment were separated from the membrane by centrifugation. The fluorescent peptides were purified using an LDC-Milton Roy computer-driven HPLC system as described by Kirley, et al. (1984). The purified fluorescent fractions were then sequenced on an Applied Biosystems 470A gas phase Protein Sequenator.

RESULTS AND DISCUSSION

Pretreatment with Iodoacetate

Estimates of binding stoichiometry based on absorbance at 492 nm indicates that there are 2 to 3 moles of 5-IAF bound per mole of αβ protomer. To increase the specificity of 5-IAF labeling, Na,K-ATPase was treated with iodoacetate before fluorescent labeling to block nonspecific sulfhydryl modification. Like 5-IAF, iodoacetate does not inhibit enzymatic activity.

Pretreatment of the enzyme with 25 mM iodoacetate for 30 min results in 1) a decrease in binding stoichiometry of 5-IAF from 15.5 to 8.8 nmoles mg^{-1}; 2) a decrease in the fluorescent intensity per mg protein of about one third; and 3) an enhancement of the conformationally dependent signals of up to 50% (Figure 1).

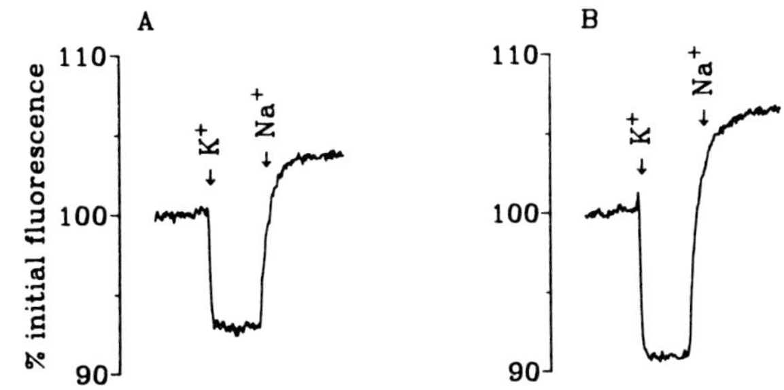

Figure 1. Fluorescence responses of 5-IAF labeled Na,K-ATPase. The fluorescence of enzyme (10-20 μg/ml) in 25 mM imidazole-HCl, 0.1 mM EDTA, pH 7.5 was measured at 25°C as described in METHODS. KCl (30 μM) or NaCl (1.0 mM) were added at the arrows as indicated. A, enzyme labeled with 5-IAF only; B, enzyme pretreated with iodoacetate before 5-IAF labeling.

These results suggest that iodoacetate reacts with one or more sulfhydryl residues which would normally be labeled by 5-IAF. Furthermore, these blocked residues are not reporter sites, since the remaining fluorescence signal has enhanced conformational responses as compared to the control.

Purification and Sequence Analysis of the Fluorescent Tryptic Fragments

Na,K-ATPase that had been labeled with 5-IAF after pretreatment with iodoacetate as outlined above was treated with trypsin as described in METHODS. The supernatant fraction after trypsinolysis was further treated with 10 mM iodoacetate for 10 min in order to prevent inter-peptide crosslinking through disulfide bond formation. After lyophilization, the tryptic supernatant was dissolved in 0.1% trifluoroacetic acid and applied to a reverse phase C_{18} column (Figure 2).

The fluorescence chromatogram shows six major fluorescent peaks, indicated as I through VI. The six peaks were collected and repurified on C_4 and phenyl columns. Of the six

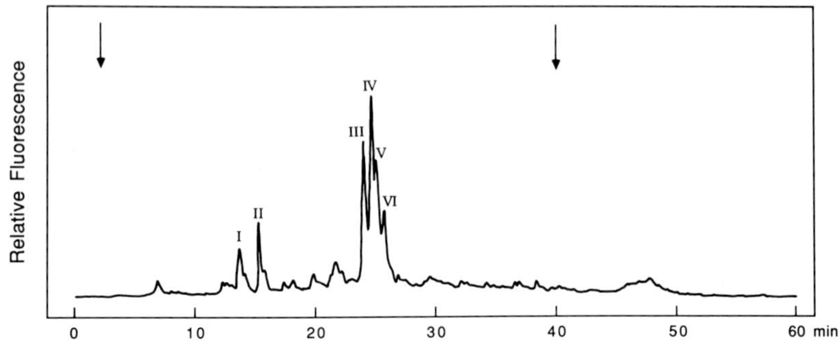

Figure 2. HPLC chromatogram of the supernatant resulting from 2 hrs of tryptic digestion of 15 mg dog kidney Na,K-ATPase sequentially treated with iodoacetate and 5-IAF. The arrows show the extrema of the 15-35% linear acetonitrile gradient. Fluorescence was monitored by exciting at 440 nm, with emission at 500-700 nm.

peptides, three (I,III,IV) yielded sequences (Table 1).

TABLE 1. Amino acid sequence analysis of peptides I, III and IV. The analysis was carried out as described in METHODS. Component X was not identified by the sequencing; and is assumed to be cys (see text).

Peptide	Sequence
I	- X -gly-ser-val-lys
III	X -ile-glu-leu- X - X -gly-ser-val-lys
IV	(cm)cys-ile-glu-leu-(cm)cys - X -gly-ser-val-lys

The three other peaks did not yield a sequence and may therefore be nonproteinaceous derivatives of 5-IAF. The residue designated as X is identified as cysteine based on the following: 1) cysteine is destroyed during the sequencing procedure, 2) alkyl halides have high sulfhydryl selectivity and 3) strong homology with known sequences is observed (see below). The sequence identified centers around the only vicinal cysteine residues in the alpha subunit of Na,K-ATPase. Peptide I appears to be a truncated form derived from non-specific cleavage between the vicinal cysteine residues. It is clear from peptide IV that the 5-IAF labeled cysteine is the second of the two vicinal cysteines, since the other cysteines are carboxymethylated either by the prior- or post-iodoacetate treatment. Peptide III could be labeled at any of the 3 cysteines in the sequence.

The decapeptide cys-ile-glu-leu-cys-cys-gly-ser-val-lys corresponds to residues 452 through 461 in the primary structure of the lamb α-subunit, with the substitution of leu for val_{455} (Table 2).

TABLE 2. Sequence comparison of the 5-IAF-labeled peptide with several Na,K-ATPases. The asterisk indicates the position of the IAF label. Sequences are aligned to show the maximum homology between peptides, and nonhomologous residues are underlined. Data from this work (dog); Shull et al., 1985 (lamb); Kawakami et al., 1986 (human); Ovchinnikov et al., 1986 (pig); Kawakami et al., 1985 (torpedo).

Source		Sequence (* above second cys)	
dog kidney		cys-ile-glu-leu-cys-cys*-gly-ser-val-lys	
lamb kidney	452	cys-ile-glu-<u>val</u>-cys-cys-gly-ser-val-lys	461
human (HeLa)	457	cys-ile-glu-leu-cys-cys-gly-ser-val-lys	466
pig kidney	452	cys-ile-glu-leu-cys-cys-gly-ser-val-lys	461
torpedo	459	cys-ile-glu-leu-cys-cys-gly-ser-val-<u>ser</u>	468

This region is highly conserved among the Na,K-ATPases which have been sequenced. It is located in the putative main cytoplasmic domain, between the phosphorylation site and the fluorescein isothiocyanate site.

It is noteworthy that a site has been identified on the sarcoplasmic reticulum Ca-ATPase that is labeled by both iodoacetamide and IAEDANS. This site is at the first of the only vicinal cysteine residues in that sequence (Yoshita and Kawakita, 1987).

This research was supported by NSF Grant DM8602627

and USPHS Grant NS05430 (to MS), a Pharmaceutical Manufacturers Association Foundation Advanced Pre-doctoral Fellowship (to PAT), and NIH Grant #P01 HL 22619 [CORE 2] (to TLK and ETW).

REFERENCES

Castro, J and Farley, RA (1979). Proteolytic fragmentation of the catalytic subunit of the sodium and potassium adenosine triphosphatase. J Biol Chem 254:2221-2228.

Jorgensen, PL (1974). Purification and Characterization of (Na+K)-ATPase. Biochim Biophys Acta 356:36-52.

Kapakos, JG and Steinberg, M (1982). Fluorescent labeling of (Na^{+}+K^{+})-ATPase by 5-iodoacetamidofluorescein. Biochim Biophys Acta 693:493-496.

Kawakami, K, Noguchi, S, Noda, M, Takahashi, H, Ohta, T, Kawamura, M, Nojima, H, Nagano, K, Hirose, T, Inayama, S, Hayashida, H, Miyata, T, and Numa, S (1985). Primary structure of the α-subunit of *Torpedo california* (Na^{+}-K^{+})ATPase deduced from cDNA sequence. Nature 316:733-736.

Kawakami, K, Ohta, T, Nojima, H, and Nagano, K (1986). Primary structure of the α-subunit of human (Na^{+}-K^{+}) ATPase deduced from cDNA sequence. J Biochem 100:389-397.

Kirley, TL, Wallick, ET, and Lane, LK (1984). The amino acid sequence of the fluorescein isothiocyanate reactive site of lamb and rat kidney Na^{+} and K^{+}-dependent ATPase. Biochem Biophys Res Comm 125:767-773.

Ovchinnikov, YuA, Modyanov, NN, Broude, KE, Petrukhin, KE, Grishin, AV, Arzamazova, NM, Aldanova, NA, Monastyrskaya, GS, and Sverdlov, ED (1986). Pig kidney (Na^{+}-K^{+})ATPase. Primary structure and spatial organization. FEBS Letters 201:237-245.

Shull, GE, Schwartz, A, and Lingrel, JB (1985). Amino-acid sequence of the catalytic subunit of the (Na +K)-ATPase deduced from a complementary DNA. Nature 316;691-695.

Tyson, PA and Steinberg, M (1987). Characterization of the 5-iodoacetamidofluorescein (5-IAF) binding site on the Na,K-ATPase. The Pharmacologist 29:222.

Yamashita, T and Kawakita, M (1987). Reactive sulfhydryl groups of sarcoplasmic reticulum ATPase. J Biochem 101:377-385.

The Ion Pumps: Structure, Function,
and Regulation, pages 45–50

IDENTIFICATION OF TWO CATALYTIC SUBUNITS OF NA^+/K^+ ATPase IN RAT VENTRICLE.

D. Charlemagne, M. Poyard, P. Oliviero, and
E. Mayoux.
INSERM U 127, Hôpital Lariboisière, 41 Bd
de la Chapelle 75010 Paris, France.

INTRODUCTION

In adult rat heart, the inotropic effect of ouabain has been demonstrated to be related to a high-and a low-affinity component (Lelièvre et al, 1986). In vitro studies on rat heart sarcolemmal preparations have shown the presence of high- and low- affinity ouabain binding sites related to a biphasic dose response curve of the inhibition by ouabain of the Na^+/K^+ ATPase (Erdmann et al 1980, Mansier and Lelièvre, 1982) (Table 1).

However, SDS gel electrophoresis of the sarcolemmal membranes allowed the detection of only one form of the α catalytic subunit of the Na^+/K^+ ATPase (Charlemagne et al, 1986). According to the presence of at least three m-RNAs in rat brain, encoding for three distinct chains (Shull et al 1986), the question arises whether the single band detected so far could be the comigration of two or more catalytic subunits. The denaturation of the sarcolemmal proteins by reduction and alkylation show that two distinct forms of the α catalytic subunit of the Na^+/K^+ ATPase are resolved by SDS gel electrophoresis in rat heart. These two forms originate from cardiomyocytes (Charlemagne et al, 1987).

TABLE 1

Characteristics of the two sites of Na^+/K^+ ATPase in rat heart.

		High affinity 10^{-8} M	Low affinity 10^{-5} M
Ouabain binding	Kd	2,9	1,8
	% HA/LA	10-20	80-90
Na/K ATPase inhibition	Ic50	2	0,7
	% HA/LA	40-60	60-40
Inotropic effect	Ec50	2	2
	% HA/LA	25-40	60-75
Isoform		S	F

METHODS

- Rat cardiac myocytes were isolated according to Eckel et al (1983).
- Sarcolemmal vesicles from rat left ventricle and from isolated myocytes were purified by the method of Mansier et al (1983).
- Microsomal fractions from rat kidney and brain were partially purified according to Sweadner (1979).
- SDS gel electrophoresis were performed according to Laemmli as modified by Sweadner (1979) on 6 % acrylamide gel (8 x 5 cm-Bio Rad Protean II). After solubilisation in 2 % SDS, 10 mM DTT, 10 mM EDTA, Tris-HCl pH 6.8 for 30 min at 37°C, electrophoresis was run at a constant power of 3 watts.
- Reduction and alkylation treatments were carried out on membrane proteins (50 ug) sedimented by centrifugation for 1 min at 100, 000 x g (Airfuge, Beckman Instrument). The sediment was treated for 15 min at 37°C with DTT (10-100 mM) in 8M urea, 150 mM Tris-HCl pH 8.8, 0,1 mM EDTA, at a concentration of 3 mg of protein/ml. Alkylation was performed with iodoacetamide (0.12 M) for 30 min at 37°C. The pH was adjusted to 6.8 and SDS was added for 30 min at 37°C.

Specific immunological detection of α and β subunits of Na^+/K^+ ATPase was performed on proteins electrophoretically transferred to a nitrocellulose sheet after electrophoresis. Blots were saturated in PBS, 5 % low fat milk for one hour at 37°C then incubated with anti-lamb Kidney Na^+/K^+ ATPase antiserum (1/1000), rabbit anti IgG coupled to biotin (1/500 Amersham) and streptavidin-peroxydase (1/300, Amersham) in PBS 0.5 % low fat milk, 0.05 % Tween 20, each for one hour at 37°C. Between incubations, nitrocellulose sheets were washed four times (30 min). Revelation was performed in 100 ml of 20mM Tris-HCl pH 7.5, 0.5 M NaCl, with 60 mg of 4-chloro-1-naphthol in methanol.

RESULTS

Under limited denaturation conditions (solubilisation in SDS 2 % and reduction by DTT), the α catalytic subunit of the Na^+/K^+ ATPase appeared as a single band of 96 KD in kidney, heart and myocytes and as two bands of 96 KD and 100 KD in neonatal brain (Fig. 1, lanes 1).

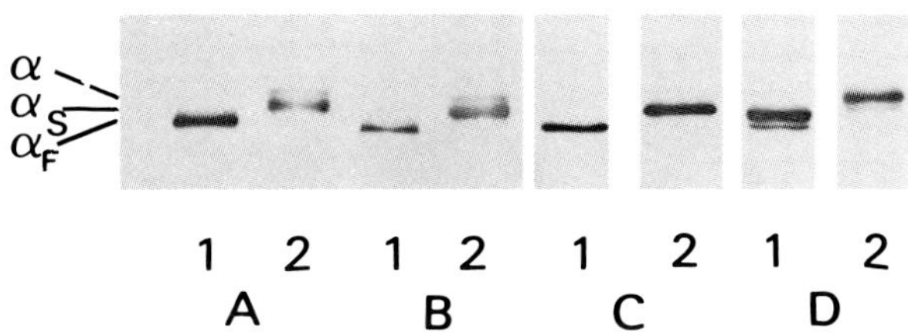

Figure 1. Comparison of the α catalytic subunits of Na^+/K^+ ATPase from rat heart (A), isolated myocytes (B) kidney (C), neonatal brain (D). Samples were treated with SDS (lanes 1) or reduced and alkylated before SDS solubilisation (lanes 2).

As shown in Fig. 1, lanes 2, the enhanced denaturation of the protein samples by reduction of the disulfide bonds and alkylation of the sulfhydryl groups allowed the separation of two bands for the catalytic subunits of Na^+/K^+ ATPase from rat heart, myocytes and brain whereas only one component was observed in the kidney enzyme.

The apparent molecular weight of the α subunits was increased by this treatment. Under reduction and alkylation the Mr 96,000 subunit from kidney migrated as Mr 100,000 (α_F) and the Mr 96,000 from rat heart migrated as Mr 100,00 (α_F) and 103,000 (α_S).

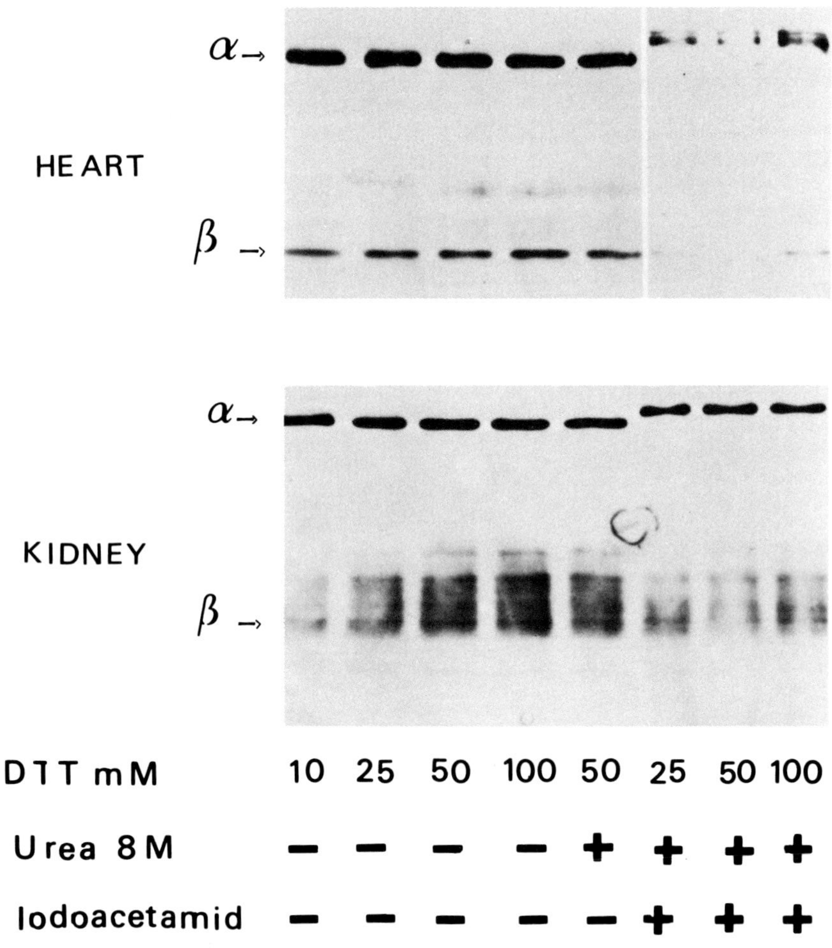

Figure 2. Denaturation of the α and β subunits of Na^+/K^+ ATPase in heart (A) and kidney (B). Effect of urea, DTT and iodoacetamid.

As shown in Fig 2, reduction even at high concentration of DTT and denaturation in 8M urea did not influence the mobility of the α subunits in kidney or heart. Alkylation was necessary to resolve the two subunits from rat heart whereas under the same conditions and even at 100 mM DTT only one subunit was observed in kidney. Moreover, this treatment was only effective when the protein concentration was less than 4.5 mg of protein/ml. Note that the mobility of the β subunit was not influenced by this treatment.

It seems likely that the α_F form was that of the low affinity form from kidney (Matsuda et al, 1984). Further support for this conclusion arises from the observation that the α^+ form is predominant in neonatal rat brain (Fig. 1D) as previously shown by Matsuda et al (1984) and that the α_S form is also predominant in neonatal rat brain under our experimental conditions. Thus this α_S form might be the ouabain sensitive α^+ form.

CONCLUSION

Reduction and alkylation of the sarcolemmal proteins allow the electrophoretic detection of two forms of the catalytic subunit of the Na^+/K^+ ATPase from rat heart myocytes.

Due to a higher denaturation than with SDS only, the apparent mobility of the α forms, from brain, kidney and heart is reduced by this treatment. Although increased from Mr 96,000 to Mr 100,000 the molecular weights determined here are far below their exact values ($\simeq$ 112, 000) according to the amino-acid sequence deduced from their cDNA clones (Shull 1986). Separation of the two isoforms might be related to known differences in the reactivity of the sulfhydryl groups (Matsuda, 1984) and to the resulting unfolding of the isoforms.

It seems likely that the α_F form (the only one present in kidney) is the form of low affinity for ouabain while the α_S form is the form of high affinity for ouabain. As shown in Table 1, these two forms well account for

i) The high and low affinity ouabain binding sites.

ii) The biphasic dose response curve of inhibition by ouabain of the Na^+/K^+ ATPase.

iii) The two components of inotropic responsiveness to ouabain of rat heart.

REFERENCES

Charlemagne D, Maixent JM, Preteseille M, and Lelièvre LG (1986). Ouabain binding sites and $(Na^{+} + K^{+})$-ATPase activity in rat cardiac hypertrophy : expression of the neonatal forms. J. Biol. Chem. 261:185-189.

Charlemagne D, Mayoux E, Poyard M, and Oliviero P, and Geering K (1987). Identification of two isoforms of the catalytic subunit of Na^{+},K^{+}-ATPase in myocytes from adult rat heart. J. Biol. Chem. 262: 8941-8943.

Eckel J, Pandalis G, and Reinfauer H. (1983) Insulin action on the glucose transport system in isolated cardiocytes from adult rat. Biochem. J. 212:385-392.

Erdmann E, Philipp G, and Scholz R (1980). Cardiac glycoside receptors, $(Na^{+}+K^{+})$-ATPase activity and force of contraction in rat heart. Biochem. Pharmacol. 29:3219-3229.

Lelièvre LG, Charlemagne D, Mouas C, and Swynghedauw B (1986). Respective involvements of high-and low-affinity digitalis receptors in the inotropic response of isolated rat heart to ouabain. Biochem. Pharmacol. 35:3449-3455.

Mansier P, Charlemagne D, Rossi B, Preteseille M, Swynghedauw B, and Lelièvre LG (1986). Isolation of impermeable inside-out vesicles from and enriched sarcolemma fraction of rat heart. J. Biol. Chem. 258:6628-6635.

Matsuda T, Heitaroh I, and Cooper JR (1984). Specific inactivation of (+) molecular forms of $(Na^{+}+K^{+})$-ATPase by pyrithiamin. J. Biol. Chem. 259:3858-3863.

Shull GE, Greeb J. and Lingrel JB (1986). Molecular cloning of three distinct forms of the Na^{+},K^{+} ATPase subunit from rat brain. Biochemistry, 25:8125-8132.

Sweadner KJ (1979). Two molecular forms of $(Na^{+}+K^{+})$- ATPase in brain. Separation and difference in affinity for strophantidin. J. Biol. Chem. 254:6060-6067.

The Ion Pumps: Structure, Function, and Regulation, pages 51–56

HIGH YIELD INCORPORATION OF THE NEUROSPORA PLASMA MEMBRANE H+ATPase INTO PROTEOLIPOSOMES : LIPID REQUIREMENT AND SECONDARY STRUCTURE OF THE ENZYME BY IR SPECTROSCOPY

Erik Goormaghtigh, Jean-Marie Ruysschaert and Gene A. Scarborough°
Université Libre de Bruxelles CP206/2, Service de Chimie Physique des Macromolécules aux Interfaces, B1050 Brussels. °Department of Pharmacology, School of Medecine, University of North Carolina , Chapel Hill NC27514, USA.

INTRODUCTION

Among the ATPases characterized by an aspartyl-phosphoryl-enzyme intermediate, the proton ATPase in the plasma membrane of *Neurospora crassa* is one of the most thoroughly studied. Recently we have shown that monomers of the ATPase are able to catalyze efficient proton translocation in asolectin liposomes (Goormaghtigh et al., 1986). Our previously described reconstitutions were carried out by using a large excess of lipid (i.e. 1 µg of purified enzyme to 2 mg of asolectin (Scarborough and Addison, 1984)). The complex composition of asolectin and the excess required to achieve the reconstitution preclude the use of spectroscopic techniques for studies of the structure of the ATPase so reconstituted, as well as studies of the role of the lipid phase on the ATPase activity. In the present communication, we describe a new procedure for reconstituting the proton ATPase that largely circumvents these problems. This procedure is then applied to study the lipid requirement for the ATPase activity and the structure of the protein in its lipid environment by infrared (IR) spectroscopy.

METHODS

- Preparation and assay of the ATPase were carried out as described (Smith and Scarborough, 1984)
- Reconstitution:Multilamellar liposomes were formed by

dispersion of 3 mg of phospholipids in 150 ul of buffer (10 mM Mes, 50 mM CH3COOK, pH 6.8 with Tris, or, for the IR experiments, 5mM Mes, 0.25% trehalose (w/v), pH 6.8 with Tris) + 4 µl of 0.1 M EDTA + 40 µl of 10% Na deoxycholate. This mixture was shaken until clear. 70 ug of purified ATPase (250 ug for the IR experiments) were then added and the reconstitution was achieved by passing the sample through a 27 cm long G50 Sephadex column (diameter=1.5 cm) at a flow rate of 2 ml/min.

- IR spectroscopy:The application of infrared Attenuated Total Reflection (ATR) spectroscopy to membrane research has been reviewed by Fringeli and Günthard (1981). Spectra were recorded with a Perkin-Elmer infrared spectrophotometer 983G. The internal reflection element was a Germanium ATR plate (50x20x2 mm) with an aperture angle of 45° yielding 25 internal reflections. Spectra were recorded at a nominal resolution of 2 cm-1. The ATPase-containing liposomes were spread and dried on a Germanium internal reflection element under a N2 stream and then rehydrated by flushing the chamber in which the dried sample was sealed with D2O-saturated N2 for 6 hours. Liposomes prepared in the same way but in the absence of the ATPase were subjected to an identical treatment and their contribution was subtracted from the spectra of the proteoliposomes prior to any subsequent treatment.

- Secondary structure determination : Fourier self-deconvolution was applied (Kauppinen et al, 1981) to increase the resolution of the spectra in the amide I' region which is the most sensitive to the secondary structure of proteins. In order to quantify the area of the different components of amide I' revealed by the self-deconvolution, a least square iterative curve fitting was performed to fit Lorentzian line shapes to the spectrum between 1700 and 1600 cm-1. The proportion of a particular structure was computed to be the sum of the area of all the fitted Lorentzian bands having their maximum in the frequency region where that structure occurs divided by the area of all the Lorentzian bands having their maximum between 1689 and 1615 cm-1. The frequency regions for the different structures are as follows : 1661-1647.5 cm-1 : alpha-helix; 1689-1682 + 1637.5-1615 cm-1 :beta-sheet; 1644.5-1637.5 cm-1 : "random"; 1682-1661 cm-1 : "turns". Both the "random" and "turns" structures were added and referred to as "unordered" structures.

RESULTS and DISCUSSION

When the ATPase was reconstituted as described in "Methods" using dimyristoylphosphatidylcholine (DMPC), dioleylphosphatidylcholine (DOPC) or egg PC, very little activity was recovered. However, addition of a negatively charged lipid yielded much higher specific activities (10-11 µmol/min/mg protein). The increase in the specific activity of the ATPase as a function of the DOPC/cardiolipin (CL) ratio is reported in Figure 1.

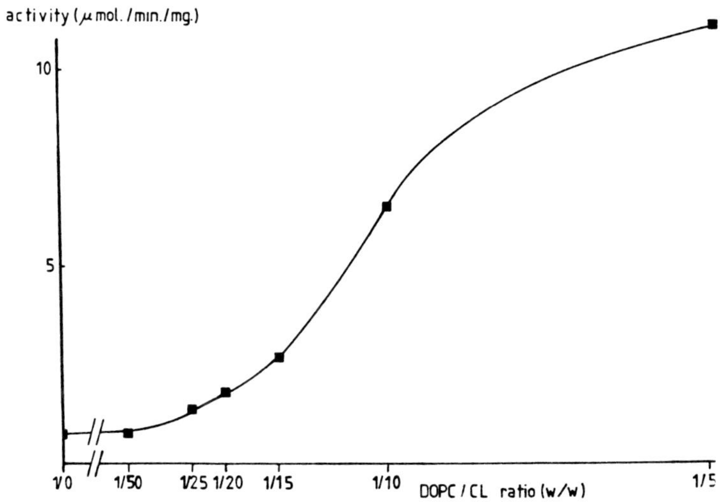

Figure 1 : Activity of the ATPase when reconstituted in the presence of dioleylphosphatidylcholine (DOPC) containing increasing amounts of cardiolipin (CL). The assays were carried out in the presence of nigericin (Scarborough and Addison, 1984)

Clearly, the negatively charged lipid is essential for full activity of the enzyme. It is worth noting that in the reconstitution procedure used, a random insertion of the ATPase molecules in the bilayer might occur, while in our previous reconstitutions (freeze-thaw) it was shown that at least 91% of the ATPase molecules were oriented with the ATP side toward the outside of the liposome (Scarborough and Addison, 1984). This could explain, at least in part, why the specific activity of the ATPase in these studies is about 1/2 to 1/3 of the specific activity measured previously. The altered lipid composition might also affect the ATPase specific activity. Importantly, the specific activity measured in the presence of nigericin is

3 times higher than the specific activity measured its absence (not shown). This indicates that at least 2/3 of the active ATPase molecules are properly inserted in the bilayer and able to pump protons towards the inner compartment of the liposome.

At this stage of the experiments it was still possible that the negatively charged CL was necessary for the insertion of the ATPase into the liposomes but not for its enzymatic activity. In order to test this hypothesis, we added increasing amounts of adriamycin (ADM) to the proteoliposomes (80% DOPC, 20% CL w/w). Figure 2 shows that an inhibition of the ATPase activity appears at ADM concentrations between 0.01 and 0.001 mM.

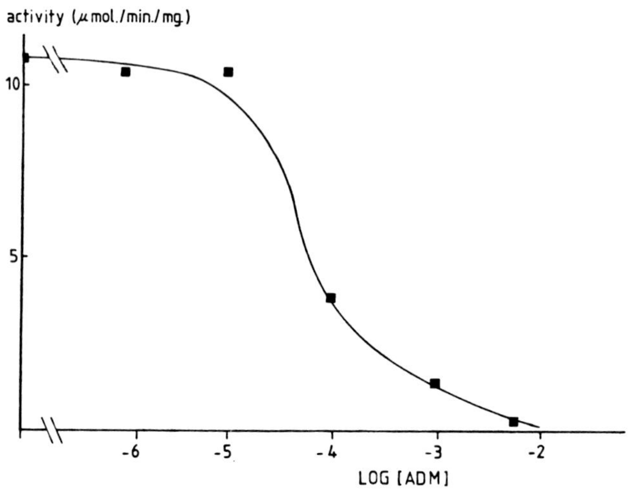

Figure 2 : Activity of the ATPase reconstituted in liposomes (80% DOPC, 20% CL (w/w)) measured in the presence of increasing amounts of adriamycin (ADM).

We have previously shown that ADM interacts specifically with CL, but not with neutral phospholipids, and that this interaction alone is responsible for the inhibition of CL-dependent enzymes such as mitochondrial cytochrome c oxidase or complex I-III. The inhibition of other enzymes by modification of their CL environment by ADM has also been shown (for a review see Goormaghtigh and Ruysschaert, 1984). The remarkable property of ADM to inhibit CL-dependent enzymes without interacting directly with the enzymes themselves indicates that the negatively charged lipid is indeed required for ATPase activity.

The secondary structure of the reconstituted ATPase was next studied by infrared spectroscopy. We first demonstrated that, provided that trehalose is present, full enzyme activity is retained when the proteoliposomes are dried under a nitrogen flow and then rehydrated by the addition of water (not shown). We were therefore able to prepare multibilayers spread on the surface of germanium crystals by dehydration of the samples (Fringeli and Günthard, 1981) followed by rehydration in the presence of D2O-saturated N2 without loss of enzyme activity. Spectra of the ATPase recorded under these conditions are reported at various stages of the deconvolution along with their curve fitting in Figure 3. The percentage of alpha-helix

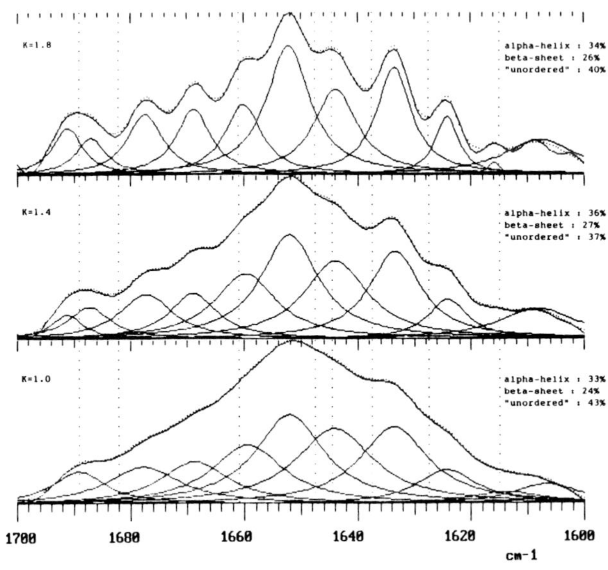

Figure 3 : IR spectrum of the ATPase in the amide I' region presented at different stages of the deconvolution. The resolution enhancement factor K has been defined by Kauppinen et al (1981). The vertical dotted lines delimit the frequency regions listed in "Methods". The dotted spectrum is the sum of the lorentzian bands. The input parameters for the curve fitting were determined for K=1.8 and reused for K=1.4 and K=1.0.

structure reported in Figure 3 (about 35%) is only slightly less than that reported for the sarcoplasmic reticulum calcium ATPase for which values between 38% and 53% have been reported (Williams et al, 1986; Csermely et al, 1987) but is in good agreement with the value of 36% determined by CD for the same enzyme (Hennessey and Scarborough, J. Biol. Chem. in the press). Additional data about the secondary structures of ATPases can be found in the paper by N.M. Green in this volume.

REFERENCES

Csermely P, Katopis C, Wallace BA, Martonosi A. (1987). The E1-E2 transition of Ca2+-transporting ATPase in sarcoplasmic reticulum occurs without major changes in secondary structure. Biochem. J. 241:663-669.

Fringeli UP , Günthard HH (1981). Infrared membrane spectroscopy. in "Membrane Spectroscopy" ed E Grell 270 -332, Springer-Verlag.

Goormaghtigh E, Chadwick C, Scarborough GA (1986). Monomers of the Neurospora plasma membrane H+ATPase catalyze efficient proton translocation. J. Biol. Chem. 261:7466 -7471.

Goormaghtigh E, Ruysschaert JM (1984). Anthracycline glycoside-membrane interactions. Biochim. Biophys. Acta 779:271-288.

Kauppinen JK, Moffat DJ, Cameron DG, Mantsch HH (1981). Noise in Fourier self-deconvolution. Applied Optics 20:1866-1879.

Scarborough GA, Addison R. (1984). On the subunit composition of the Neurospora plasma membrane H+ATPase. J. Biol. Chem. 259:9109-9114.

Smith R, Scarborough GA (1984). Large-scale isolation of the Neurospora plasma membrane H+ATPase. Anal. Biochem. 138:156-163.

Williams RW, McIntyre JO, Gaber BP, Fleischer S. (1986). The secondary structure of calcium pump protein in light sarcoplasmic reticulum and reconstituted in a single lipid component as determined by raman spectroscopy. J. Biol. Chem. 261:14520-14524.

The Ion Pumps: Structure, Function,
and Regulation, pages 57–66

PROTON-ATPases: UNIVERSAL CATALYSTS IN BIOLOGICAL ENERGY CONSERVATION.

G. Schäfer, J. Weber, H. Tiedge* and M. Lübben

Institute of Biochemistry, Medical University of Lübeck, 2400 Lübeck 1, (FRG), and *Center for Neurobiology and Behaviour, Columbia University New York, (USA).

The chemiosmotic principle of primary energy conservation has been recognized almost ubiquitously, using various mechanisms to store energy as an ion-gradient, normally of H^+. Consequently, the question arises whether or not the catalytic devices secondarily coupling the dissipation of this proton gradient to ATP-synthesis are universally appearing unique catalysts with regard to structure and function.

The best investigated examples of such H^+-ATPases are those from mammalian mitochondria, spinach-chloroplasts and escherichia coli. They are composed of a membrane residing proton-channel Fo, and a large peripheral protein complex F1. The latter typically contains five types of subunits α(55 kd), ß(51,5kd), γ(30kd), δ(15kd) and ε(5,6kd) (values of molecular mass: beef heart), and catalyzes either formation or hydrolysis of ATP, depending on the direction of proton flow across the membrane. In its solubilized form F1 is capable only of net ATP-hydrolysis.

Whereas a great deal of information has been accumulated on either of the constituent parts Fo and F1 (for review see Senior, and Wise, 1983; Vignais, and Satre, 1984; Hatefi, 1985; Pedersen, and Carafoli, 1987) the mechanism of coupling between proton transport and ATP-hydrolysis (resp.-synthesis) is still unknown. As a common denominator these F1-ATPases share some properties, discriminating them from other types of membrane bound ATPases, namely: the proton channel Fo is specifically inhibited by DCCD; the catalytic F1-moiety contains multiple copies of the major subunit polypeptides α and ß in a stoichiometry of 3:3; it exhibits multiple nucleotide binding sites with n=3 as a minimum number; it does not form a phopshorylated intermediate during the catalytic cycle.

From numerous labeling studies using covalently reacting nucleotide analogs agreement has been achieved, that all binding sites are located on or at interfaces of α- and ß-subunits (Schäfer, and Dose, 1984; Vignais, and Lunardi, 1985). Moreover, with the beef heart enzyme it has been shown that a total of six nucleotides can be bound simultaneously (Weber et al., 1985; Kironde FAS, and Cross RL, 1986). Out of these, three nucleotides interact with rapidly reversible sites and are exchanged during catalysis, ultimately leading to a functional model known as the "binding change mechanism" (Kayalar et al., 1977; Cross et al., 1984).

As documented by equilibrium binding and by kinetic studies, the three reversible sites do not behave equivalently under all conditions (Tiedge et al., 1982; Cross, and Nalin, 1982; Lunardi et al., 1986). Thus, taking into account that only one gene is coding for either the α- or the ß-polypeptides, respectively (Senior, and Wise, 1983) a conceptual paradoxon becomes immediately obvious: On a genetic basis each αß-pair, considered to comprise a catalytic unit in the oligomeric complex, should be functionally equivalent. On the other hand, an apparent functional asymmetry of the nucleotide binding domains suggests that these may be different. For an escape from this conflict it has to be solved, whether or not the three αß-pairs are permanently maintaining an invariant asymmetric structure; an assumption in direct contradiction to the idea of a perpetual and sequential involvement of three a priori equivalent catalytic centers as introduced by the "binding change mechanism"(Cross et al., 1984).

Our previous findings on nucleotide binding to beef heart F1 (Tiedge et al., 1982) are consistent with a model assuming strong negative cooperativity of binding between three a priori equivalent sites. This has been corroborated by kinetic studies on the mutual displacement of different nucleotide analogs from F1 (Tiedge, and Schäfer, 1986) and is perfectly in line with positive cooperativity of catalysis (Grubmeyer, and Penefsky, 1981; Penefsky, 1985).

The binding data of table 1 are illustrating that several nucleotides bind to the first accessible site of previously nucleotide depleted F1 with a higher affinity than to the remaining two reversible sites; others, however, can apparently occupy all three sites with equal affinity making them functionally indistinguishable. Therefore, a

Table 1: Typical dissociation constants of nucleotides at the three reversible binding sites of beef heart F1-ATPase.

Nucleotide	n_1=1	n_2=2
ADP +Mg2+	50-80 nM	1-3 µM
ADP -Mg2+	0.6-0.8 µM	8-10 µM
ATP -Mg2+	0.3-0.5 µM	7-10 µM
N-ADP +Mg2+	50-100 nM	50-100 nM
N-ATP +Mg2+	120 nM	120 nM
ANA-ADP +Mg2+	200 nM	200 nM
AMP-PNP +Mg2+	18 nM	1 µM
TNP-ADP +Mg2+	<< 1 nM	< 1 nM

(N- = 3'-O-naphthoyl-; ANA- = 3'-O-(5-azido-naphthoyl)-; TNP- = 3'-O-2,4,6-trinitro-phenyl-)

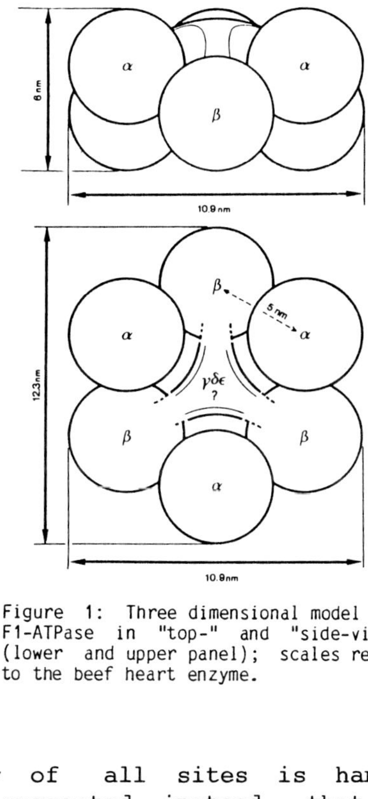

Figure 1: Three dimensional model of F1-ATPase in "top-" and "side-view" (lower and upper panel); scales refer to the beef heart enzyme.

permanent intrinsic asymmetry of all sites is hardly conceivable; it is rather suggested instead, that an asymmetry of binding is induced by the nature of the specific ligand. Though unlikely, such a conclusion might result accidently when exclusively drawn on the basis of thermodynamic binding data. However, careful examination of physical parameters using 3'-O-naphthoyl-AD(T)P as a fluorescent probe revealed, that no differentiation of its three binding sites with respect to their environment (polarity, mobility, accessibility, energy transfer between neighbouring groups) was possible, and no inhomogeneity of sites could be detected (Weber et al., 1987). Including available information from the literature these data provide further support to our earlier conclusion (Tiedge, and Schäfer, 1986) that known F1-ATPases expose three a priori "equipotent" reversible nucleotide binding domains which, however, become non-equivalent by an "induced asymmetry" imposed to the enzyme by cooperative binding of the nucleotides themselves. Obviously some nucleotide analogs are capable to overcome this cooperativity.

This conclusion is thought to apply also to F1-ATPases from other genera and species though certain distinctions -for example of regulatory susceptibility- may result from differences in structure and function of their smaller polypeptide constituents (γ,δ,ε), while the principle function and interaction of the catalytic αß-assemblies is alike. This view is supported also from comparison of structural properties on different levels of complexity.

Primary sequence data are now available for F1-ATPases from a variety of species, revealing an extraordinary degree of homology between ß-subunits (Walker et al., 1984; Walker et al., 1985) and a high homology also between α and ß per se; homology is less pronounced between corresponding smaller subunits. Regarding quarternary structure there is ample evidence that all F1-ATPases exhibit an $\alpha_3\beta_3$ stoichiometry whereas the smaller subunits (γ-ε) are present only as single copies.

Despite F1 has been crystallized from rat liver and beef heart mitochondria (Amzel, and Pedersen, 1978; Walker, Lutter, 1987), the resolution achieved hitherto with X-ray analysis is insufficient for comparison of various ATPases or for a spatial assignment of individual polypeptides within the oligomeric enzyme. The first three-dimensional model providing such information has been derived with the beef heart enzyme from molecular electron microscopy in our laboratory (Tiedge et al., 1983). Figure 1 shows the model including the cross-dimensions and the juxtaposition of the large subunits α and ß. Controlled tilting of individual molecules has been employed to derive spatial information from the changes of their projected images. Basic symmetry elements for the location of the large peripheral subunits are two trigonal layers of globular masses; these are stacked on top of each other and twisted by 60°. This typical "top view" of the molecule gives rise to pseudo-hexagonal projections also confirmed by others (Boekema et al., 1986; Lücken et al., 1987). The side view makes obvious that the central mass, most probably composed of the smaller polypeptides γ-ε, is slightly displaced out of the central plane of the molecule towards one surface. Its actual shape appears variable and can not be determined exactly, due to the limited resolution of the method of about 18 nm. By decoration with monoclonal antibodies against the α- and ß-subunits an assignment of their position became accessible as depicted also from figure 1. An identical distribution could be directly demonstrated for the enzymes from spinach chloroplasts (Tiedge et al., 1985) and from e.coli (Lünsdorf et al., 1984); by analogy it also appears likely for the respective enzymes from other sources (see fig.2). Clearly, the small subunits may introduce an element of asymmetry and/or of variability. Present only as single copies these can not be distributed in a symmetric way, and equivalent contacts to only one kind of the symmetrically arranged α- or ß-subunits can not exist. Therefore a question mark is put at the respective part of our model. Averageing methods when applied to "top views" lead others to propose a permanent asymmetric location of these subunits relative to α and ß (Boekema et al., 1986). Such a conclusion is subject, however, to serious objections. First, all available methods including X-ray diffraction are observing an object under fixed static conditions. Secondly, in electron microscopic imaging the tilting angle and submersion into the stain should be known for each individual molecule before averaging and sorting images into specific classes. Finally, on the basis of dynamic nucleotide interactions as discussed above it appears conceivable that F1-molecules may not be confined to an invariant asymmetric state but rather may assume time averaged symmetry during a squence of catalytic cycles as envisaged by the proposed "binding change" mechanism (Cross et al., 1984).

While on the basis of sequence homologies a surprising degree of identity between various organelle and procaryotic F1 ATPases has been found, we extended the search to the level of quarternary structure. Figure 2 shows a gallery of F1-ATPases in top-view projections. Without further explanation it becomes obvious that all of these enzymes are constructed according to a unique building plan (c.f.fig.1), exhibiting the typical pseudohexagonal array of the peripheral globular masses and therefore presumably

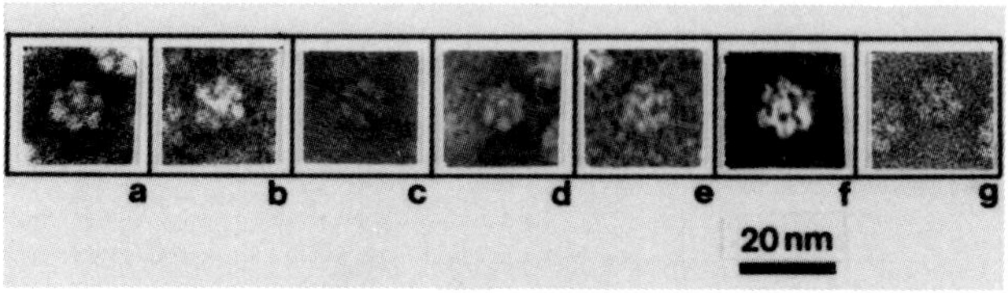

Figure 2: Gallery of F1-ATPases from various sources; selected top-view projections of negatively stained enzyme molecules: a) CF1 from spinach (Tiedge H et al., 1985: Proc Natl Acad Sci 82:7874-7878); b) bovine heart F1 (Tiedge H et al., 1983:Eur J Biochem 132:37-45); c) E.coli F1 (Lünsdorf H et al.,1984: J Membr Biol 173: 131-136); d) F1 from PS 3 (Yoshida M, 1986: unpublished); e) Wolinella succinigenes F1 (Bokranz M et al., 1985: Biochim Biophys Acta 810: 84-93); f) Clostridium thermoaceticum F1 (Mayer et al., 1986: J Bacteriol 166:1128-1130); g) F1 from Sulfolobus acidocaldarius (Lübben M et al., 1987:unpublished).

also possess an identical stoichiometry of their large subunits. One of these enzymes has been recently isolated from the thermoacidophilic archaebacterium sulfolobus acidocaldarius (Lübben, and Schäfer, 1987; Lübben et al., 1987) which was demonstrated to use a proton driven chemiosmotic mechanism for ATP synthesis (Moll et al., 1987; Moll, 1987). Actually, a DCCD-binding chloroform soluble 6kd polypeptide could be extracted from the membane which may resemble the well known proteolipid of Fo from eucaryotes and e.coli.

Interestingly, an antiserum raised against the ß-subunit of the archaebacterial ATPase revealed a remarkable immunological crossreactivity versus the ß-subunits of known procaryotic and eucaryotic F1-ATPases as well as against polypeptides of closely similar molecular mass from membrane- or whole-cell-lysates of other archaebacteria (Lübben et al., 1987). Table 2 gives a list of hitherto tested crossreacting species and as an important information clearly demonstrates the close immunological relation of the respective polypeptides. Under phylogenetic aspects these immunological data together with the structural analogies strongly suggest not only that F1-type ATPases are universally found in nature, but also may have evolved from a common ancestor. Though sequence data of

Table 2: Examples of crossreactivity of anti-ß serum against the ATPase from Sulfolobus acidocaldarius.

Sulfolobus acidocaldarius	reference
Sulfolobus solfataricus	+++
Staphylothermus marinus	++
Acidianus brierleyi	+++
Acidianus infernus	+++
Pyrodictium occultum	+++
Halobacterium halobium	+
Halobac. saccharov.-F1	++
Methanobac. thermoautotr.	++
Methanolobus tindarius	++
Methanosarcina barkeri	++
Escherichia coli-F1	+
Propiogenium modestum-F1	++
Spinach chloroplasts-F1	++
Mammalian mitochondria-F1	++

Crossreactivity measured against membrane lysates or purified enzymes, asigned as F1. For details see (Lübben et al., 1987).

archaebacterial ATPases are to be determined, this conclusion is supported by comparative analysis of primary sequences from various other F1 ATPases (Walker et al., 1984; Walker et al., 1985) especially of the ß-subunits which are believed to contain the catalytic sites (Cross et al., 1987; Gromet-Elhanan Z, Khanashvili D, 1984). Besides homologies even consensus sequences could be determined at the insertion sites of covalently reacting substrate analogs. Since within these sequence stretches also a short consensus structure can be found between F1-ATPases and other cation transport ATPases (Pedersen, and Carafoli, 1987), it may be hypothesized that both types of ATPases evolved from the same progenotic gene segment coding for adenine nucleotide binding.

REFERENCES:

Amzel LM, Pedersen PL (1978) Adenosine triphosphatase of rat liver mitochondria; crystallization and X-ray diffraction studies of the F1-component of the enzyme. Biochim Biophys J Biol Chem 253:2067-2069.

Boekema EJ, Berden JA, van Heel MG (1986) Structure of mitochondrial F1-ATPase studied by electron microscopy and image processing. Acta 851:353-360.

Cross RL, Nalin CM (1982) Adenine nucleotide binding sites on beef heart F1-ATPase. J Biol Chem 257:2874-2881

Cross RL, Cunningham D, Taura JK (1984) Binding change mechanism for ATP synthesis by oxidative phosphorylation and photophosphorylation. Curr Top Cell Regul 24:335-344.

Cross RL Cunningham D, Miller CG, Xue Z Zhou J, Boyer PD (1987) Adenine nucleotide binding sites on beef heart F1-ATPase; photoaffinity labeling of β-subunit Tyr-368 at a noncatalytic site and β Tyr-345 at a catalytic site. Proc Natl Acad Sci (USA) 84:5715-5719.

Gromet-Elhanan Z, Khanashvili D (1984) Characterization of two nucleotide binding sites on the isolated reconstitutively active β-subunit of the FoF1-ATP-synthase. Biochem 23:1022-1028.

Grubmeyer C, Penefsky HS (1981) The presence of two catalytic sites on beef heart mitochondrial adenosine triphosphatase. J Biol Chem 256: 3718-3727, and 3728-3734.

Hatefi Y (1985) The mitochondrial electron transport and oxidative phosphorylation system. Ann Rev Biochem 54:1015-1069.

Kayalar C, Rosing J, Boyer PD (1977) An alternating site sequence for oxidative phosphorylation suggested by measurement of substrate binding patterns and exchange reaction inhibitions. J Biol Chem 252: 2486-2491.

Kironde FAS, Cross RL (1986) Adenine nucleotide binding sites on beef heart F1- ATPase. Conditions that affect occupancy of catalytic and noncatalytic sites.. J Biol Chem 261:12544-12459.

Lübben M, Lünsdorf H, Schäfer G (1987) The Plasma membrane ATPase of the thermoacidophilic archaebacterium Sulfolobus acidocaldarius; purification and immunological relationships to F1-ATPases. Eur J Biochem 167:211-219.

Lübben M, Schäfer G (1987) A plasma-membrane associated ATPase from the thermoacidophilic archaebacterium Sulfolobus acidocaldarius. Eur J Biochem 164:533-540.

Lücken U, Bork T, Remington SJ, Capaldi RA (1987) Electron microscopy and image processing of stained two-dimensional arrays obtained from e.coli F1 and F1Fo preparations. Biophys J 51:242a.

Lünsdorf H, Ehrig K, Friedl P, Schairer HU (1984) Use of monoclonal antibodies in immuno electron microscopy for determination of subunit stoichiometry in oligomeric enzymes. J Mol Biol 173:131-136.

Lunardi J, Klein G, Vignais PV (1986) Interaction between aurovertin and adenine nucleotide binding sites on mitochondrial F1-ATPase and the isolated β-subunit. J Biol Chem 261:5350-5354.

Moll R, Lübben M, Schäfer G (1987). in preparation -

Moll R (1987). Protonenbewegungen an der Membran des thermoacidophilen Archaebakteriums Sulfolobus acidocaldarius. Thesis, Univ of Hannover.

Pedersen PL, Carafoli E (1987) Ionmotive ATPases. I.Ubiquity, properties, and significance to cell function. II. Energy coupling and work output. Trends Biochem Sci 12:146-150, 186-189.

Penefsky HS (1985) Energy dependent dissociation of ATP from high affinity catalytic sites of beef heart mitochondrial adenosine triphosphatase. J Biol Chem 260:14728-13734, and 13735-13741.

Schäfer HJ, Dose K (1984) Photoaffinity crosslinking of the coupling factor 1 from micrococcus luteus by arylazido-8-azido-ATP. J Biol Chem 259:15301-15306.

Senior AE, Wise JG (1983) The proton-ATPase of bacteria and mitochondria. J Membrane Biol 73:105-124.

Tiedge H, Lücken U, Weber J, Schäfer G (1982) High-affinity binding of ADP and of ADP analogues to mitochondrial F1-ATPase. Eur J Biochem 127:291-299.

Tiedge H, Schäfer G, Mayer F (1983) An electron microscopic approach to the quarternary structure of mitochondrial F1-ATPase.Eur J Biochem 132:37-45.

Tiedge H, Lünsdorf H, Schäfer G, Schairer HU (1985) Subunit stoichiometry and juxtaposition of the photosynthetic coupling factor 1: Immunoelectron microscopy using monoclonal antibodies. Proc Natl Acad Sci (USA) 82:7874-7878.

Tiedge H, Schäfer G (1986) Site-site interaction on mitochondrial F1-ATPase; functional symmetry of the high affinity binding sites.Biol Chemistry (Hoppe-Seyler) 367:689-694.

Vignais PV, Satre M (1984) Recent developments on structural and functional aspects of the F1-sector of H^+-linked ATPases. Mol Cell Biochem 60:33-70.

Vignais PV, Lunardi J (1985) Chemical probes of the mitochondrial ATP-synthesis and translocation. Ann Rev Biochem 54:977-1014.

Walker JE, Saraste M, Gay NJ (1984) Nucleotide sequence, regulation and structure of ATP synthase. Biochim Biophys Acta 768:164-200.

Walker JE, Fearnley IM, Gay NJ, Gibson BW, Northrop FD, Powell SJ, Runswick M J, Saraste M, Tybulewicz VLJ (1985) Primary structure and subunit stoichiometry of F1-ATPase from bovine mitochondria. J Membrane Biol 184:677-701.

Walker JE, Lutter R (1987). personal communication, unpublished.

Weber J, Lücken U, Schäfer G (1985) Total number and differentiation of nucleotide binding sites on mitochondrial F1-ATPase; an approach by photolabeling and equilibrium binding studies.Eur J Biochem 148:41-48.

Weber J, Rögner M, Schäfer G (1987) Novel Approaches towards characterization of the high-affinity nucleotide binding sites on mito- chondrial F1-ATPase by the fluorescence probe 3'-O-(1-naphthoyl)adenosine di- and triphosphate. Biochim Biophys Acta 892:30-41.

The Ion Pumps: Structure, Function, and Regulation, pages 67–74

STRUCTURE OF THE ATP-SYNTHASE FROM CHLOROPLASTS AS REVEALED FROM BIOCHEMICAL STUDIES AND ELECTRON MICROSCOPY

Peter Gräber, Petra Fromme, Günter Schmidt and Egbert Boekema *
Max-Volmer-Institut, Technische Universität Berlin, Strasse des 17. Juni 135, 1000 Berlin 12
* Fritz-Haber-Institut, Faradayweg 4-6, 1000 Berlin 33, FRG

The ATP-synthase from chloroplasts, CF_oF_1, can be isolated, purified and reconstituted into liposomes (Pick and Racker, 1979) and this reconstituted CF_oF_1 shows practically the same rate of ATP synthesis and ATP hydrolysis as in its natural thylakoid membrane (Schmidt and Gräber, 1987a,b). This highly active preparation has been used for biochemical and electron microscopic studies. First, we have studied the subunit composition: Fig. 1 shows an SDS-gel of CF_1 (lane 1) with the well-known subunits α, β, γ, δ, ε and of CF_oF_1 showing additionally the subunits I, II, III, and IV. Two phenomena require some discussion:

1. Subunit IV is only weakly stained by Coomassie Blue. Therefore, in earlier work this subunit has been considered an impurity. However, this subunit can be stained strongly by silver. We have, therefore, isolated this subunit by quantitative gel electrophoresis and a partial amino acid sequence analysis has been carried out from the N-terminus (Fromme et al., 1987a). Comparison of this sequence with the amino acid sequence deduced from the DNA sequence of the corresponding chloroplast gene shows that the isolated protein is identical with the gene product of the plastid gene AtpI from spinach (Hennig and Herrmann, 1986) and pea (Cozens et al., 1986). However, the first 18 amino acids are missing, implying a post-translational processing. Subunit IV is homologue to subunit a from E. coli and according to hydrophopathy plots it has 4 or 5 transmembrane helices.

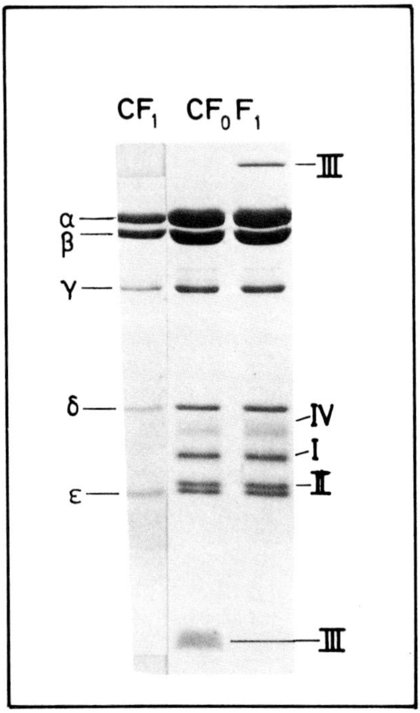

Figure 1. SDS-gel of CF_1 (lane 1), CF_oF_1 (lane 2) incubated 3 min at 95 °C and CF_oF_1 (lane 3) incubated 20 min at 25 °C. 15 µg protein for lane 1, 25 µg protein in lane 2 and 3, all stained with Coomassie Blue.

2. Subunit III ("DCCD-binding protein") can be observed at the molecular mass of 8 kDa only when CF_oF_1 is heated in 2% SDS-buffer to 95 °C for 5 min before it is subjected to gel electrophoresis (lane 2). If CF_oF_1 is incubated in the same buffer at room temperature, a new band at the apparent molecular mass 100 kDa is observed (lane 3). We have isolated this 100 kDa band by quantitative SDS-gel electrophoresis (Fromme et al., 1987b). Upon incubation of the 100 kDa protein in 2% SDS for 5 min at 95 °C, the 100 kDa band disappears and now only a band at 8 kDa is observed. Partial amino acid sequence analysis of the 100 kDa protein from the N-terminus reveals that the first five amino acids are identical with those reported in literature for subunit III. From these results it is clear that the 100 kDa protein is a multimeric form of subunit III and it contains at least 12 subunits III. Because of its exceptional stability, we assume that this multimeric form represents a stable native substructure of CF_oF_1 in the thylakoid membrane.

Fig. 2 shows electronmicrographs of negatively stained samples of the 100 kDa protein (top) with our interpretation of the structure and definition of different parameters (bottom)(Fromme et al., 1987b). The 100 kDa subunit III complex forms string-like structures with up to 35 units arranged along the string axis. The diameter of the strings varies, indicating that one ("monomer"), two ("dimer") or three ("trimer") complexes can form the

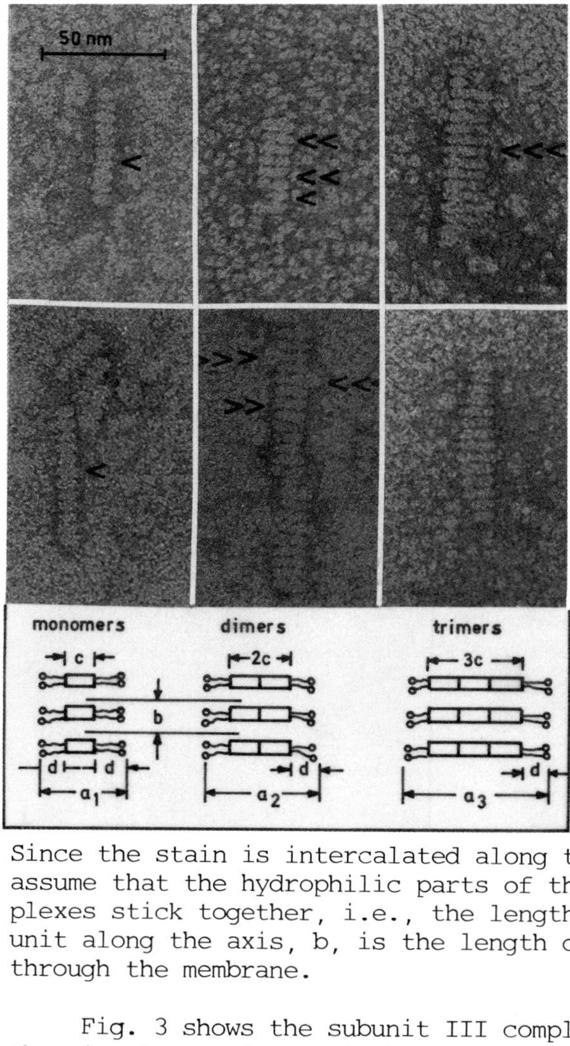

Figure 2. Electronmicrographs of the isolated 100 kDa complex of subunit III (negatively stained with uranylacetate). Top: gallery of monomers, dimers and trimers. Bottom: Presentation of the arrangement of the complexes in the string and dcfinition of parameters.

repeating units of the string. We assume that the detergent (SDS) substitutes the lipid environment of CF_o; i.e., it is located at the hydrophobic outer surface of the complex as indicated in Fig. 2, bottom. Since the stain is intercalated along the string axis, we assume that the hydrophilic parts of the subunit III complexes stick together, i.e., the length of one repeating unit along the axis, b, is the length of subunit III through the membrane.

Fig. 3 shows the subunit III complex after increasing the signal-to-noise ratio by a computerized image analysis (van Heel and Frank, 1981). The diameter of the strings, a, is plotted in Fig. 4B versus the number of complexes in the string perpendicular to the string axis. The slope of this curve gives the increase in length per complex III. We assume that there is no detergent between the aggregated complexes. With this assumption the slope in Fig. 4B gives directly the diameter of the complex III; i.e., a = 6.2 nm. Extrapolation to zero gives the length of the detergent at

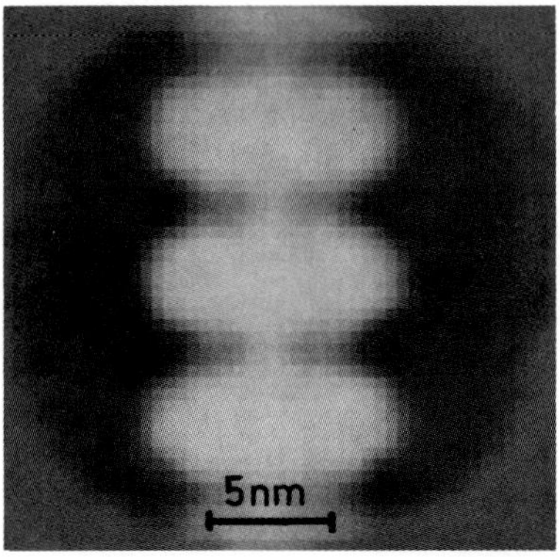

Figure 3. Electronmicrograph of the subunit III complex with increased signal-to-noise ratio by image processing, using 96 digitized images.

both sides of the complex. It results 4.4 nm in accordance with the length of SDS (2.2 nm).

Detergent solubilized CF_oF_1 forms also string-like structures in which the hydrophobic parts of CF_o are aggregated and the F_1 parts are arranged on alternating sides along the string (Boekema et al., 1987, 1988). Fig. 4A shows the length of the string as function of the number of attached F_1 parts. The slope of the curve gives the increase in length per F_1 part. We assume that there is no detergent between the aggregated F_o parts, and with this assumption the slope gives directly the diameter of the F_o part. It results 6.2 nm for the diameter of F_o (Boekema et al. 1988). This is identical with the diameter found for the subunit III complex.

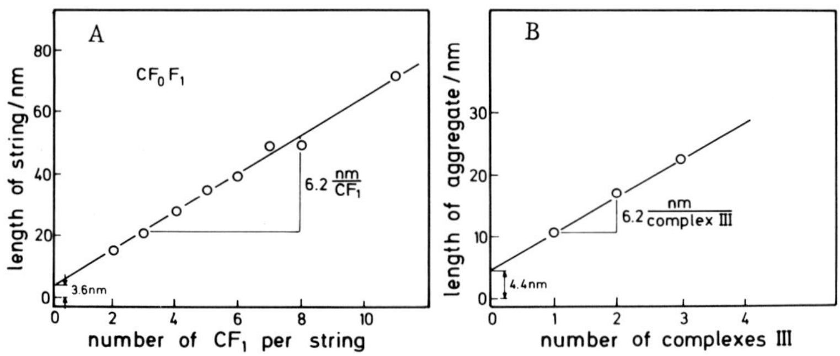

Figure 4. A: Length of the string of CF_oF_1 aggregates as a function of the number of attached CF_1 parts. The slope gives the diameter of one CF_o part.
B: Length of the aggregate of the subunit III complex perpendicular to the string axis as a function of the number of complexes. The slope gives the diameter of one subunit III complex.

One can now try to construct a model for the subunit arrangement in CF_o based on the following considerations:

1. From the amino acid sequence and from the polarity profile of subunit III it has been concluded that the protein traverses the membrane in a hair pin-like structure with two α-helical segments (Sebald and Hoppe, 1981).
2. From the amino acid sequence and from the polarity profile of subunit IV and I it has been concluded that subunit IV has at least 4 transmembrane helices and subunit I one transmembrane helix (Hennig and Herrmann, 1986). Similar data for the nuclear encoded subunit II are not published yet.
3. The diameter of an α-helix is between 0.7 and 0.9 nm.
4. The subunit III complex has the same diameter (6.2 nm) as the complete CF_o part.

Fig. 5 shows a top view of two hypothetical arrangements of helices (each helix is depicted as a circle) in the F_o part, assuming a diameter of 0.9 nm for each helix. The subunits III must be arranged in such a way that for the complex III and CF_o the same diameter is obtained. Two of the possible structures are shown because of their highly symmetrical arrangement: In model a the subunits III are arranged in two non-identical positions. In the middle of the structure there is place for one helix, possibly from subunit I. Four helices from subunit IV and hypothetically two helices from subunit II are distributed at the outside. This model is similar to the one proposed by Sebald et al., 1987. In model b all subunits III are arranged in identical positions in one ring. In this case the center can be filled with 7 helices from subunit I, II and IV as indicated. This model is similar to the one proposed by Cox et al., 1986.

Since there is no agreement on the stoichiometry of subunits I, II, III, and IV and the amino acid sequence of subunit II is not known, the model is highly speculative. Additionally, if we assume a diameter of 0.7 nm for one helix, the number of possible helices in CF_o would increase considerably. Therefore, this model should be regarded as a minimal structure of F_o showing the high complexity of this part of the ATP-synthase.

On the basis of electronmicroscopic data one can also try to construct a model for the subunit arrangement in

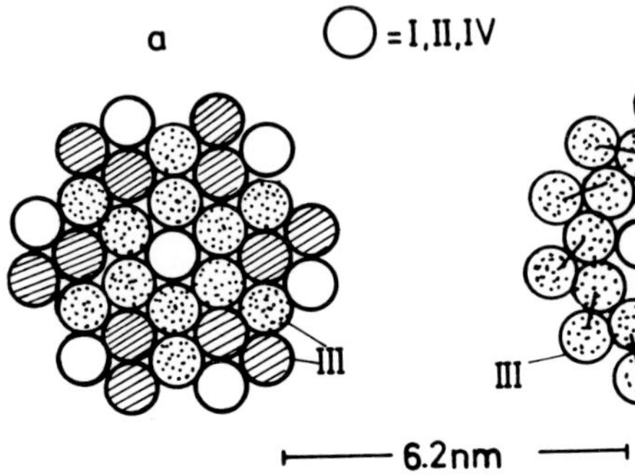

Figure 5. Two hypothetical arrangements of helices in the F_o part of the ATP-synthase from chloroplasts (top view). Each helix is represented by a circle. The figure is drawn to scale, assuming a diameter of 0.9 nm per helix. Helices from subunit III are dotted or shaded.

CF_oF_1:

1. The 100 kDa subunit III complex has a membrane spanning length of 6.1 nm, whereas, CF_o has a length of 8.3 nm (Boekema et al., 1988). This implies that the hydrophilic parts of subunits I, II and IV are located on the top (i.e. facing the CF_o part) or at the bottom (or on both sides) of the subunit III complex.
2. The CF_1 part forms a flattened trigonal antiprism (11.5 x 11.5 x 8.5 nm) with two layers of the subunits α and β (Tiedge et al., 1983; Tsuprun et al., 1984; Boekema et al., 1987) and the α-subunits are located towards the F_o part (McCarty and Hammes, 1987; Gao and Bäuerlein, 1987). The small subunits γ, δ, and ε are located asymmetrically in the center of the α and β subunits (Boekema et al., 1987).
3. The membrane part of CF_o is connected to the hydrophilic F_1 part by a stalk (length 3.7 nm, diameter 2.7 nm) (Boekema et al., 1987, 1988). Presumably, the stalk is formed mainly from subunit I (which has a large hydrophilic domain) and subunit δ (which is known to have an elongated shape).

From these results and considerations a hypothetical model of the subunit arrangement can be constructed which is shown in Fig. 6. The model is drawn to scale as derived from our electronmicroscopic data.

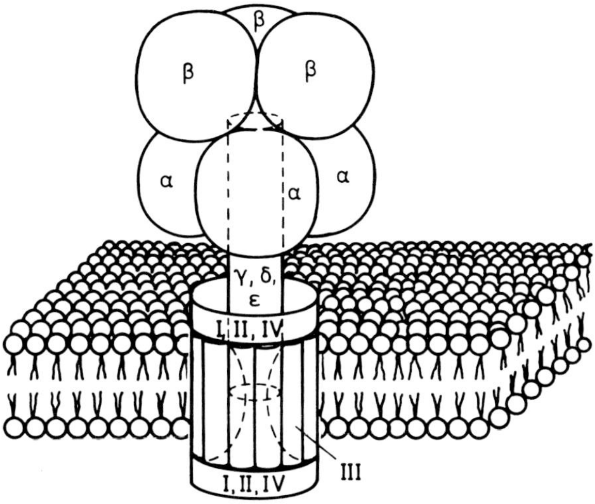

Figure 6. Hypothetical scheme of subunit arrangement in CF_oF_1. The shape is drawn to scale according to the electronmicroscopic data

ACKNOWLEDGEMENTS

We thank Prof. B. Rumberg for stimulating this work and Prof. E. Zeitler for continuous support. This work was supported by a grant from the Deutsche Forschungsgemeinschaft (Sfb 312).

REFERENCES

Boekema EJ, van Heel M, Gräber P (1987). Structure of the ATP-synthase from chloroplasts studies by electron microscopy and image processing. This meeting.

Boekema EJ, Schmidt G, Gräber P, Berden JA (1988). Structure of the ATP-synthase from chloroplasts and mitochondria studied by electron microscopy. Z. Naturforsch. 42, in press

Cox GB, Fimmel AL, Gibson F, Hatch L (1986). The mechanism of ATP synthesis: a reassessment of the functions of the b and a subunits. Biochim. Biophys. Acta 849:62-69

Cozens AL, Walker JE, Phillips AL, Huttly AK, Gray JC (1986). A sixth subunit of ATP synthase, an F_o component, is encoded in the pea chloroplast genome. EMBO Journal 5:217-222

Fromme P, Gräber P, Salnikow J (1987a). Isolation and Identification of a fourth subunit in the membrane part of the chloroplast ATP-synthase. FEBS Lett. 218:27-30.

Fromme P, Boekema EJ, Gräber P (1987b). Isolation and Characterization of a supra-molecular complex of subunit III of the ATP-synthase from chloroplasts. Z.Naturforsch. 42c:117-123.

Gao Z, Bäuerlein E (1987). Identifying subunits of ATP synthase TF_oF_1 in contact with phospholipid head groups. FEBS Lett. 223:366-370.

Hennig J, Herrmann RG (1986). Chloroplast ATP synthase of spinach chloroplasts contains nine nonidentical subunit species, six of which are encoded by plastid chromosomes in two operons. Mol. Gen. Genet. 203:117-128.

McCarty RE, Hammes GG (1987). Molecular architecture of chloroplast coupling factor 1. TIBS 12:234-237.

Pick U, Racker E (1979). Purification and reconstitution of the N,N'-Dicyclohexylcarbodiimide-sensitive ATPase complex from spinach chloroplasts. J. Biol. Chem. 254:2793-2799.

Sebald W, Hoppe J (1981). On the structure and genetics of the proteolipid subunit of the ATP synthase complex. Curr. Top. Bioenerg. 12:1-64.

Sebald W, Weber H, Hoppe J (1987). In: Structure and Function of Energy Transducing Systems (T. Izawa and S. Papa, eds.), Japan Sci. Soc. Press, Tokyo, pp. 215-224. Conformational stability of the proteolipid oligomer of the H-conducting F_o of the ATP-synthase.

Schmidt G, Gräber P (1987a). The rate of ATP synthesis catalyzed by reconstituted CF_oF_1 liposomes: dependence on ΔpH and $\Delta\Psi$. Biochim. Biophys. Acta 890:392-394.

Schmidt G, Gräber P (1987b). The rate of ATP hydrolysis catalyzed by reconstituted CF_oF_1-liposomes. Z. Naturforsch. 42c:231-236.

Tiedge H, Schäfer G, Mayer F (1983). An electron microscopy approach to the quaternary structure of mitochondrial F_1 ATPase. Eur. J. Biochem. 132:37-45.

Tsuprun VL, Mesyanzhinova IV, Kozlov IA, Orlova EV (1984). Electron microscopy of beef heart mitochondrial F_1-ATPase. FEBS Lett. 167:285-290.

van Heel M, Frank J (1981). Use of multivariate statistics in analyzing the images of biological macromolecules. Ultramicroscopy 6:187-194.

The Ion Pumps: Structure, Function, and Regulation, pages 75–80

STRUCTURE OF THE ATP-SYNTHASE FROM CHLOROPLASTS STUDIED BY ELECTRON MICROSCOPY AND IMAGE PROCESSING

Egbert Boekema, Marin van Heel and Peter Gräber

Fritz-Haber-Institut der Max-Planck Gesellschaft Faradayweg 4-6, D-1000 Berlin (E.B and M.v.H) and Max Volmer Institut, Technische Universität Berlin (P.G)

INTRODUCTION

ATP synthesis/hydrolysis is carried out by a large membrane-bound enzyme (ATP-synthase), which is similar in different types of organelles (chloroplasts, mitochondria, bacteria). It consists of a membrane integrated part, F0, which is supposed to act as a proton channel through the membrane and a hydrophilic part, F1, which contains the nucleotide binding sites. The F1 part is connected by a thin stalk to the F0 part.

The hydrophilic part of the ATP-synthase from chloroplasts, CF1, is composed of 5 different subunits: alpha, beta, gamma, delta and epsilon. One CF1 molecule contains 3 copies of alpha and beta, and one of gamma and one or possibly two copies of delta and epsilon.

An arrangement of the larger subunits was proposed on the basis of electron microscopy of crystals and single molecules. These results show that the F1 ATP-synthase is built up from two layers of large subunits alpha and beta in the form of a flattened trigonal antiprism with dimensions of about 11.5 x 11.5 x 8 nm (Tsuprun et al., 1984; Tiedge et al., 1985, Boekema et al., 1986).

Computer image analysis of single molecules in combination with multivariate statistical techniques were developed for analyzing mixed populations of images (Van Heel and Frank 1981; Van Heel and Stöffler-Meilicke 1985). The characteristic views, present in the population of images, are determined with an automatic classification scheme and the resulting class averages represent noise-reduced projection images of the molecule under investigation. In an earlier study we applied these techniques to MF1 molecules from beef heart mitochondria (Boekema et al. 1986). In the current study we investigate CF1, isolated from spinach chloroplasts, using the same techniques.

Exact information on the shape and dimensions of the holoenzyme FOF1 is scarce. Isolated CFOF1 ATPase forms string-like aggregates in which the CFO parts stick together (Mullet et al. 1981). We have investigated such strings in order to determine the dimensions of CFO.

RESULTS AND DISCUSSION

A typical micrograph showing molecules with a clear subunit structure is shown in Fig. 1. From 21 of such negatives 3300 images were selected for image processing. The selected images were pretreated, aligned and submitted to an eigenvector-eigenvalue data compression procedure ("correspondence analysis"). The images were partitioned into (arbitrarely) 16 classes, which shows the image variation as present in the data set (fig 2). All classes show hexagonal views. The rare but characteristic side views seen Tsuprun (1984), which point to a flattening of the hexagonal antiprism, are so rare in our data set that no reliable analysis was possible. However, if CF1 was prepared in a different way, more of such side views could be seen (see also Fig 3).

Some classes (N,O,P) represent molecules in a "flattened hexagonal projection". These classes appear smaller in one direction; Some outer subunits appear slightly smaller and this is accompanied by a displacement of the central mass in the direction of the smaller looking outer subunits. Measurements on the distances in the classes showed that the flattened hexagonal projections originate from an uneven stain distribution around part of the molecules, rather than from specific tilting.

The other differences within the hexagonal view mainly represent the effects of different stain accumulation: high stain in I,K,M, low stain in A-D. Other classes might represent molecules oriented differently on the carbon support (face-up and face-down, C,D versus E-G). The centre of several classes is more or less split into a smaller and a larger part (A,B,C).

The features of the hexagonal views of CF1 and MF1, (Boekema et al. 1986) can be compared. In both cases, the arrangement of the large outer subunits is practically

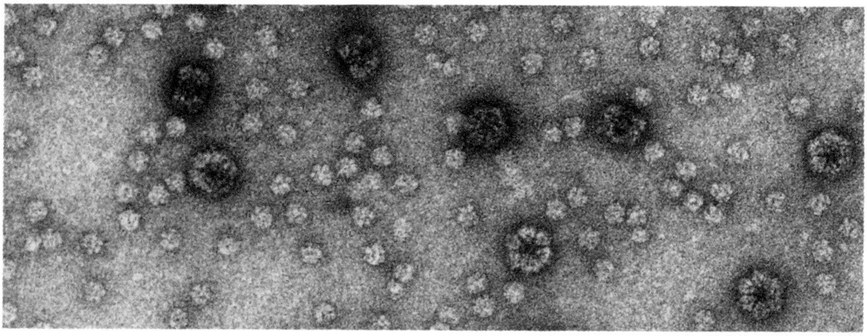

Fig. 1. A micrograph showing F1 molecules used for the analysis of single particles. The specimen was mixed with giant erythrocruorin molecules during the staining with with 1% uranyl acetate. In this way a better stain distribution could be obtained easily.

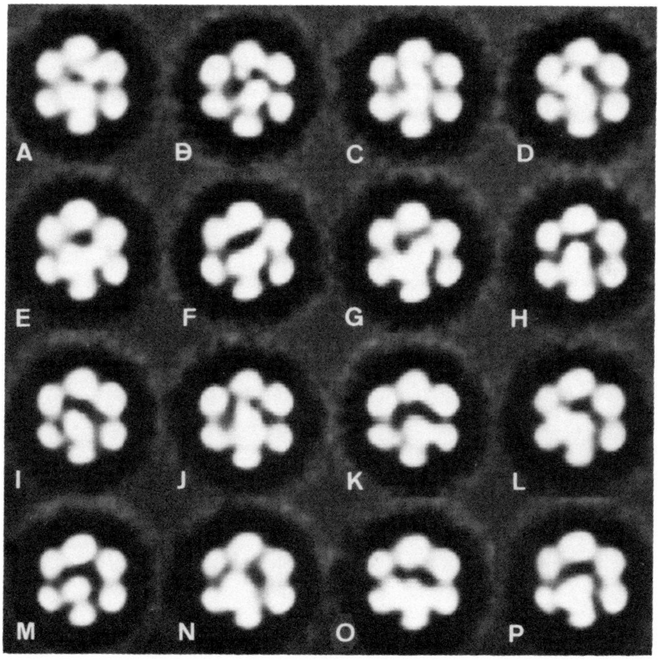

Fig. 2. Molecular projections of CF1 ATP–synthase as determined by multireference alignments and classification of 3300 images.

the same and the views differ mainly in the center: the central mass of CF1 is larger than in MF1 and is frequently split into two parts, a feature that was never observed in MF1. The central mass with the shape of the letter V, that is placed asymmetrically in the center of the en-

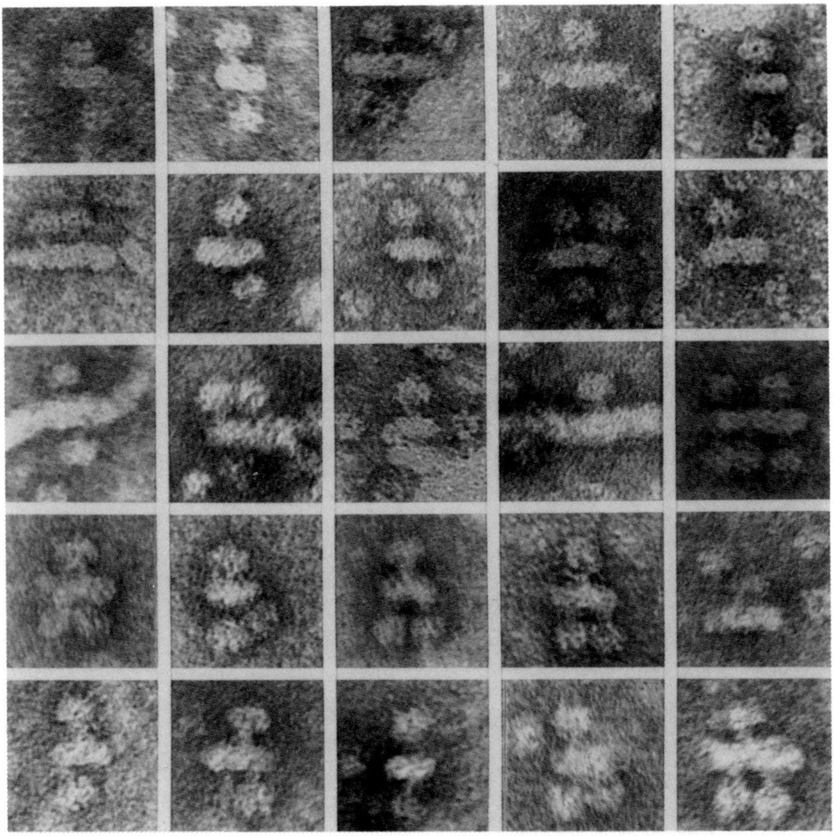

Fig. 3. A gallery of CFOF1 ATP-synthase strings (upper three rows) and MFOF1 strings (lower two rows). The diameter of FO in the direction of the strings was calculated by dividing the string length of well-prepared strings by the number of attached F1 parts. Diameters of about 6.2 and 6.4 nm were found for CFO and MFO, respectively. Note that in most images the longest dimension of F1 is parallel to the string (F1 in "side view" position).

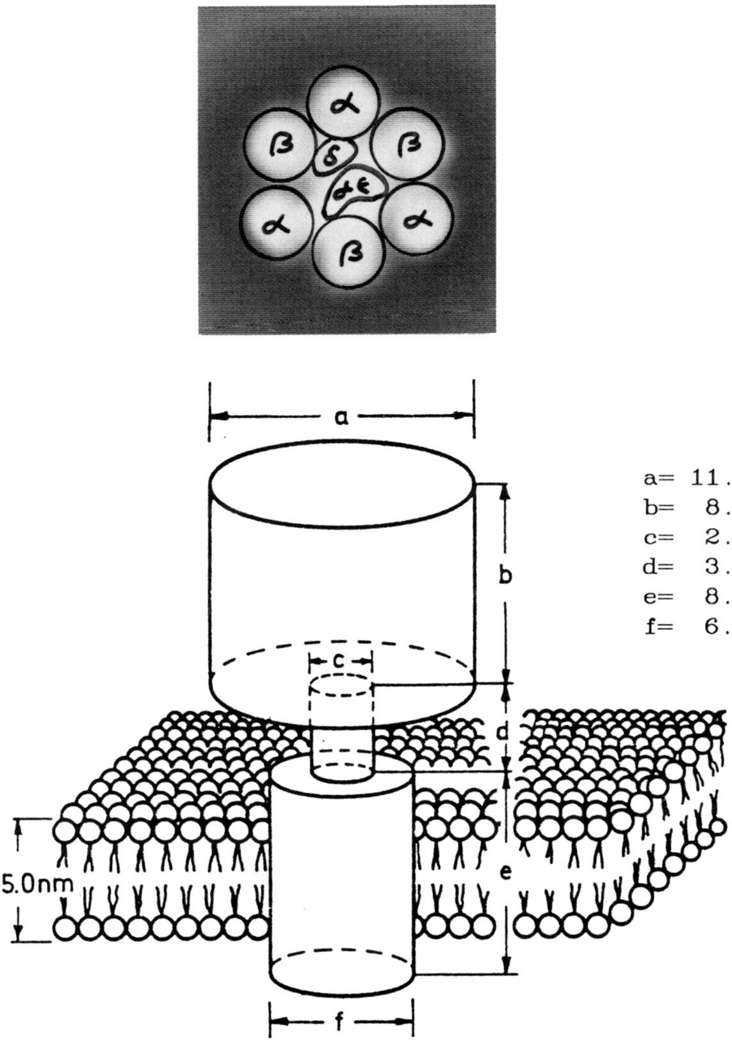

Fig. 4. The shape of ATP–synthase measured from FOF1 in strings and isolated F1, and scheme for the F1 subunit arrangement.

zyme, is more or less similar in both F1s. The second smaller part of the central mass was detected in CF1 only. The analyzed MF1 preparation lacks the OSCP subunit, which is equivalent to delta of CF1. Therefore, the V-shaped structure observed in both cases possibly is the gamma- and epsilon-subunit and the second mass the the CF1 delta subunit. Fig. 4 shows a model based on these considerations and on the results of others. This scheme is supported by recent microscopic studies on MF1 using a ferritin label (Tsuprun et al. 1987), from which was concluded that gamma/epsilon are asymmetrically oriented.

Fig. 3 shows a gallery of FOF1 strings. From such strings, in which the FO parts come together, the dimensions of FO and the stalk could be measured. The diameter of FO in the membrane was found to be only 6.2 nm. It could be calculated from dividing the string lengths through the number of attached F1 parts. The measurements lead to a model showed in Fig. 4.

REFERENCES

Boekema, EJ, Berden, JA and van Heel, MG (1986) Structure of mitochondrial F1-ATPase studied by electron microscopy and image processing. Biochim. Biophys.Acta 851, 353-360

Mullet, JE, Pick, U and Arntzen, CJ (1981) Structural analysis of the isolated chloroplast coupling factor and the N,N'-dicyclohexylcarbodiimide binding proteolipid. Biochim. Biophys. Acta 642, 149-157

Tiedge, H, Lünsdorf, H, Schäfer, G and Schairer, HU (1985) Proc. Natl. Acad. Sci. USA 82, 7874-7878

Tsuprun, VL, Mesyanzhinova, IV, Kozlov, IA and Orlova, EV (1984) Electron microscopy of beef heart mitochondrial F1-ATPase. FEBS Letters 167, 285-290

Tsuprun, VL, Mesyanzhinova, IV., Milgrom, YM and Kalashnikova, TY (1987) Electron-microscopic studies on location of SH-groups in mitochondrial F1-ATPase using a ferritin label. Biochim. Biophys. Acta 892, 130-137

Van Heel, M. and Frank, J. (1981) Use of multivariate statistics in analysing the images of biological macromolecules. Ultramicroscopy 6, 187-194

Van Heel, M. and Stöffler-Meilicke M. (1985) Characteristic views of E. coli and B. stearothermophilus 30S ribosomal subunits in the electron microscope. EMBO J. 4, 2389-2395

The Ion Pumps: Structure, Function, and Regulation, pages 81–86

DISSOCIATION-RECONSTITUTION EXPERIMENTS WITH NBD-MODIFIED F_1: SUPPORT FOR THE PRESENCE OF TWO CATALYTIC β-SUBUNITS

J.A.Berden, P.Nieboer and A.F.Hartog

Laboratory of Biochemistry of the University of Amsterdam, B.C.P. Jansen Institute, Plantage Muidergracht 12, 1018TV Amsterdam, The Netherlands

INTRODUCTION

After it had been established that mitochondrial F_1 contains 3 α- and 3 β-subunits Stutterheim et al. (1981), the discussion on the number of catalytic sites kept going on, but the proposal of three catalytic sites was highly favoured, expecially after the kinetic analysis of Gresser et al. (1982). The binding studies with *p*-fluorosulphonyl-5'-adenosine (Esch and Allison, 1979) and the data of Cross and Nalin (1982), indicative for three nonexchangeable binding sites, were thought to fit with the presence of three cooperative catalytic sites and three non-exchangeable non-catalytic sites. Our own data obtained with photolabelling studies, disagreed with such proposal and we concluded that F_1 contains 2 catalytic sites on β-subunits, 2 exchangeable non-catalytic sites, probably regulatory, on α-β interfaces and 2 non-exchangeable sites, one on the third β-subunit and one on the third α-β interface (van Dongen and Berden, 1986, Van Dongen et al., 1986) . The three tightly bound nucleotides, present on isolated F_1, were thought to represent one nucleotide on a catalytic site and two on the non-exchangeable sites. The recent data of Kironde and Cross (1986) as well as the data on the binding of dicyclohexylcarbodiimide and the structure of F_1 (Boekema et al.,1986) are in agreement with our proposal.

A completely different conclusion has been reached by Wang on the basis of dissociation-reconstitution experiments. (Wang et al. 1986)The proposal, however, of only one catalytic site for ATP hydrolysis seems to contradict many other data and since the experimental approach seemed to be a valid one, we performed experiments similar to those of Wang.

MATERIALS AND METHODS

Bovine heart mitochondria F_1-ATPase was isolated according to the method described by Knowles and Penefsky (1972). Prior to use, F_1 was precipitated with $(NH_4)_2SO_4$ and dissolved in buffer A, containing 50 mM Hepes-KOH (pH 7.0), 2 mM EDTA, 5 mM ATP and 25% (v/v) glycerol. $(NH_4)_2SO_4$ was removed by means of a centrifuge column packed with Sephadex G-50 and equilibrated with buffer A. Protein concentrations were determined with the Biorad protein assay. The starting concentration of F_1 was about 6 mg/ml with a specific activity of 140 μmol/min per mg.

Pyruvate kinase (in glycerol), lactate dehydrogenase (in glycerol), Hepes, NADH, ATP and phospho*enol*pyruvate were obtained from Boehringer Mannheim. NBD-Cl was produced by Serva, dithiothreitol by Sigma, and Sephadex G-50-Coarse by Pharmacia. 2-azido-ATP was synthesized as described previously (Van Dongen et al. 1986).

ATP hydrolysis was assayed with a regenerating system.

Binding of NBDCl was performed in the dark at 20 ^{0}C in buffer A containing 75 μM NBDCl . The level of inhibition was determined relative to the activity obtained in the presence of 125 μM dithiothreitol.

Binding of 2-N-ATP to a catalytic site was performed as described (Van Dongen and Berden, 1987), with the modification that illumination was performed with a xenon-flash unit. Because of the slow transition of the tetrazolium form into the azido-form of the compound, (Czarnecki, 1984) the illumination (10 flashes with intervals of 15 s) was repeated after 30 and 60 min.

Non-denaturating polyacrylamide gel electrophoresis was as described in (Knowles and Penefsky, 1972), using slabgels instead of tubegels.

RESULTS AND DISCUSSION

Dissociation and Reconstitution The procedure for dissociation and reconstitution as described (Wang et al. , 1986) was slightly modified to obtain reproducible results. Centrifuge columns were pre-centrifuged at 2000 rpm and samples were centrifuged at 2400 rpm. The incubation in 3 M LiCl was stopped after 100 s and the activity after reconstitution was measured not earlier then 15 min. after removal of LiCl. With this procedure the β-subunits of F_1 and NBD-F_1 were fully dissociated and 70% of the original activity was restored.

Statistics In order to evaluate the results of the experiments, some statistical calculations were made. The degree of inhibition was considered to be the same as the amount of label (NBD) on the F_1. The assumptions being made were:
(i) labelled and unlabelled β-subunits behave the same upon treatment with LiCl;
(ii) at reconstitution the β-subunits integrate randomly into F_1.

If we consider the β-subunits of F_1 to have different functions in the enzyme, we can call them β-1. β-2 and β-3. The β-1 subunit in this calculation is the subunit were the label is incorporated in the enzyme. The total amount of label is called X (mol of label/ mol of F_1). Inactivation after reconstitution is expressed by r. At total inhibition r=1 and at zero inhibition r=0. Before LiCl-treatment there are only two kinds of F_1 possible:
(i) β-1, β-2, β-3 = X, O, O (X=labelled, O=unlabelled); and
(ii) β-1, β-2, β-3= O, O, O
After reconstitution many more kinds of F_1 are possible (see Table 1)

Table 1.

X=0			X=1			X=2			X=3		
β-1	β-2	β-3	β-1	β-2	β-3	β-1	β-2	β-3	β-1	β-2	β-3
O	O	O	X	O	O	X	X	O	X	X	X
			O	X	O	O	X	X			
			O	O	X	X	O	X			

The amount of labelled β-subunits is 1/3*X and the amount of unlabelled β-subunits is 1-1/3*X. The expected values for r were calculated for three kinetic models:
(i) one catalytic site:

$$r=1/3\ X$$

(ii) Two cooperative catalytic sites

$$r=2(1-1/3\ X)^2*1/3\ X + 3(1-1/3\ X)(1/3\ X)^2 + (1/3\ X)^3$$

(iii) Three cooperative catalytic sites:

$$r=3(1-1/3\ X)^2*1/3\ X + 3(1-1/3\ X)(1/3\ X)^2 + (1/3\ X)^3$$

These values are plotted in fig. 1.

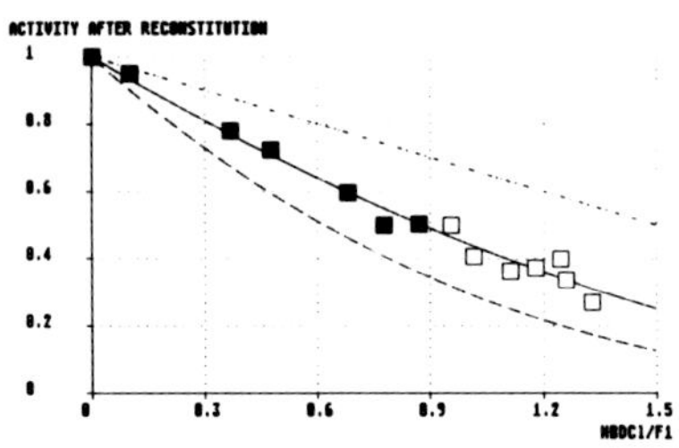

Figure 1.

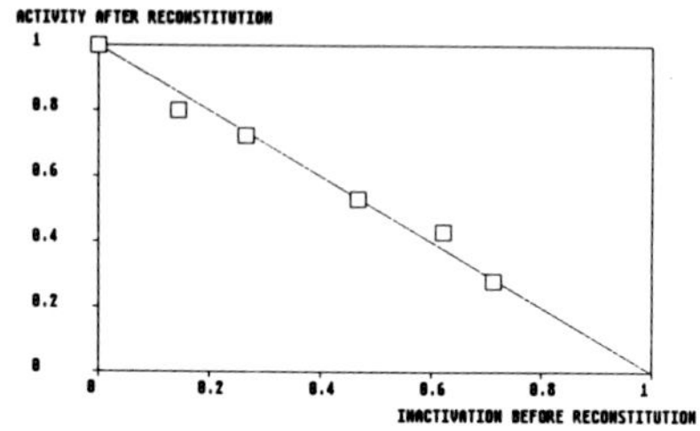

Figure 2.

Fig. 1. Reactivation of NBD-modified F_1. F_1 was incubated with NBDCl and samples were treated with LiCl, desalted as described in Materials and Methods, and the ATPase activities after reconstitution were measured. The activity in the presence of dithiothreitol was taken as the 100% value. (■). All samples were incubated again with NBDCl and the dissociation-reconstitution procedure was repeated (□). The curves - ·· - ·· -, ——— and - · - · - are theoretical curves on the basis of random incorporation of the dissociated β-subunits into F_1 and the involvement of three, two or one β-subunit, respectively, in the catalytic process of ATP hydrolysis.

Fig. 2. Reactivation of F_1 modified by 2-nitreno-ATP. F_1 (4.1mg/ml) was incubated with 1mM 2-azido-ATP in medium B and after 10 min not bound 2-azido-ATP was removed by repeated centrifuge column gel filtration. The obtained F_1 preparation, containing 1 mol 2-azido-AT(D)P per mol F_1 (or the tetrazolium form of the compound) was illuminated to various degrees resulting in the formation of F_1 modified by 2-nitreno-AT(D)P. The partially inactivated samples were treated with LiCl and desalted, and the activities were measured before and after the dissociation-reconstitution procedure.

Dissociation and Reconstitution of NBD-modified F_1 Binding of NBDCl to F_1 was followed spectrophotometrically and by measuring the activity. On the basis of earlier data it was assumed that inactivation is proportional with the binding of 1 mol NBDCl/mol of F_1.

When after dissociation followed by reconstitution, the samples were again treated with NBDCl, the time-course of the binding (inactivation) was identical with the time-course of the original binding indicating that inactivation was again due to binding of NBDCl to one catalytic site. Fig. 1 shows the results of a typical experiment. As a control for full activity, both unmodified F_1 and the activities after treatment with dithiothreitol were used, giving identical values within 2%. The open symbols in the figure represent the data obtained when the samples after the first dissociation reconstitution step were treated again with NBDCl and then subjected to the dissociation-reconstitution procedure for a second time. It is clear that all points are in agreement with the theoretical curve for random incorporation of β-subunits into F_1 with two of the three β-subunits being involved in cooperative catalysis. The data of Wang (Wang et al., 1986) also fit with this curve.

Modification of F_1 with 2-N-ATP Since 2-azido-ATP can be bound tightly to a catalytic site (van Dongen and Berden, 1987), we also repeated the dissociationassociation procedure with F_1 partially modified and inactivated with 2-nitreno-AT(D)P, covalently bound to a catalytic site. As can be seen in Fig. 2, the dissociation-reconstitution procedure did not result in reactivation, in contrast to that with NBD-modified F_1. The reason for this lack of reactivation was found when samples taken after 100 s of incubation in the dissociation medium were put on polyacrylamide gels in the absence of SDS. With unmodified F_1 the band of native F_1 had largely disappeared, while the band at the position of the isolated β-subunits was very intense, but with modified F_1 the residual F_1-band was about proportional to the level of modification. It is concluded from this experiment that covalent binding of 2-nitreno-ATP to a catalytic site protects the enzyme against dissociation and that, as suggested previously, bound adenine nucleotides contribute significantly to the stability of the mitochondrial F_1.

REFERENCES

Boekema, E.J., Berden, J.A. and Van Heel, M.G. (1986). Structure of mitochondrial F_1-ATPase studied by electron microscopy and image processing Biochim. Biophys. Acta 851, 353-360

Cross, R.L. and Nalin, C.M. (1982). Adenine nucleotide binding sites on beef heart F_1-ATPase J. Biol. Chem. 257, 2874-2881

Czarnecki, J.J. (1984). Tautomerism of 2-azidoadenine nucleotides effect on enzyme kinetics and photoaffinity labelling. Biochim. Biophys. Acta 800, 41- 51

Van Dongen, M.B.M. and Berden, J.A. (1986). Demonstration of two non exchangeable non-catalytic and two cooperative catalytic sites in isolated bovine heart mitochondrial F_1, using the photoaffinity labels [2-^{3}H]8-azido-ATP and [2-^{3}H]8-azido-ADP Biochim. Biophys. Acta 850, 121-130

Van Dongen, M.B.M., De Geus, J.P. Korver, T., Hartog, A.F. and Berden, J.A. (1986). Binding and hydrolysis of 2-azido-ATP and 8-azido-ATP by isolated mitochondrial F_1: characterisation of high-affinity binding sites. Biochim. Biophys. Acta 850, 359-368

Van Dongen, M.B.M. and Berden, J.A. (1987). Exchange and hydrolysis of tightly bound nucleotides in normal and photolabelled bovine heart mitochondrial F_1-ATPase Biochim. Biophys. Acta 893, 22-32

Esch, F.S. and Allison, W.S. (1979). On the subunit stochiometry of the F_1-ATPase and the sites in it that react specifically with *p*-fluorosulphonylbenzyl-5'-adenosine. J. Biol. Chem. 254, 10740-10746

Gresser, M.J., Myers, J.A. and Boyer, P.D. (1982). Catalytic site cooperativity of beef heart mitochondrial F_1 adenosine triphosphatase. J. Biol. Chem. 257, 12030-12038

Kironde, F.A.S. and Cross, R.L. (1986). Adenine nucleotide-binding sites on beef heart F_1-ATPase. J. Biol. Chem. 261, 12544-12549

Knowles, A.F. and Penefsky, H.S. (1972). The subunit structure of beef heart mitochondrial F_1-ATPase J. Biol. Chem. 247, 6617-6623

Stutterheim, E., Henneke, M.A.C. and Berden J.A. (1981). Subunit composition of mitochondrial F_1-ATPase isolated from *Saccharomyces Carlsbergensis*. Biochim. Biophys. Acta 634, 271-278

Wang, J.H., Joshi, V. and Wu, J.C. (1986). Geometric isomers of covalently labeled mitochondrial F_1-adenosinetriphosphatase with different properties Biochemistry 25, 7996-8001

The Ion Pumps: Structure, Function, and Regulation, pages 87–92

MITOCHONDRIAL ATP SYNTHASE: ROLE OF METAL BINDING IN STRUCTURE AND FUNCTION.

Noreen Williams*, Joanne Hullihen, and Peter L. Pedersen
Johns Hopkins University School of Medicine, Baltimore, MD 21205

INTRODUCTION

The mitochondrial ATP synthase is comprised of two major components, F_1 and F_0 (see Pedersen and Carafoli,1987a,b); Hatefi,1985 for reviews). The F_0 component is a multisubunit complex although the subunit structure is not identical throughout phylogeny. Minimum subunit composition for the _E. coli_ F_0 is three distinct subunits while for the beef heart enzyme there appear to be greater than eight subunit types. The F_0 or some component(s) of the F_0 functions as a channel directing protons to the F_1 portion of the ATPase complex. The F_1 component is composed of five distinct subunits with a stoichiometry of $\alpha_3\beta_3\gamma_1\delta_1\epsilon_1$. The soluble F_1 headpiece is capable of hydrolysis and, under highly defined conditions (Sakamoto,1984;Kandpal,_et al_,1987), synthesis of ATP in the presence of divalent cation. Under physiological conditions the divalent cation present is Mg^{++} although other divalent cations may support hydolysis and synthesis at even higher rates (Williams,_et al_,1987). Previous reports from this laboratory have established a structural asymmetry in the F_1 portion of the molecule which appears to be due to uniqueness of one of the three $\alpha\beta$ subunit pairs. This uniqueness is the result of the interaction of the smaller subunits, $\gamma\delta\epsilon$, with only one of the $\alpha\beta$ pairs of F_1. This asymmetry has been demonstrated both in structural studies such as cold dissociation (Williams,_et al_,1984), by cross-linking (Lotscher and Capaldi,1984), and in functional studies with divalent cation and nucleotide binding (Williams,_et al_,1987 and Pedersen,_et al_,in press). The work presented here

*Current Address: USUHS,Dept. of Biochem.,Bethesda,MD 20814

supports those studies and also demonstrates the presence of a stability associated divalent cation binding site not previously seen.

Earlier work from this laboratory demonstrated the presence of a number of divalent cation binding sites on the soluble F_1-ATPase. When the enzyme is isolated it contains a single non-exchangeable metal site, which appears to reside on an α subunit or on an $\alpha\gamma$ complex. This divalent cation cannot be removed from the F_1 without a loss of structural integrity (Williams, et al,1987). The condition which resulted in partial vacancy of this site was overnight incubation at room temperature in the presence of 50 mM Tris-HCl and 50 mM EDTA (pH 7.6). No conditions were found which allow recovery of activity from the resulting precipitated enzyme. The enzyme which possesses a single occupied divalent cation site also contains an occupied adenine nucleotide site. However, the divalent cation site does not appear to correlate directly with an adenine nucleotide binding site since all adenine nucleotides are easily removed by several different treatments (repeated ammonium sulfate precipitations, cold denaturation or incubation in 50 mM Tris-HCl, pH 7.5).

A second divalent cation exists on the enzyme as isolated by the above described procedure. It, unlike the first divalent cation site, is easily removed by a one hour incubation in 50 mM Tris-HCl and 50 mM EDTA, pH 7.6 (or overnight in 250 mM KPi and 50 mM EDTA, pH 7.6). This divalent cation binding site has been identified by the use of radioactive Co^{2+}. (Co^{2+} was chosen as the divalent cation for these radioactive assays since it most closely mimics Mg^{2+} in all measured activities with the ATPase.) The presence or absence of the adenine nucleotide was shown to have no effect on the binding of Co^{2+} to this site, indicating that although this site may also bind nucleotide it is not required for cation binding. It is of interest however that the enzyme contains approximately two mols of both divalent cation and adenine nucleotide as isolated.

In the presence of the ATP analog, AMP-PNP, a total of 2.8 mol of Co^{2+} were bound per mol of F_1 rather than the 1 mol Co^{2+} seen in the absence of adenine nucleotide. In the reversed experiment where radioactive AMP-PNP was employed, it was determined that only approximately 2 mol AMP-PNP are bound in the absence of divalent cation. The results from these experiments taken together would suggest that one metal and one adenine nucleotide can bind to a single site independently. Two other sites can be occupied either by adenine nucleotide alone or by a nucleotide-metal complex. These two sites cannot be occupied by divalent cation alone. These results are summarized in Table 1. Scatchard analysis of AMP-PNP binding also supports the presence of two distinct types of sites. The first site binds more tightly with a K_D of 1.3 uM. The second type of site, of which there appears to be two present on the enzyme, has a lower affinity. A separate ADP binding site also exists on the F_1 and is not affected by the presence of AMP-PNP or Co^{2+}.

TABLE 1. DIVALENT CATION AND ADENINE NUCLEOTIDE BINDING TO F_1 UNDER VARIOUS CONDITIONS

DIVALENT CATION	ADENINE NUCLEOTIDE	EXPERIMENTAL CONDITIONS
1st	not known	present in isolated enzyme treated with EDTA 1 hr
2nd	may be associated but not required	present in isolated enzyme Co^{2+} binding $\pm$ AMP-PNP
3rd	required for binding	only seen in the presence of AMP-PNP
4th	required for binding	only seen in the presence of AMP-PNP

The current study examines a fifth divalent cation site not previously reported. This site appears to promote inactivation of the soluble F_1-ATPase but is masked in the presence of the F_0 component of the complex.

RESULTS

The F_1-ATPase is shown to be inhibited when incubated in the presence of only divalent cation. Half maximal inhibition occurs in less than 30 min (0.32 mg/ml protein with 5 mM Mg^{2+} and 50 mM Tris-HCl, pH 7.5). The site(s) involved in this inactivation do not show the same specificity or kinetic properties as any of the other divalent cation binding sites previously reported (Table 2). The apparent K_m for $MgCl_2$ of the sites involved in ATPase activity is 33 times less than that of the inactivation site (4 μM vs 135 μM).

Table 2. Specificity of Divalent Cation Binding Sites on F_1

Divalent Cation Site	Substrate Specificity
ATPase Activity	$Zn^{2+} > Mn^{2+} > Mg^{2+} > Co^{2+} >> Ca^{2+} > Sr^{2+}$
Inactivation Site	$Mn^{2+} > Zn^{2+} > Ca^{2+} > Co^{2+} > Mg^{2+} > Sr^{2+}$

Incubation of the F_1-ATPase with $MgCl_2$ results in formation of a precipitate. Following a 4 h incubation at room temperature in the presence of 5 mM $MgCl_2$ (2.32 mg/ml in Tris buffer, pH 7.5) the F_1 had lost 90% of the initial ATPase activity. The incubation was centrifuged to separate the sediment and supernatant fractions. The results of that analysis indicate that the sediment contains only the α subunits and some of the γ subunits. The supernatant contained the β, δ, ϵ and the remaining fraction of the γ subunit. It should be noted that the loss in ATPase activity parallels the loss in α and γ subunits into the precipitate fraction.

The divalent cation inactivation caused by binding at this unique site is not protected by the substrate or products of the ATPase reaction or by the divalent cation chelator, EDTA. Protection experiments were performed by including one of the following in 1 h incubations of F_1 with $MgCl_2$: ADP, KP_i, ATP, ITP, GTP and EDTA. None of these compounds showed any ability to protect F_1 from inactivation. By contrast the ATPase activity was reduced only by 50% over <u>4 h</u> when incubated with $MgCl_2$ in the presence of 200 mM KP_i. The same incubation with $MgCl_2$ in Tris buffer results in 90% inactivation of ATPase activity. This is the only compound tested which partially prevented inactivation.

Table 3 shows that the inhibition caused by divalent cation binding to this site is <u>completely</u> prevented by the association of the F_1 moiety with F_0. When F_1 is rebound to F_1 depleted inner membrane vesicles (urea particles) no inhibition by $MgCl_2$ (2 h) is found. Control experiments showed no protection in the presence of $MgCl_2$ with any component of the reconstitution mixture (BSA, mannitol, sucrose). Therefore this protection is due to the specific interaction of the F_1 with the F_0.

Table 3. Protection By F_0 Against $MgCl_2$ Induced Inhibition of the Soluble F_1-ATPase

Incubation Conditions	% Loss in ATPase Actvity
F_1 alone	5
F_1 plus $MgCl_2$	70
F_1 + urea particles	3
F_1 + urea particles plus $MgCl_2$	6

DISCUSSION

Results from earlier divalent cation binding studies by this and other laboratories (Senior,1979,1980, Gruys,et al,1985, Daggett,et al,1985, Williams,et al,1987) have yielded a complex scheme of binding sites on the soluble F_1-ATPase. Here we have reported the presence of still another divalent cation binding site, one which may prove to be significant in the interaction of F_1 with F_0 in the membrane. Figure 1 summarizes these results. We distinguish three classes of divalent cation binding sites. Class I consists of three exchangeable sites which are localized on the β subunits (Kironde and Cross,1987). This class in fact is composed of two subsets. Subset A is the unique divalent cation binding site which does not require the presence of adenine nucleotide to be filled (Table 1, 2nd site). The other two exchangeable sites are only found with divalent cation when adenine nucleotide is present (Table 1, 3rd and 4th sites). Class II is represented by a single non-exchangeable divalent cation binding site which is on either an α or $\alpha\gamma$ complex (Table 1, 1st site). This site is only removed with a loss of functional and structural integrity of the F_1. The Class III site which is discussed in this report and in Pedersen,et al (1987) is a unique site which when occupied by divalent cation renders the soluble F_1 inactive. The F_0 component of the H^+-ATPase when added back to F_1 however protects against this inactivation.

This last class of divalent cation site may be related to or identical with a hysteretic site (inhibitory nucleotide binding site) described by DiPietro,et al (1980) since Mg^{2+} plus either ATP or ADP appears to inhibit at this site as well. On the basis of the results from our experiments it appears that when divalent cation (with or without ATP or ADP) occupies this site a subunit dissociation appears to occur with the α and some of the γ subunits being released from the F_1. The function of this site in the mitochondrion may represent some control mechanism for the inhibition of F_1 molecules that have been released from the membrane.

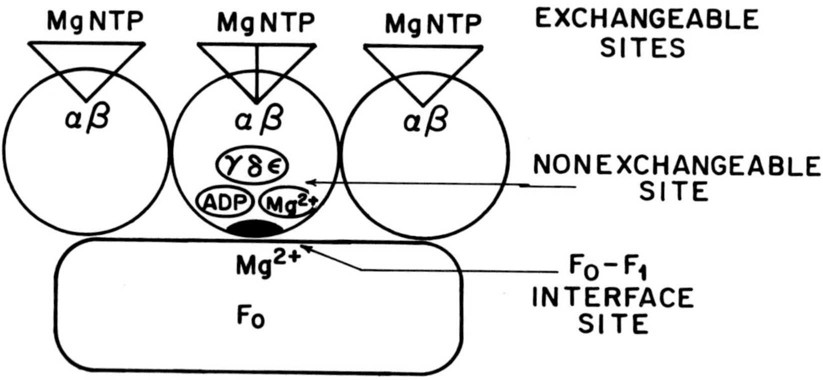

Figure 1. Model for Divalent Cation Binding Sites on F_1.

REFERENCES

Amzel LM, Pedersen PL (1983). Ann Rev Biochem 52:801.
Daggett SG, Gruys KJ, Schuster SM (1985)J Biol Chem 260:6213.
DiPietro A, Penin F, Godinot C, Gautheron DC (1980) Biochem 19:5671.
Gruys KJ, Urbauer JL, Schuster SM (1985)J Biol Chem 260:6533.
Hatefi Y (1985) Ann Rev Biochem 54:1015.
Kandpal RP, Stempel KE, Boyer PD (1987) Biochem 26:1512.
Kironde FA, Cross RL (1987) J Biol Chem 262:3488.
Lotscher HR, Capaldi RA (1984) Biochem Biophys Res Commun 121:331.
Pedersen PL, Carafoli E (1987) Trends in Biochem Sci 12:146.
Pedersen PL, Carafoli E (1987) Trends in Biochem Sci 12:186.
Pedersen PL, Williams N, Hullihen J (1987)Biochem (in press).
Sakamoto J (1984) J Biochem 96:475.
Senior A E (1979) J Biol Chem 254:11319.
Senior A E, Richardson L V, Baker K, Wise J G (1980) J Biol Chem 255:7211.
Williams N, Hullihen J, Pedersen P L (1984) Biochem 23:780.
Williams N, Hullihen J, Pedersen P L (1987) Biochem 26:162.

The Ion Pumps: Structure, Function, and Regulation, pages 93–98

IMAGE ANALYSIS OF THE *ESCHERICHIA COLI* LACTOSE PERMEASE MOLECULE IN FILAMENTOUS ARRAYS

Jade Li and Phillip Tooth

Medical Research Council Laboratory of Molecular Biology, Cambridge, England

INTRODUCTION

The *E.coli* lactose permease catalyzes an obligatory co-transport of 1 mole of proton with 1 mole of galactoside across the cytoplasmic membrane (Kaback, 1986; Overath and Wright, 1983). The *lac y* gene coding for the permease has been cloned and sequenced, allowing the polypeptide sequence to be deduced (Buchel et al., 1981). Of 417 residues, 71% are made up by non-polar amino acids, making the lactose permease one of the most hydrophobic membrane proteins known. The cloned permease has been over-expressed in the *E. coli* T206 strain, where it amounts to 10-15% of the protein in the cytoplasmic membrane (Teather et al., 1980). From these membranes the permease has been purified in detergents, and reconstituted into lipid vesicles in a functional state (Newman et al., 1981; Wright and Overath, 1984). By probing the surfaces of the cytoplasmic membrane and the reconstituted vesicles, it was found that both the N- and C-termini of the polypeptide are located on the cytoplasmic surface, and the polypeptide is also accessible to monoclonal antibodies and proteases from the periplasmic surface (see above reviews). However, freeze-etching electron microscopy showed that the lactose permease molecule does not protrude significantly outside the bilayer surface (Costello et al., 1984). Based on the CD and Raman spectroscopy evidence that 70-85% of the residues are in α-helical conformation, and on hydropathy plots of the sequence (Kyte and Doolittle, 1982), secondary structure models have been proposed dividing the sequence into 12 or 14 relatively hydrophobic trans-membrane segments, linked by short hydrophilic peptides (Forster et al., 1983; Vogel et al., 1985).

In order to determine the three-dimensional structure of lactose permease so necessary to the understanding of its function, we have prepared ordered arrays of the lactose permease molecules in the absence of large excess of lipids and detergents. Analysis of these arrays by electron microscopy determined the dimensions of the permease molecule in the membrane plane, and revealed that this protein comprises two domains separated by a solvent-accessible cleft.

METHODS

The lactose permease was purified from the cytoplasmic membrane of the T206 strain by Ecteola chromatography in the detergent dodecyl-ß-maltoside (DM; Wright and Overath, 1984). Further purification on the CM-52 column and assays for protein, phospholipids, and detergent concentrations, as well as for the substrate-binding activity have been described elsewhere (Li and Tooth, 1987). Details of the preparation of filaments, electron microscopy, and procedures for image analysis are given in the same reference.

RESULTS

The lactose permease preparation eluted from the CM-52 column contained a minimal amount of phospholipids, i.e. 4-5 moles per mole of the permease, and it contained free DM (unbound to the protein) at the CMC of this detergent. This preparation of lactose permease remained soluble and retained the substrate-binding activity as measured by equilibrium dialysis assay (Li and Tooth, 1987).

Dialysis of the CM-52 purified lactose permease against a solution containing 20 mM MES or Tris buffer, pH 6 to 8, 0.1 mM EDTA, and 3 mM NaN_3, but without added detergent, further reduced the free DM concentration to below the CMC and caused the lactose permease to aggregate. Under three conditions, ordered aggregates in the form of sheets of filaments lying side by side (Fig. 1) were obtained. These conditions were: low temperature (4°C) during dialysis, a protein concen-tration above 0.1 mg/ml, and the presence of 0.1 M of KCl or NH_4Cl which promoted lateral association of the filaments. Filaments up to several µm in length have been observed, and within small areas of the sheets, such as boxed off in Figure 1, two-dimensional order can be demonstrated by optical diffraction or by computing the Fourier transform from electron micrographs.

Figure 2 shows a representative collection of the filamentous arrays. The parallel filaments are well delineated by the negative stain. Along the filament the staining also has a periodic character. Moreover, the repeat unit in the filament is bisected by a stain-

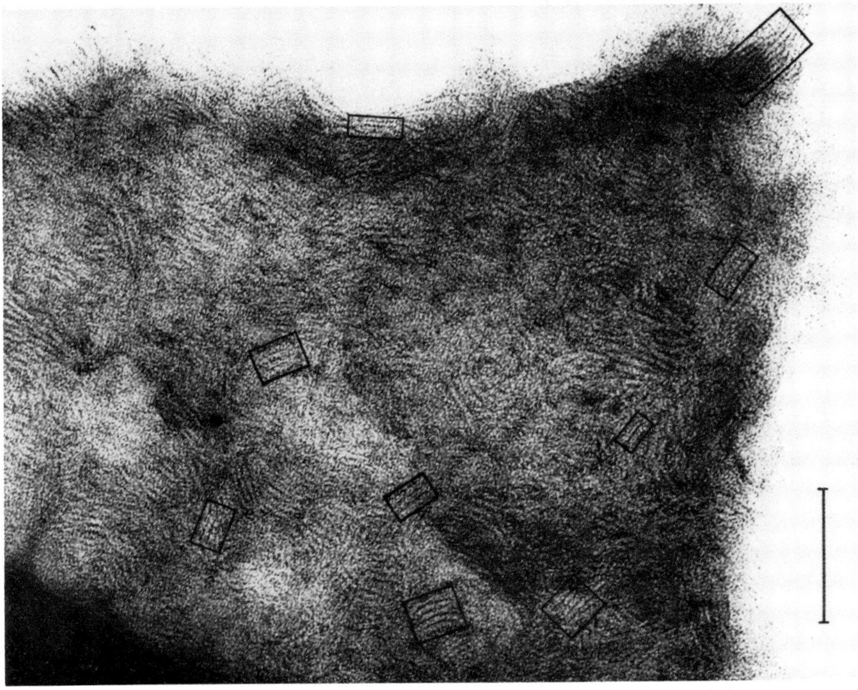

Fig. 1: Electron micrograph of a sheet formed by the lateral association of lactose permease filaments, negatively stained with 1% uranyl formate. The boxed areas show two-dimensional order. Scale bar: 1000 Å.

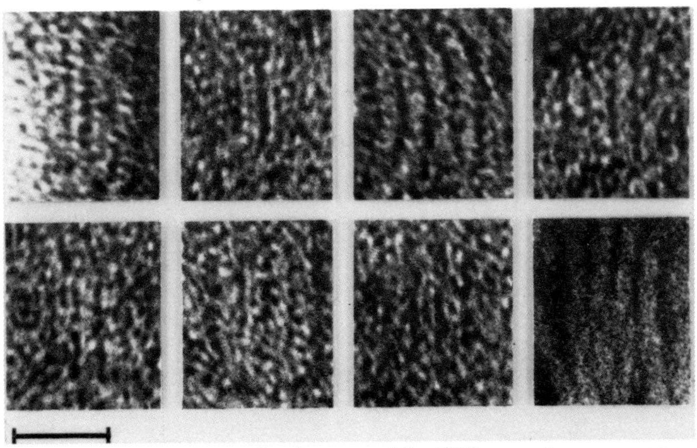

Fig. 2: A sample of the small filamentous arrays subjected to image analysis. Note the stain-penetrated clefts on the midline of the filaments. Scale bar: 200 Å.

penetrated cleft, so the filaments appear double-stranded. The optical and computed transforms of such arrays showed one or two orders of equatorial spots due to the interfilament repeat, which was 50.7+-1.6 Å averaged over 23 arrays. They also showed a near-axial spot due to the periodicity along the filaments. This was 28.4+-1.6 Å averaged over 15 arrays.

DISCUSSION

When the free detergent concentration in the lactose permease sample was reduced by dialysis to below the CMC, filamentous arrays were formed to internalize the hydrophobic molecular surfaces that would otherwise be exposed to the aqueous environment. The lactose permease molecules had a ribbon-like arrangement in the filaments, rather than a helical arrangement, as indicated by the common orientation of the cleft in subunits in the same filament. The computed transforms of the arrays had a handedness, implying that the subunits of the array point to the same face of the sheet, and not alternately up and down. This is best accounted for by adjacent molecules having the same orientation as they would in the natural membrane.

The projected area per subunit was 1440 $Å^2$, defined by the filament width of 50.7+-1.6 Å and the axial repeat of 28.4+-1.6 Å. This size is within 10% of the estimated cross-sectional area of the molecule, which is 1322-1487 $Å^2$ based on the molecular weight of 46504, a partial specific volume of 0.77 cm^3/g for membrane proteins, and a membrane thickness of 40-45 Å. The area to molecular weigh ratio can be compared, and agrees well, with that calculated for bacteriorhodopsin, another very hydrophobic membrane protein with a high helical content. The latter has an experimentally determined cross-sectional area of about 800 $Å^2$ (Henderson and Unwin, 1975) and a molecular weight of 26000. The agreement is consistent with the subunit of the filamentous arrays being the lactose permease monomer seen in the face-on view as if it were held in the membrane. Thus, the lactose permease molecule has an elongated cross-section of 51 Å by 28 Å in the membrane plane .

The observation that a stain-penetrated cleft bisects the projected area of the subunits strongly suggests that the lactose permease molecule is comprised of two globular domains of similar size. This cleft would appear to be accessibe from the aqueous compartments at the membrane surfaces. Very likely the galactoside binding site is located within this cleft.

The organization into two domains may be correlated with internal sequence repeats between the N- and C-terminal halves of the polypeptide. We can identify two oligopeptide segments within the N-terminal half which are homologous in tandem with two segments at corresponding positions within the C-terminal half. These segments are: (68)DKLGLR(73)...(136)SNFE(139), and, (284)NRIGGK(289)...(339)SQFE(342). Sequences adjacent to these segments can also be aligned such that the corresponding residues from the N- and C-halves are similar in hydrophobicity. Moreover, the homologous segments found here contain the motif RXG(R)R, which is strongly conserved in a family of sugar transporters and is internally duplicated within each sequence at similar positions (Maiden et al., 1987). These proteins and the lactose permease thus appear to have evolved from an ancestral transporter that had undergone an internal gene duplication. This implies that despite the limited sequence homology between N- and C-halves of the present protein, the folding of the two halves of polypeptide can be rather similar, so that each half folds into one of the two globular domains which are made up of similarly arranged secondary structure elements. One consequence of the two domains having similar structure is that the lactose permease polypeptide must traverse the membrane by a multiple of 4 times. Therefore secondary structure models predicting 12 trans-membrane helices are plusible, whereas those predicting 14 are not.

In summary, by placing the lactose permease molecule in ordered arrays of filaments lying side by side, we were able to determine that it has an elongated cross-section of 51 Å by 28 Å in the membrane plane, and that it is made up of two globular domains separated by a cleft accessible from the membrane surface. The two- domain organization may be correlated with a 2-fold internal repeat in the polypeptide sequence.

REFERENCES

Buchel DE, Gronenborn B, Muller-Hill B (1981). Sequence of the lactose permease gene. Nature 283:541-545.

Costello MJ, Viitanen P, Carrasco N, Foster DL, Kaback HR (1984). Morphology of proteoliposomes reconstituted with purified *lac* carrier protein from *Escherichia coli*. J Biol Chem 259:15579-15586.

Forster DL, Boublik M, Kaback HR (1983). Structure of the *lac* carrier protein of *Escherichia coli*.

J Biol Chem 258:31-34.
Henderson R and Unwin PNT (1975). Three-dimensional model of purple membrane obtained by electron microeoscopy. Nature 257:28-32.
Kaback HR (1986). Proton electrochemical gradients and active transport: the saga of *lac* permease. Ann N Y Acad Sci 456:291-304.
Kyte J, Doolittle RF (1982). A smiple method for displying the hydropathic character of a protein. J Mol Biol 157:105-132.
Li J, Tooth P (1987). Size and shape of the *Escherichia coli* lactose permease measured in filamentous arrays. Biochem 26:4816-4823.
Maiden MCJ, Davis EO, Baldwin SA, Moore DCM, Henderson PJF (1987). Mammalian and bacterial sugar transport proteins are homologous. Nature 325: 641-643.
Newman MJ, Foster DL, Wilson TH, Kaback HR (1981). Purification and reconstitution of functional lactose carrier from *Escherichia coli.* J Biol Chem 256:11804-11808.
Overath P, Wright, K (1983). Lactose permease: a carrier on the move. Trends in Biochem Sci 8:404-408.
Teather RM, Bramhall J, Riede I, Wright JK, Furst M, Aichele G, Wilhelm U, Overath P (1980). Lactose carrier protein of *Escherichia coli*. Structure and expression of plasmids carrying the *y* gene of the *lac* operon. Eur J Biochem 108:223-231.
Vogel H, Wright JK, Jahnig F (1985). The structure of the lactose permeae derived from Raman spectroscopy. EMBO J 4:3625-3631.
Wright JK, Overath P (1984). Purification of the lactose:H^+ carrier of *Escherichia coli* and characterization of galactoside binding and transport. Eur J Biochem 138:497-508.

The Ion Pumps: Structure, Function, and Regulation, pages 99–104

FREEZE-FRACTURE ELECTRON MICROSCOPY STUDY OF BACTERIO-RHODOPSIN OLIGOMERISATION.

Jean-Louis Rigaud, Tadeusz Gulik-Krzywicki and Michel Seigneuret.

Dept of Biology, CEN Saclay (J.L.R and M.S) and Centre de Genetique Moleculaire, C.N.R.S (T.G-K) Gif sur Yvette, 91190, France.

INTRODUCTION

One of the most striking structural feature of the purple membrane of Halobacterium halobium is the organisation of bacteriorhodopsin (BR) in a two dimensional hexagonal lattice of protein trimers (Henderson and Unwin, 1975 ; Blaurock, 1975). This ability to cristallize in two dimensions appears to be an intrinsic property of the BR polypeptide. Indeed, when the purified protein is forced to aggregate in reconstituted liposomes, e.g by using low lipid to protein ratios or by cooling the lipids below their phase transition temperature, the hexagonal lattice spontaneously forms (Cherry et al, 1978 ; Heyn et al, 1981). Although several intrinsic proteins tested in this regard are known to aggregate under similar conditions, in no other case than BR is such a cristalline arrangement formed (Swanson et al, 1980 ; Kusumi and Hyde, 1982 ; Chen and Hubbell, 1973). Thus a detailed study of the BR aggregation process may be important, in particular to know whether BR can exist in the membrane as limited and well defined states of association which may represent intermediates in the cristallisation process. To this end an improved freeze-fracture electron microscopy procedure has been developped and applied to the study of the association of BR in large proteoliposomes reconstituted by reverse-phase evaporation with egg lecithin (Rigaud et al, 1983).

MATERIALS AND METHODS.

BR proteoliposomes were cryoprotected with glycerol (30% w/w) and rapidly frozen in liquide propane, using conventional Balzer's gold planchetts. The fracturing was performed at - 150°C with a nitrogen cooled knife, for the conventionally frozen samples. The replication of the fractured surfaces was performed in the direction of fractures using Tungsten-Tantalum (W-Ta) alloys in four to six steps, each lasting a few seconds and separated by about 10 second periods during which the partly shadowed, fractured surfaces were allowed to cool. In order to avoid contamination of the fractured surfaces, the samples were protected, between each shadowing step, by a liquid nitrogen cooled knife.

RESULTS AND DISCUSSION

High magnification freeze-fracture electron micrographs of BR-proteoliposomes reconstituted at high lipid to protein ratio are shown in Fig.1 together with views of pure lipid vesicles. As opposed to pure lipid material (Fig.1A), the BR-liposomes clearly show the presence of intramembrane particles (Fig.1B) randomly dispersed ou both concave and convex fracture faces. The absence on particles in the background ice indicates that all the protein is incorporated into the proteoliposomes during the reconstitution (this was observed for all lipid to protein ratios tested). A quantitative analysis of the intramembrane particle size distribution (Fig.1D) indicates that the average diameter of the particles is about 4.5nm, which is very close to the dimensions of a monomeric BR (2.5x3.5x4.5nm) as determined by electron, microscopy of the purple membrane (Henderson and Unwin, 1975). This clearly suggests that BR is monomeric at high lipid to protein ratios.

The first result of this study is that our improved freeze-fracture-shadowing procedure can provide a relatively good estimation of the size of intrinsic membrane proteins. The choice of BR for testing this procedure was particularly appropriate, since it is one of the smallest intrinsic membrane proteins (M.W.27K). With such a small protein, the relative enlargement due to metal deposition may be expected to be rather large, leading to an important overestimation of its size. The results shown here demonstrate, however, that the average diameter of the intramembrane particles corresponding to the BR monomers (4.5nm) is only some 1-1.5nm larger than the average molecular size of BR. It is very likely that for larger intrinsic membrane proteins, this relative enlargement will be even smaller. This situation contrasts with the conventional freezefracture procedure, which leads to a mean size of intramembrane particles of the order of 10nm (see Fig.1C and 1D), i.e. an enlargement of about 6-7nm in comparison to the true molecular diameter.

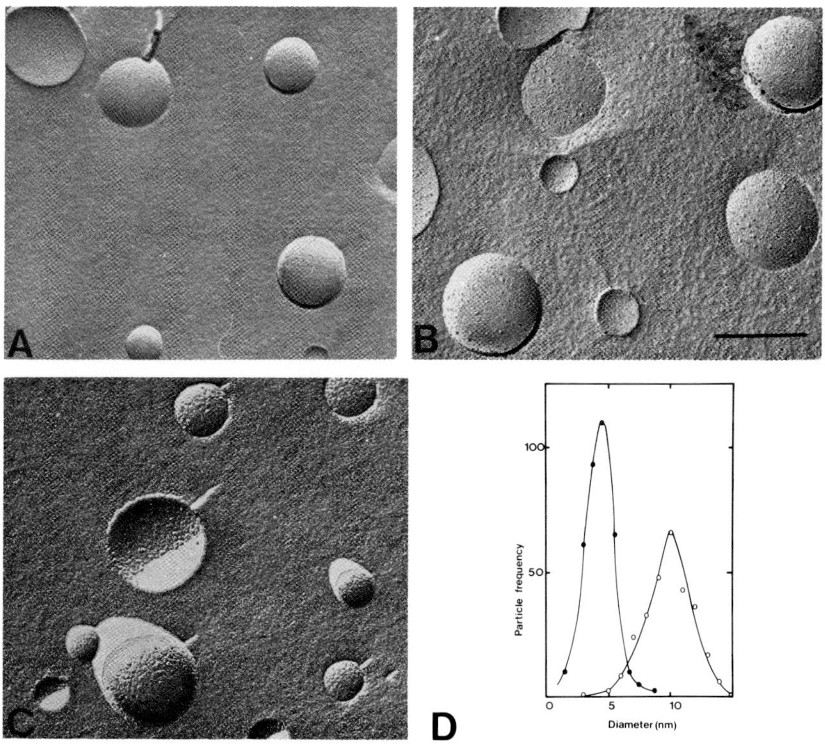

Figure 1 : A, pure liposomes. B, bacteriorhodopsin proteoliposomes at a lipid to protein ratio of 40 (w/w). C, the same preparation as in B, but freeze-fractured in a conventional way and shadowed with Pt-C.D, size distribution of intramembrane particles measured on bacteriorhodopsin proteoliposomes such as those shown in B and C, (closed circles, improved freeze-fracture procedure ; open circles, conventional freeze-fracture procedure). The bar in B represents 200nm.

BR-proteoliposomes of different protein to lipid ratios, from 80 to 1 w/w were analyzed by freeze-fracture : the striking observation is that the size of the intramembrane particles increases with the increasing protein content. This is clearly seen on the histograms of particle size distributions, such as those shown in Fig.2 At high (40 w/w) lipid to proteins ratios this distribution is symmetric and centered at about 4.5nm. At the intermediate ratios (20-3 w/w) the distribution is bimodal, probably indicating the appearence of rather well defined larger particles. At the lowest lipid to protein ratio tested (1w/w) the particle size distribution is again almost symetric but now it is centered at about 7.5nm. Lower lipid to protein ratios were not tested since these resulted in incomplete incorporation and partial bleaching of BR.The simplest explanation of these data is to assume that, at high lipid to protein ratios BR is monomeric. When the protein content increases above a given lipid to protein ratio the BR monomer starts to self associate into a rather well defined oligomer, as can be deduced from the presence of a bimodal particle size distribution, instead of a broadening of such a distribution which would be expected for the random aggregation of BR molecules. This self association process seems to be completed at a lipid to protein ratio of about 1 w/w.

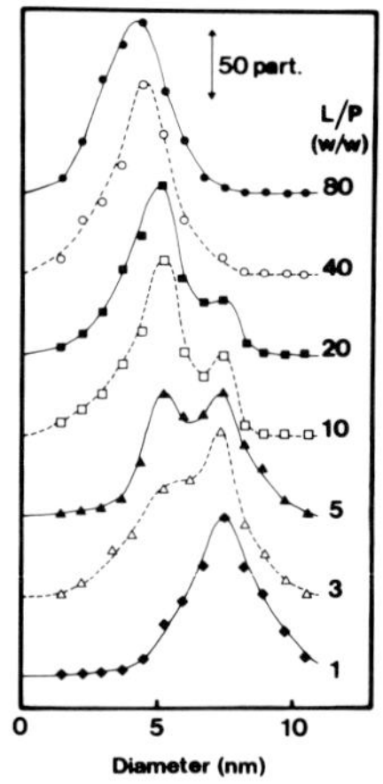

Figure 2 : Size distribution of intramembrane particles in bacteriorhodopsin proteoliposomes as a function of lipid to protein ratio (w/w).

Independent information on the aggregation properties of BR within proteoliposomes can be obtained from measurement of the surface density of intramembrane particles as a function of the lipid to protein ratio. The estimation of particle densities takes advantage of the presence in each preparation of a certain number of large vesicles (500nm) displaying almost flat central regions. Counting the particles on such large surfaces gives the most accurate particle density values. These values are compared in Table 1 with the expected densities of BR molecules, calculated for corresponding lipid to protein ratios.

TABLE 1 : Intramembrane particle density as a function of lipid to protein ratios.

Lip/BR (w/w)	Lip/BR (M/M)	Calculated BR surface density ($BR/\mu m^2$) (*)	Measured Particle density ($BR/\mu m^2$)
80	2 500	1 150	500
40	1 250	2 240	1 200
20	625	4 400	2 000
10	312	8 500	3 500
5	155	15 800	3 500
3	93	24 400	3 600
1	31	51 000	7 500

(*) The surface density of BR was calculated by assuming a surface area of 875 $Å^2$ for a BR molecule and of 70 $Å^2$ for phospholipid.

At high lipid to protein ratios, the two densities are equal within experimental error (after dividing the calculated density by 2 to take into account the fact that particles distribute equally between concave and convex fracture faces). This reinforces the conclusion that the particles observed at high lipid to protein ratios correspond to BR monomers.

For lipid to protein ratios of 20 (w/w) and below, the measured values of the particle densities deviate progressively from the expected ones, confirming that BR aggregation is taking place. The degree of BR oligomerization obtained from the ratio of the experimental to the expected particle density values for the lowest lipid to protein ratio tested here is 3.4.

CONCLUSION

Several intrinsic membrane proteins reconstituted with lipids above the order-disorder transition tempéra-ture have been shown to aggregate at low lipid to protein ratios. Usually such effects occur at lipid to protein ratios of the order of 100-200 (mole/mole) and it results in the segregation of protein patches. The BR oligomeri-zation process described here appears to be of different nature. Indeed, it begins at very high lipid to protein ratio, below 40 (w/w) (i.e for about 1400 lipids/pro-tein). Furthermore it is limited to the formation of BR trimers or tetramers and does not lead to protein segregation even when very low (1w/w, i.e. about 30 lipids/protein) lipid to protein ratios are reached.

Such proporties may be related to the formation of the BR hexagonal lattice during the biogenesis of the purple membrane in Halobacterium : indeed the first step may be the formation of such oligomers which aftewards segregate in an hexagonal lattice when a very high local protein concentration is reached. Thus, formation of BR patches would occur in a sequential and progressive way rather than by a rapid all or nothing phase separation. This could explain why a two-dimensional cristalline arrangement is formed when BR segregates, unlike what is observed with other membrane proteins.

REFERENCES

Blaurock AE (1975). J Mol Biol 93 : 139-158

Chen YS and Hublell W (1973). Exp Eye Res 17 : 517-532

Cherry RJ, Muller U, Henderson R and Heyn MP (1978). J Mol Biol 121 : 283-298

Henderson R and Unwin PNT (1975). Nature 257 : 28-32

Heyn MP, Cherry RJ and Dencher MA (1981). Biochemistry 20 : 840-849

Kusumi A and Hyde SJ (1982). Biochemistry 21 : 5978-5983

Rigaud JL, Bluzat A and Buschlen S (1983). Biochem Bio phys Res Commun 111 : 373-382

Swanson MS, Quintanilha AT and Thomas DD (1980). J Biol Chem 255 : 7494-7502

The Ion Pumps: Structure, Function,
and Regulation, pages 105–112

MOLECULAR CHARACTERIZATION OF A UNIQUE ANION PUMP: THE ArsA PROTEIN IS AN ARSENITE(ANTIMONATE)-STIMULATED ATPase.

Barry P. Rosen, Ulrich Weigel, Cyrus Karkaria, and Preeti Gangola.
Department of Biochemistry, Wayne State University School of Medicine, Detroit, MI 48201

INTRODUCTION

Every ion pump reported to date translocates cations except for two: halorhodopsin, a light-driven chloride pump (Lanyi, 1987), and an arsenite(antimonate) pump proposed to be an oxyanion translocating ATPase (Chen et al., 1986). The latter is a plasmid-encoded system which, when expressed in *Escherichia coli*, produces resistance to the oxyanions arsenate, arsenite and antimonate. Extrusion as a mechanism of resistance is a frequent strategy evolved by plasmids (Mobley and Summers, 1987). The energy for arsenical transport is most likely ATP (Mobley and Rosen, 1982, Silver and Keech, 1982, Rosen and Borbolla, 1984). Two genes, *arsA* and *arsB*, are required for resistance to arsenite and antimonate (Chen et al.,1985), and the anion pump is composed of two proteins, the 63 kDa hydrophilic ArsA protein and the hydrophobic 45.5 kDa ArsB protein (Chen et al., 1986). The ArsB protein, which is found in the inner membrane of *E. coli* (San Francisco et al., 1987), is predicted to be the anion-conductive component. The ArsA protein is found in the cytosol and on the membrane (Mobley et al., 1983), and membrane binding requires the ArsB protein (L. Tisa and B.P. Rosen, unpublished). Two nucleotide binding sites have been predicted in the ArsA protein based on the predicted amino acid sequence (Chen et al., 1986). In this paper we discuss the evidence that the ArsA protein binds ATP and has ATPase activity. We postulate that that the oxyanion pump is composed of a catalytic subunit, the ArsA protein, and an anion translocation subunit, the ArsB protein.

RESULTS

The ArsA protein was isolated from a strain of *E. coli* carrying the recombinant plasmid pUM3, which constitutively produces it in large amounts (Mobley et al., 1983). The ArsA protein was purified from the cytosol by sequential chromatography on columns of Q Sepharose, red agarose and Superose. The degree of purification was determined by SDS gel electrophoresis (Fig. 1). The sequence of the first 12 residues of the purified protein determined to be identical to the predicted sequence of Met-Gln-Phe-Leu-Gln-Asn-Ile-Pro-Pro-Tyr-Leu-Phe (Chen et al., 1986).

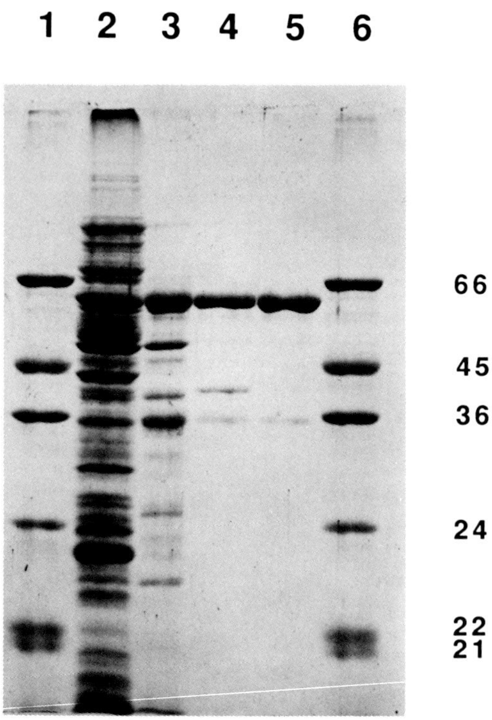

Fig. 1. **Purification of the ArsA protein.** The ArsA samples from each purification step column were analyzed by SDS gel electrophoresis. A 10% polyacrylamide gel was used with Coomassie blue staining. Lanes 1 and 6: mixture of proteins standards; lane 2: 30 µg cytosol; lane 3: 6 µg pooled fractions from Q Sepharose; lanes 4-5: 1 µg each of pooled fractions from red agarose and Superose columns.

The ArsA protein has two short regions of sequence homology with the members of a family of nucleotide binding proteins (Walker et al., 1982). The regions of homology are in the predicted nucleotide binding domains, suggesting that the ArsA protein has two nucleotide binding regions. Columns with coupled triazine dyes, which bind to proteins with such nucleotide binding folds, are used for affinity chromatography of these proteins (Thompson et al., 1975). The binding of the ArsA protein to a red agarose column (Fig. 1) supports the idea that the protein binds nucleotides.

To demonstrate direct binding of nucleotides the ArsA protein was photolabeled with α-[^{32}P]ATP by exposure to ultraviolet light (Yue and Schimmel, 1977, Biswas and Kornberg, 1984, Banks and Sedgwick, 1986). Unlabeled ATP and ADP were competed with radioactive ATP for photolabeling, but AMP did not (Fig. 2). Other nucleotides, including GTP, UTP, or CTP, did not compete even at a 20-fold excess over labeled ATP. ATP binding did not require arsenite or antimonate.

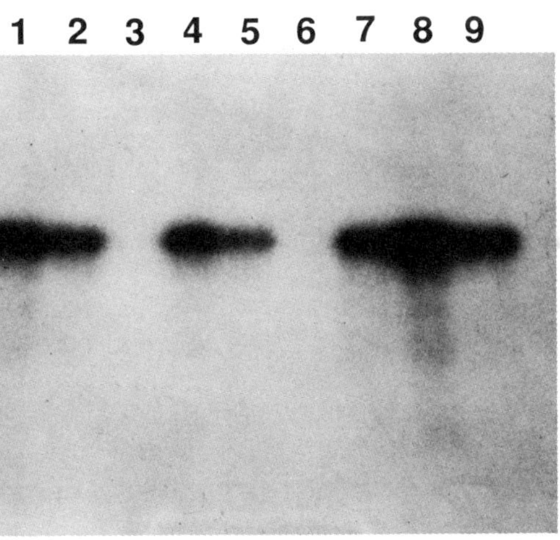

Fig. 2. **ATP photolabeling of the ArsA protein.** Photolabeling was performed by a modification of the method of Yue and Schimmel, 1977. Unlabeled nucleotides were added as competitors at the following concentrations: lane 1, no addition; lane 2, 10 μM ATP; lane 3, 100 μM ATP, lane 4, 1 μM ADP; lane 5, 10 μM ADP; lane 6, 100 μM ADP; lane 7, 1 μM AMP; lane 8, 10 μM AMP; lane 9, 100 μM AMP. Samples were analyzed on a 15 % SDS polyacrylamide gel.

The ArsA protein could bind the fluorescent ATP analog 2',3'-o-(2,4,6)trinitrophenyl adenosine triphosphate (TNP-ATP), as measured by the enhancement of TNP-ATP fluorescence produced by the addition of protein (Grubmeyer and Penefsky, 1981, Mas and Colman, 1985). Binding saturated with an 2 μM apparent Kd (Fig. 3). TNP-ATP binding did not require arsenite or antimonate and was competed only by ATP or ADP. Binding required $Mg2^{+}$.

The purified ArsA protein catalyzed ATP hydrolysis. In contrast to nucleotide binding, hydrolysis was dependent on the presence of antimonate or arsenite. The maximal velocity was 5- to 10-fold greater with arsenite and 20- to 50-fold greater with antimonate, and the concentration of either oxyanion required for half-maximal stimulatory concentration was 0.1 mM (Fig. 4). No oxyanion stimulated hydrolysis of GTP, CTP or UTP was observed.

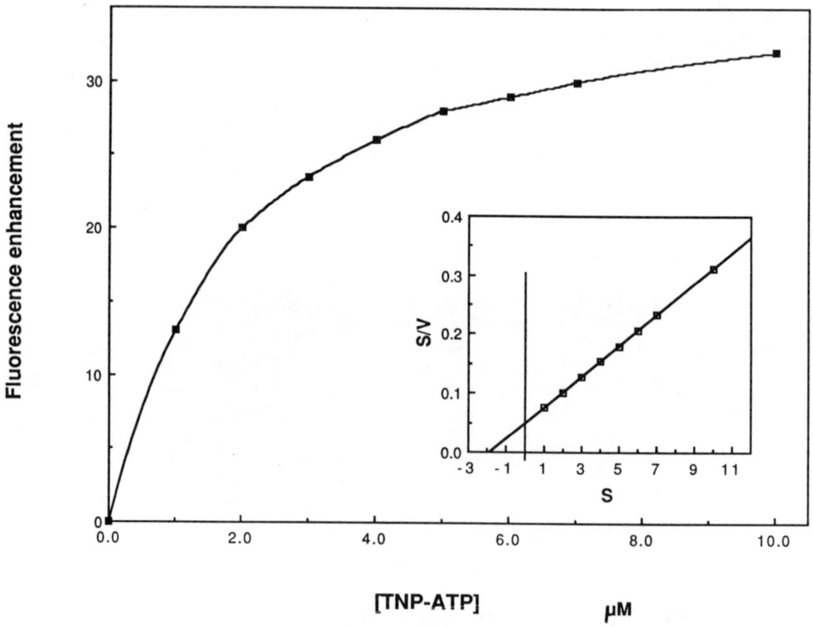

Fig. 3. **Binding of TNP-ATP to the ArsA protein.** Binding of TNP-ATP was quantified according to the method of Mas and Colman, 1985. Increasing amounts of TNP-ATP were added to a solution with or without 5 μM of purified ArsA protein. The values shown represent the difference between those two sets of data.

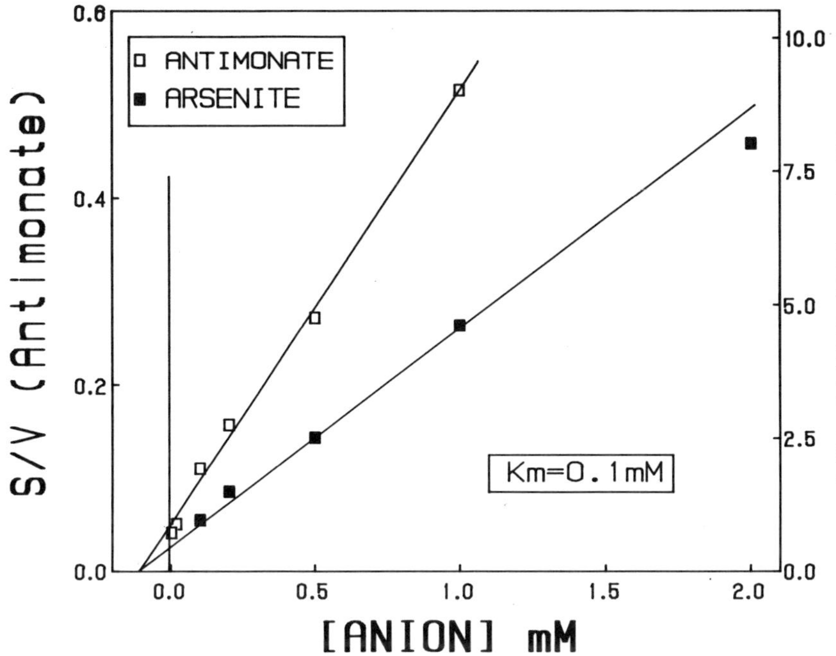

Fig. 4. **ATPase activity of the purified ArsA protein.** ATPase activity of the purified ArsA protein was assayed in the presence of varying concentrations of sodium arsenite (■) or potassium antimonyl tartrate (□). Liberated phosphate was measured by the method of Lanzetta et al., 1979.

DISCUSSION

The oxyanion pump encoded by the arsenical resistance operon of plasmid R773 is unique in that all other ATP-translocating ATPases thus far reported pump cations and not anions. The nucleotide sequence of the arsA gene shows that it is the product of a gene duplication, so that the first half of the arsA gene and first half of the ArsA protein are both homologous to the other half of gene and protein. Thus the protein has two potential nucleotide binding sequences (Chen et al., 1986). The binding of such proteins to triazine dye columns was utilized for purification of the ArsA protein. The purified protein was shown to binding adenylate nucleotides by two methods: UV photolabeling with α-[^{32}P]ATP and fluorescence enhancement of TNP-ATP. By both assays only ATP and ADP were active, and in neither case was an oxyanionic substrate required. The purified protein also hydrolyzed ATP, but, in contrast to binding, either antimonate or arsenite was required. Thus binding and hydrolysis of ATP are separable steps, where binding is independent of oxyanion, while catalysis requires both nucleotide and oxyanion. These results are consistent with a model of an oxyanion-translocating ATPase in which the ArsA protein is the catalytic subunit and the ArsB protein is an inner membrane protein which acts as the anion translating subunit (Fig. 5).

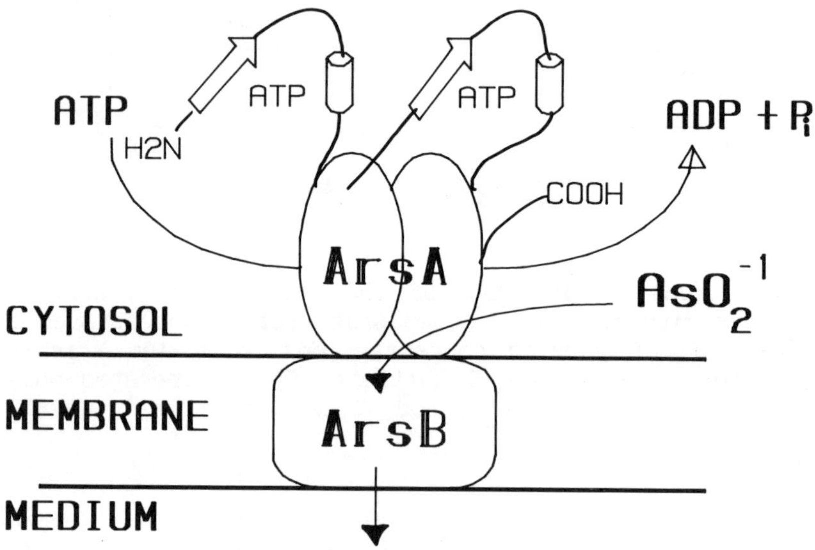

Fig. 5. Model of the arsenical pump. The integral membrane protein ArsB functions as an anion channel. The ArsA protein is an extrinsic membrane protein with ATPase activity. The complex of the ArsA and ArsB proteins is an arsenite pump. Interaction of a third subunit, the ArsC protein, (not shown) with the complex allows recognition and extrusion of arsenate (Rosen and Borbolla, 1984, Mobley et al., 1985, Chen et al., 1986)

REFERENCES

Banks, GR, Sedgwick, SG (1986). Direct ATP photolabeling of Escherichia coli RecA proteins: Identification of regions required for ATP binding. Biochemistry **25**:5882-5889.

Biswas, SB, Kornberg, A (1984). Nucleoside triphosphate binding to DNA polymerase III holoenzyme of Escherichia coli: a direct photoaffinity labeling study. J. Biol. Chem. **259**:7990-7993.

Chen, C-M, Mobley, HLT., Rosen, BP (1985). Separate resistances to arsenate and arsenite(antimonate) encoded by the arsenical resistance operon of R-factor R773. J. Bacteriol. **161**:758-763.

Chen, C-M, Misra, T, Silver, S, Rosen, BP (1986). Nucleotide Sequence of the Structural Genes for an Anion Pump: the Plasmid-Encoded Arsenical Resistance Operon. J. Biol. Chem. **261**:15030-15038.

Grubmeyer, C, Penefsky, HS (1981). Cooperativity between catalytic sites in the mechanism of action of beef heart mitochondrial adenosinetriphosphatase. J. Biol. Chem. **256**:3728-3734.

Laemmli, U. (1970). Cleavage of structural proteins during the assembly of the head of bacteriophage T4. Nature **227**:680-685.

Lanzetta, PA, Alvarez, LJ, Reinach, PS, Candia, OA (1979). An improved assay for nanomole amounts of inorganic phosphate. Anal. Biochem. **100**:95-97.

Lanyi, JK (1987) Chloride,, nitrate, and sulfate transport in bacteria. In Rosen, B.P. and Silver, S. (eds) "Ion Transport in Prokaryotes". New York: Academic Press, in press.

Mas, MT, Colman, RF (1985) Spectroscopic studies of the interactions of coenzymes and coenzyme fragments with pig heart, oxidized triphosphopyridine nucleotide specific isocitrate dehydrogenase. Biochemistry **24**:1634-1646.

Mobley, HLT, Chen, C-M, Silver, S, Rosen, BP (1983). Cloning and expression of R-factor mediated arsenate resistance in Escherichia coli. Mol. Gen. Genet. **191**:421-426.

Mobley, HLT, Rosen, BP (1982). Energetics of plasmid-mediated arsenate resistance in Escherichia coli. Proc. Natl. Acad. Sci. USA. **79**:6119-6122.

Mobley, HLT, Summers, AO (1987). Plasmid-encoded transport systems. In Rosen, B.P. and Silver, S. (eds) "Ion Transport in Prokaryotes". New York: Academic Press, in press.

Rosen, BP, Borbolla, MG (1984). A plasmid-encoded arsenite pump produces arsenite resistance in Escherichia coli. Biochem. Biophys. Res. Commun. **124:**760-765.

San Francisco, MJD, Chen, C-M, Rosen, BP (1987). Identification of the membrane component of the anion pump encoded by the arsenical resistance operon of R-factor R773. In Bronner, F, Peterlik, M (eds): "Calcium and Phosphate Transport Across Biomembranes", New York, Alan R. Liss, in press.

Silver, S, Keach, D (1982). Energy-dependent arsenate efflux: the mechanism of plasmid mediated resistance. Proc. Natl. Acad. Sci. USA **79:**6114-6118.

Silver, S, Budd, K, Leahy, KM, Shaw, WV, Hammond, D, Novick, RP, Willsky, GR, Malamy, MH, Rosenberg, H (1981). Inducible plasmid-determined resistance to arsenate, arsenite and antimony (III) in Escherichia coli and Staphylococcus aureus. J. Bacteriol. **146:**983-996.

Thompson, ST, Cass, KH, Stellwagen, E (1975). Blue dextran-sepharose: an affinity column for the dinucleotide fold in proteins. Proc. Natl. Acad. USA **72:**669-672.

Walker, JE, Saraste, M, Runswick, MJ, Gay, NJ (1982). Distantly related sequences in the α- and β-subunits of the ATP synthase, myosin, kinases and other ATP-requiring enzymes and a common nucleotide binding fold. EMBO J. **1:**945-951.

Yue, VT, Schimmel, PR (1977). Direct and specific photochemical cross-linking of adenosine 5'-triphosphate to an aminoacyl-tRNA synthetase. Biochemistry **16:**4678-4684.

The Ion Pumps: Structure, Function,
and Regulation, pages 113–120

THEORETICAL STUDIES ON THE FORMATION AND PROPERTIES OF BUNDLES OF α-HELICES AND THEIR APTITUDE TO FORM ION-CHANNELS.

Alberte Pullman

Laboratoire de Biochimie Théorique associé au CNRS, Institut de Biologie Physico-Chimique, 13 rue Pierre et Marie Curie, 75005 Paris, France.

INTRODUCTION

One of the most favored hypotheses concerning the structure of the membrane-spanning part of ion-transport proteins is the existence of bundles of "hydrophobic" α-helical polypeptide segments apt to include channels, or at least pathways, for ions. The generalization of this hypothesis was largely influenced by the model built for bacteriorhodopsin on the basis of low-resolution maps of electron density (Henderson and Unwin, 1975), model which represents the polypeptide chain as folded into seven transmembrane α-helical segments, roughly aligned along the edges of a distorted heptagonal prism and closely packed together laterally. Although no structural evidence of comparable accuracy is available for other ion-transport proteins, models based on bundles of α-helices have been actively developed, in particular for the acetylcholine receptor (see Guy and Hucho, 1987 for an updated summary), the sodium channel (Caterall, 1985) and references therein), the anion exchanging band 3 protein of the red cell membrane (Jay and Cantley, 1986), ion pumps like the Na^+, K^+-ATPase (Ovchinnikov, Y.A., 1987), etc, using as building blocks hydrophobic (or sometimes amphipatic) segments detected in the available polypeptide sequences by hydropathy analysis. A recent X-ray analysis of the three-dimensional structure of the photosynthetic reaction center complex of *Rhodopseudomonas Viridis* (Deisenhofer *et al.*, 1985) has put into evidence the existence of eleven essentially hydrophobic α-helical transmembrane segments. Even though, in the photosynthetic system, these helices appear to have essentially an architectural role and do not stricto sensu, participate in ion channel formation, their detection is the first clear-cut evidence of the existence of such segments in the membrane-spanning part of a large protein and provides an encouragement to pursue the search for hydrophobic helical segments in membrane proteins and the building of models for their fonctionnal utilization.

Without discarding the possibility of other structures to be involved in the translocation of ions, the "bundle" hypothesis is appealing. As it stands, however, it raises a number of questions due to the fact that relatively little is known on the aggregation properties of hydrophobic α-helices and their ability to form channels apt to transport ions, since most of the available (experimental and theoretical) knowledge on protein structure and folding comes from the study of soluble proteins. In the hope of contributing to the elucidation of some of the problems involved, we have in our Laboratory (Furois-Corbin and Pullman, 1986a, 1986b, 1987a, 1987b, 1988), undertaken a systematic study of the packing of appropriate hydrophobic segments.

The methodology consists in optimizing the total energy of the system of molecules in interaction, including simultaneously the interaction energies and the conformational energy variations in the global optimization. The energies are computed as sums of atom-atom interactions over all the atoms, using appropriate potential functions and including all the components of the energy, electrostatic, repulsion, dispersion, polarization and torsion (the repulsion-dispersion component corresponds to the van der Waals-London energy and is often called the Lennard-Jones term). In the energy optimizations of packages of 2 or more helices, one helix is fixed, while each of the others are allowed simultaneously six degrees of freedom with respect to the first. The side chain dihedral angles are at the same time allowed to vary, while the backbones are generally frozen in a standard α-helical conformation except in particular cases, e.g. for instance in a recent comparative study of alamethicin and trichorzianine (Furois-Corbin and Pullman, 1988).

Among the problems considered have been :

a) the pairing properties of hydrophobic α-helices (structure, stability, energy components commanding the structure, influence of the nature of the amino acids, etc) ; b) the possibilities of obtaining stable bundles for different numbers and orientations of the helices, the effect of bulky side chains at the interfaces and of polar side chains in various positions ; c) the capability of a purely hydrophobic bundle to accomodate and "channel" a cation or/and water ; d) the possible role(s) of polar side chains in the helices. Let us consider two of them.

THE INTERACTION PROPERTIES OF HYDROPHOBIC α-HELICES

In order to delineate the main characteristics of the interaction of hydrophobic helices, the simplest hydrophobic amino acid, alanine, was used to build α-helices of $(L\text{-}Ala)_n$ of increasing length from n=6 to 26, and to study the formation of stable couples by energy optimization. It was found that i) for all n the most stable arrangement is a nearly antiparallel structure. ii) after n=13 the structure is practically constant with a tilt of about 15 degrees between the axes of the helices and an equilibrium distance of about 7.9 Å between them. The packing mode is reminiscent of (but not identical to) the "knobs into holes" insertion mode developed on geometrical grounds, where the methyl groups of one helix fit into the holes made by methyls on its partner, allowing a close approach of the axes (about 7.9 Å in the constant structure) : for the short helices, there can be only a few such imbrications, the effects of the ends are more important than for the long helices. After n=13 the essential pattern of the interaction is well set and the packing mode becomes constant. The key to this behavior is in fact in the evolution of the main components of the binding energy of the pair upon increasing n (Fig. 1) : the total interaction energy of the pair (curve a) increases regularly with the number of amino acids per helix. The mutual polarization energy of the helices (represented by the difference between curves a) and b) is small and practically constant. The two other components (electrostatic and Van der Waals-London respectively) are both favorable, but vary differently upon increasing n : until about n=12, the two terms increase similarly and contribute equally to the total interaction ; after this value the electrostatic term tends towards an asymptot while the Van der Waals-London interaction continues to increase regularly so that its variation commands the slope of the total energy variation.

The fact that the Van der Waals-London energy is negative (favorable) and increases regularly with n is a reflection of the hydrophobic nature of the side chains in interaction : methyl groups in interaction give rise to an attractive dispersion (London) energy which dominates over the corresponding repulsion (Van der Waals) term. The resulting interaction increases thus with the number n of methyl groups at the interface of the helices. The more hydrophobic the residues are at the interface, the larger will be this effect. This is indeed found upon replacement of the alanines by leucines on the faces in contact of the two helices (Furois-Corbin and Pullman, 1987b).

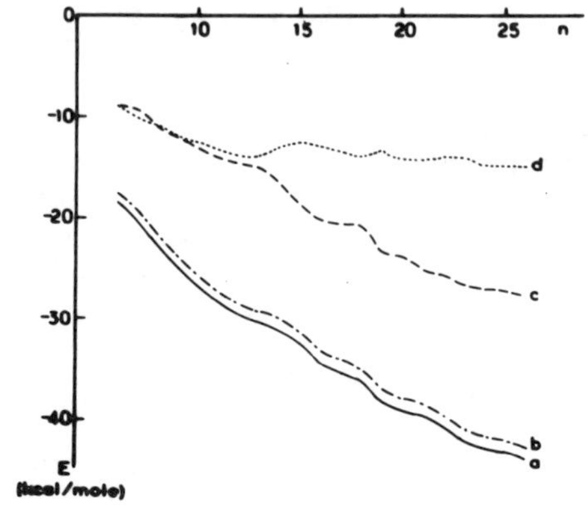

Figure 1. Evolution, in terms of n, of the optimal energy and of its components for a pair of α-helices of $(L\ ala)_n$. Curves a, b, c, d are respectively : the total energy, the same minus the polarization term, the repulsion-dispersion term and the pure electrostatic term.

At the same time, the lengthening of the helices does not increase indefinitely their electrostatic interaction. This is sometimes overlooked in reasoning based on the notion of dipole-dipole interactions : it is true that the dipole moment of an α-helix increases with the number of peptide bonds, thus with n ; but this does not imply that the interaction between two α-helices increases indefinitely with their length. A qualitative idea of the asymptotic character of the electrostatic interaction can be obtained using the very simplified standard expression for the interaction of two antiparallel dipoles separated by a distance d, energy proportional to the term in parenthesis of Fig. 2 : the variation of this term becomes rapidly negligible when l increases. The limit is reached earlier in the more accurate calculation.

The above-described interplay between the components of the binding energy is a characteristic of hydrophobic helices. This has far-reaching implications for the aggregation properties of hydrophobic protein segments in membranes : in order to span the width of the lipid part of a bilayer the helices must contain at least twenty residues. At this length, the electrostatic term has reached its asymptot and the Van der Waals-London component dominates the energy, thus the aggregation. This dominance is enhanced in the presumed transmembranes segments of real proteins which comprise a large number of leucines, isoleucines,

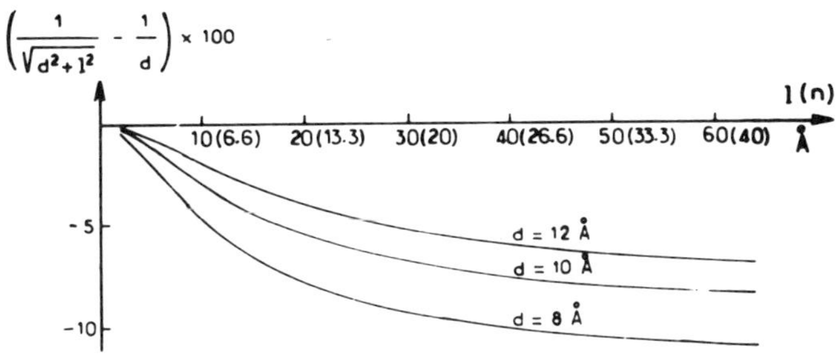

Figure 2. The asymptotic character of the interaction of two antiparallel macrodipoles at different distances.

valines, phenylalanines, instead of alanines (Furois-Corbin and Pullman, 1987b). The computations described refer to intrinsic packing properties (without inclusion of a dielectric constant). The value of the dielectric constant which would be appropriate for representing the bilayer situation is the matter of some controversy, but whatever the value adopted, it could only decrease the numerical importance of the electrostatic interaction, hence enhance the dominance of the Van der Waals-London term.

Another consequence of the energy features which govern the packing of hydrophobic helices, is the possibility of occurence of stable pairs of parallel helices. This is illustrated by the comparison given in Table I for n=14. In both cases it is the antiparallel mode of packing which is preferred, but the parallel mode is stable. The preference of α-helices for an antiparallel arrangement is often considered as the consequence of the electrostatic interactions of the net dipoles of the two helices roughly parallel to the helix axes. In fact the values of the components of the binding energy indicate that the origin of the energy preference for antiparallelism in hydrophobic helices is not due solely to an electrostatic effect but, rather, to a combination of it with the dispersion attraction : Table 1 shows the near-equality of the two main attractive energy contributions for n=14.

with a slight advantage for the second one in the antiparallel pair. In the parallel pair, the electrostatic component is unfavorable, as expected on the basis of dipole-dipole interaction but this is compensated by the second term and the polarization energy so that the pair is slightly stable. Such a stability will increase in longer helices where the electrostatic term (in that case repulsive) has reached its asymptot while the second term increases with the number of hydrophobic groups (Furois-Corbin and Pullman, 1987b).

TABLE 1. Interaction energy and its components (kcal/mole) of two α-helices in their optimal antiparallel and parallel arrangement. (a) $(L\text{-}Ala)_{14}$; (b) same with leucines at the interface.

		E_{total}	E_{elec}	$E_{rep/disp}$	E_{pol}
(a)	↑↓	−33.9	−15.9	−16.8	−1.3
	↑↑	− 0.8	14.3	−14.0	−1.2
(b)	↑↓	−33.7	−13.3	−19.3	−1.2
	↑↑	− 3.4	9.9	−12.6	−0.7

This indicates that the occurence of parallel helices in a bundle will not necessarily confer instability to the packing. Explicit calculations of the energy of packages of different sizes comprising pairs of parallel helices has indeed confirmed this feature (Furois-Corbin and Pullman, 1986b). This occurs necessarily when the bundle is made by folding a single polypeptide chain crossing the bilayer an odd number of times as in bacteriorhodopsin.

II – THE CAPABILITY OF A BUNDLE OF PURELY HYDROPHOBIC α-HELICES TO SERVE AS ION CHANNEL

One package of five poly-L-alanine helices presenting an inner hole of a size sufficient to accomodate an ion was selected to explore the possibility of a sodium ion to possibly pass through such a "channel". Placing the ion in successive planes perpendicular to the pore axis and letting it reach its optimal position in each plane, an "energy profile" is obtained giving at every step of the progression the interaction energy of the ion with the surrounding bundle (Fig. 3).

The fundamental result is that this interaction is favorable all the way through in spite of the presence of the

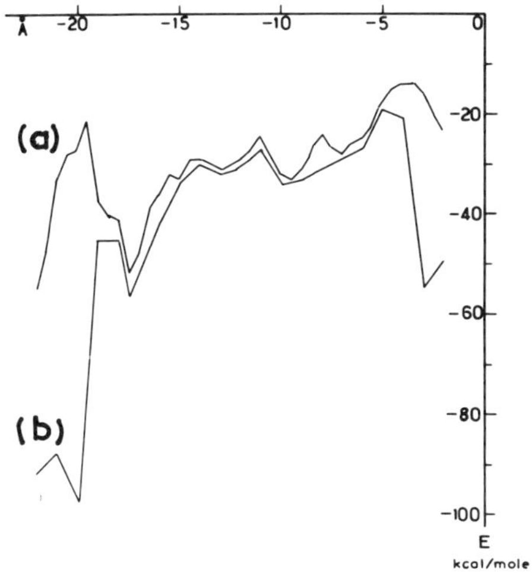

Figure 3 : The "energy profile" of Na^+ in the "channel" (made by five $(L-Ala)_{14}$ α-helices . (a) frozen, (b) helices free to reoptimize their positions.

hydrophobic methyl groups along the channel wall : the presence of the carbonyl oxygens of the peptide groups of the α-helices provide a sufficient attraction to allow the ion to find a favorable path. When the helices are allowed to readjust their positions upon the passage of the ion, the small displacements which occur are sufficient to facilitate the ion-carbonyl interactions, hence lower the profile. Similarly, the same purely hydrophobic "channel" can accomodate easily a water molecule with a favorable interaction along all its length and for the same reasons. (Furois-Corbin and Pullman, 1986b). Hence, the presence of ionized side chains thus of amphipatic helices is not necessary to provide a favorable environment for an ion or water inside a pore surrounded by α-helices and even polar non-ionized residues are not required. The channel-making properties of antibiotics like Gramicidin A confirm this point of view.

When polar residues are present inside the pore, they lower the energy profile by using the flexibility of their side chain to orient their attractive end group so as to insure the best possible interactions with the ion or with water molecules (Furois-Corbin and Pullman, 1987a). This dual characteristic – flexibility and attractive end group apt to make versatile bonds with ions and/or water (or other residues) may play a role in the dynamics of the transport whenever such residues are present. The proton translocation in bacteriorhodopsin (B. Hess, 1988) and other proton pumps, seem to utilize these properties.

REFERENCES

Caterall WA (1986). Molecular properties of voltage-sensitive sodium channels. Ann Rev Bioch 55: 953-985.

Deisenhofer J, Epp O, Miki V, Huber R and Michel H (1985). Structure of the protein subunits in the photosynthetic reaction center of rhodopseudomona Viridis at 3 Å resolution. Nature 318: 19-26.

Furois-Corbin S and Pullman A (1986a). Theoretical study of the packing of α-helices by energy minimization : effect of the length of the helices on the packing energy and on the optimal configuration of a pair. Chem Phys Letters 123: 305-310.

Furois-Corbin S and Pullman A (1986b). Theoretical study of the packing of α-helices of poly(L-alanine) into transmembrane bundles. Possible significance for ion transfer. Bioch Bioph Acta 860: 165-177.

Furois-Corbin S and Pullman A (1987a). Theoretical study of potential ion-channels formed by a bundle of α-helices : Effect of the presence of polar residues along the channel inner wall. J of Biomol Str and Dynamics 4: 589- 598.

Furois-Corbin S and Pullman A (1987b). Theoretical study of the packing of α- helices into possible transmembrane bundles. Sequences including alanines, leucines and serines. Bioch Bioph Acta 902: 31-45.

Furois-Corbin S and Pullman A (1988). Theoretical study of alamathicin and trichorzianine : conformation and pairing properties of their N terminal parts. Proteins, structure, function and genetics, in press.

Guy HR and Hucho F (1987). Trends in neurosciences 10: 318-321.

Henderson R and Unwin PTN (1975). Three-dimensional model of purple membrane obtained by electron microscopy. Nature 257: 28-32.

Hess B (1988). Bacteriorhodopsin : light-induced intramolecular proton transduction and its control, this volume.

Jay D and Cantley L (1986). Ann Rev Bioch 55: 511-538.

Ovchinnikov YA (1987). Membrane Protein Functioning. In K Yagi and B Pullman (eds.) "Ion transport through Membranes", ToKyo-New-York : Academic Press, pp 61-83

The Ion Pumps: Structure, Function,
and Regulation, pages 121–132

A TRANSISTOR MODEL FOR SODIUM FLUX ACROSS AXONAL MEMBRANES: EVALUATION OF PREDICTION CRITERIA

Jürg P. Rosenbusch
Biozentrum, University of Basel, Klingelbergstr. 70, CH-4056 Basel, Switzerland

SUMMARY

The most remarkable properties of sodium flux across axonal membranes include the very high flux rate, the distinct selectivity for sodium, and the very tight and delicately timed change of ion permeability upon an electrical stimulus. Water-filled channels, such as they exist in gramicidine, may easily account for ion flux, yet gating would have to operate through allosteric changes that convey the signal to the channel proper. Upon elucidation of the sequence of the protein catalyzing sodium flux in electric eel (Noda et al., 1984), several models have been advanced that were based on identifying channel lining elements. An alternative model (Rosenbusch, 1987) viewed sodium flux and its regulation in analogy to a field effect transistor (FET), in which a gating current (GATE) causes a drastic change in permeability such that sodium ions flow downwards their electrochemical gradient from the outside (SOURCE) to the cytoplasm (DRAIN). The sequence of the protein was therefore scrutinized for the existence of elements potentially involved in this hypothetical concept. An examination showed that a structure representing an ion lattice in the core of the protein may exist, although the criteria applied were very different from those generally adopted for the prediction of membrane protein structure. The justification as well as the significance attached to such various criteria will, of course, only be resolved once the high resolution structure will be available. Yet, as ion flux across membrane proteins is fundamental to most living processes, it appears worthwhile to examine critically the mechanisms that may differ from those most widely adopted.

THE MODEL

Ionic conduction may occur not only in solution, but also across solid-state electrolytes, provided these contain defects in crystallinity. It has been shown (Owens and Argue, 1967) that conductivity across $RbAg_4I_5$ may be nearly as high than across water. The structure analysis by X-ray solved to high resolution (Geller, 1967) revealed that such "defects" indeed exist and that they are programmed in the crystal: rubidium and iodide ions occupy precisely defined positions, while no such assignments could be made for silver ions (see Fig. 1 in Geller, 1967)). Instead, 56 binding sites per unit cell [$(RbAg_4I_5)_4$] which belong to three classes of affinities appear to be located along "pathways" trasversing the crystal. These results suggest that the 16 silver ions per unit cell move along binding sites in such pathways causing, upon application of a potential, ionic conduction. As may be estimated from the structure model (Geller, 1967), the diameter of these pathways are ~1.5Å, and are thus distinctly smaller from the 4Å channels formed by gramicidine. In it, a single file of water and cations, such as sodium, fill the channel (Wallace, 1986).

Transferring such a concept to a membrane protein structure may seem far-fetched in view of notions holding that transmembrane domains of proteins are mostly hydrophobic, a concept that has recently obtained critical support from the only membrane protein the structure of which is known at high resolution: the reaction center of phototrophic bacteria (Deisenhofer et al., 1985; Allan et al., 1987). Nonetheless, it should be borne in mind that the function of this protein is the conducting of electrons across the membrane along a redox pathway, a process which is fundamentally different from translocation of ions across a low dielectric barrier. Although rapid ion flux need not require charges (Wallace, 1986), the gating process of sodium transport invokes charge movement within the membrane domain (Hodgkin and Huxely, 1952), though the mechanism thereof is not known. If gating currents would affect the same domain of a transmembrane protein as that catalyzing ion conduction, the highly provocative question as to how low permeability is transformed to a highly permeable barrier may be closer to an understanding.

For a discussion of the assumptions made, and for the

various predictive criteria to be weighted appropriately, the model proposed (Rosenbusch, 1986; 1987; 1988) will be summarised here. Briefly, the protein is viewed as consisting of three concentric cylinders. The central core would correspond to a local ionic lattice, the charges of which would be provided by carboxylate and guanidinium (some lysyl) groups of the protein. In analogy to the rubidium-silver-iodide complex, the former would be comparable to iodide, while the latter would provide a fraction (about one half) of the cationic groups, analogous in its turn to the rubidium ions in the inorganic electrolyte. The other half or so of the cations (sodium) would bind to sites that are predetermined, though not occupied, by the protein. It might thus be analogous to the mobile silver ions in the electrolyte structure mentioned. This ionic lattice would be insulated against the hydrophobic core of the membrane by the outermost layer, or insulating mantle, which would consist of the hydrophobic segments that exist in the sequence of the sodium transducing protein of Electrophorus electricus. The intermediate layer may provide the link of these two domains, although it may have additional functions (see below). With such notions in mind, the sequence of the protein was examined. One unique configuration springs to mind immediately. There are a number of acidic protein segments in the center of the sequence, the most striking one being a polyglutamyl track (EEEEEEEEEPEE, residues 942-953) according to the sequence and nomenclature of Noda et al (1984) that will be followed throughout in this context. As the sodium salt of this carboxyl groups could exist in α-helical configuration (Holzwarth & Doty, 1965), it would span a distance of 14-18Å. Such a helix may be viewed as consisting of four tracks of carboxylate groups, each being formed of every fourth residue on the segment. These four tracks would comprise all residues of the α-helix, irrespective of its length. In the sequence of the protein from electrical eel, there exist four segments (S4), or one per homologous domain (I-IV) that contain guanidinium residues in every third position. They were originally proposed to exist as 3_{10}-helices on the cytoplasmic domain (Noda et al., 1984), a sequence that appears energetically unfavoured in proteins (Schultz & Schirmer, 1979). If the segments existed nonetheless in 3_{10}-helices, they would be highly amphipathic, with guanidinium groups located on one face of the helix located at distances of 6Å (3x2.0Å). They would thus precisely match every fourth carboxylate group on the central α-helix that would be spaced at 6Å (4x1.5Å). If four 3_{10}-helices were twisting slightly around the central α-helix, charge neutralization would occur. In such a configuration, the 3_{10}-helices may gain significantly in stability due to the electrostatic interactions involved. Yet, each carboxylate group would be close to two cationic groups, a sodium (see above) and a guanidinium residue. To these three ionic groups, a further carboxyl group could be contributed by the protein, thus completing each "elementary unit". It would have to be donated by periodic amphipathic structures, with anions at an appropriate spacing of ~6Å also. Such a configuration may exist in 2.2_7, α-helices, as 2_7-ribbons, as amphipathic β-strands, or in any other periodic structure. Prediction of such structures is not possible as several potentially amphipathic segments of the protein exist (Rosenbusch, 1987) and precise allocations are neither possible nor desirable (see below). Yet, a central sodium wire appears conceptually possible. It would consist of fixed charges provided by the proteins, with "vacancies", the coordinates of which would be determined by the protein, filled by sodium ions. These may be visualized as being located in files in intercommunicating tracks with 4-6 "elementary units" in each. At rest, the system would be self-contained. Upon an action potential, a small conformational change, such as a small translation or rotation of the central α-helix could also open up sodium sites on both sides of the membrane so that sodium ions could rush down their own electrochemical gradient. The slight change of a large number of residues in the α-helix could easily account for the six gating charges (Hodgkin & Huxley, 1952).

Of course, ion lattices have not been observed in the ever increasing number of structures of soluble proteins determined to high resolution, but then, these proteins do not catalyze vectorial ion translocation. Indeed, the densely charged core may be insulated by hydrophobic segments. From earlier models, it is clear that segments S5 and S6 and may be also S1-S3 could perform such a function. These segments may occur in an α-helical configuration, as possibly exemplified by the strictly conserved lys-glu pairs at specific positions in S2 (Noda et al., 1984).

As mentioned above, the structure of the intermediate layer is not only difficult to predict, but this may be judged rather undesirably. It would hinge upon the structure of the anionic bands that were mentioned above. Thus, if they existed as β-strands, there would in all likelihood exist more segments in the same configuration to allow intersegmental hydrogen bonding to form a β-barrel. If an intrasegmentally hydrogen bonding structure existed, such as helices or ribbons, this layer may consist of several amphipathic helices.

In this model the gating current would affect the same domain that would catalyze massive ion translocation upon a small conformational change. Intuitively, its crystalline array may appear to explain also the high selectivity of this protein. Such an assumption would tacitly imply full dehydration of sodium ions as they cross the membrane barrier, a proposal that has been discussed extensively although no conclusive evidence exists. Extensive though incomplete dehydration may indeed occur. A sodium-monohydrate might be incorporated into the ion lattice and would reduce the 3Å distance between cationic groups by a factor of two, due to the oxygen of water serving as intermediary ligand. Irrespective of these considerations, the distances involved would be realistic for ion movements.

PREDICTION CRITERIA AND THEIR SIGNIFICANCE

Since the model proposed violates most accepted prediction algorithms, these shall be dealt with at present. From the structure (Henderson and Unwin, 1975) and the sequence of bacteriorhodopsin (Ovchinnikov et al., 1979; Khorana et al., 1979), as well as manifold additional experimental evidence (Engelman et al., 1982), the conclusion emerged that the membrane spanning domains of proteins are arranged in α-helical arrays, and that they consist mostly of hydrophobic residues. As mentioned above, these notions were confirmed by the structure of photosynthetic reaction centers. They are, however, significantly different from the structure of porin, a channel-forming transmembrane protein that occurs in β-structure (Kleffel et al., 1985) with more ionizable and polar residues within the membrane domain than can be accounted for by its water-filled channels (Paul and Rosenbusch, 1985; Page and Rosenbusch, 1986; Rosenbusch, 1986). It thus appears likely that membrane domains follow not one single structural pattern, but possibly widely different ones. Porin may serve as a paradigm for the view that the core of a protein may well be "charged", and thus potentially involved in ion translocation. It should be noted that the proposal on the structure of the sodium translocating protein follows the prediction rules of Kyte and Doolittle (1982) except that it does not assume that only those segments may be transmembraneous, that are both hydrophobic and at least 20 residues long. An analogous reasoning applies to amphipathic helices (Guy, 1984; Finner-Moore and Stroud, 1984). Such structures may well be present, but the ongoing arguments on the models pertaining to the acetylcholine receptor protein (Guy and Hucho, 1987) prompt caution. Moreover it seems clear that in addition to the secondary structures already observed (α,β), any periodic structure may afford complete saturation of the hydrogen bonding potential in membrane spanning backbones. Such saturation is clearly critical as peptide bonds are rather polar. It has been suggested already by Chothia (1974) that full hydrogen bonding renders polar regions quasi-hydrophobic to their environment. The same rationale may be applied also to side chain residues (Tanford, 1972). Hydrophobicity may therefore be feigned by ion pair formation (electroneutrality) and full saturation of their hydrogen bonding potential also. The former implies that charged residues occur either paired or as single ionizable residues existing in uncharged forms with abnormal *in situ* pK-values.

As mentioned in the text, one of the most striking characteristics of these sodium "channel" protein consists in the four homologous domains with conservation of its six internal segments. There is little doubt that such homologies are functionally highly significant, and since the function is vectorial translocation across the membrane, it seems appropriate to assume that such segments span the membrane. One may, however, wonder whether this criterion is exclusive, i.e. whether no other segments may occur in the transmembrane domain. A case illustrating this may be the poorely conserved s_f and s_h segments (Greenblatt et al., 1985), also known as S6 (Guy and Seetharamulu, 1986). These segments have, or have not, been postulated to span the membrane. In contradistinction, it appears rather possible that the highly conserved S4 (Noda et al 1984; 1986; Tanabe et al 1987) provide the skeleton of the membrane moiety of the protein. But this need not rule out that nonconserved segments convey specific proteins the possibility to cope with the demands of their respective function and environment. Admittedly, this hypothesis lacks foundation in fact, but so does the opposite view.

One of the most significant criteria for any structure prediction of a protein should be its function, however. The value of models that insist on maximal charge exclusion (Engelman et al, 1980) may be contested if applied to a protein translocating charges. Similarly, a structure proposed for the axonal sodium translocating protein (Noda et al, 1984) appeared doubtful, since gating was not duly considered. It was subsequently amended (Noda et al, 1986). Finally, some other prediction criteria may be mentioned. The turn prediction based on first physical principles (Paul and Rosenbusch, 1985) seems to be compatible with the model of the sodium transducing protein suggested here (Rosenbusch, 1987). Sites of posttranslational modification have been discussed elsewhere (Greenblatt et al, 1985).

CRITIQUE OF THE MODELS PROPOSED

Only eclectic aspects shall be discussed here. The models proposed by Numa's group (Noda et al. 1984, 1986) were developed further by Catterell (1986). He suggested a sliding helix model in which a channel occurs in the center of the molecule, with four gating domains, including the S4 segment as main constituent, located more towards the periphery. The signal that depolarization has occurred would have to be conveyed as allosteric conformational change to the channel located in the center. A critical question may be raised as to the consistency with charge exclusion, mentioned above. If the ion pairings postulated were acceptable, arguments against the charged core proposed here may no longer be considered valid. Another objection to the transistor model may be the effect of tetrodotoxin binding. This inhibitor interferes with ion flux, but does not impair gating. This argument is not stringent, however, as the two processes may not be tightly coupled. Both models should therefore be subjected to further tests.

Kosower (1985) has, based on chemical logic and single group rotation, arrived independently at a model containing extensive 3_{10}-helical structure. While this is in agreement with the postulate discussed above, the spatial distribution of charges do not suggest favorable electrostatic interactions. There thus seems little reason for the proposal that anionic segments would assume a 3_{10}-helical configuration. Another reservation concerns the lack of global as well as local neutralization of electrical charges. How sodium ions may rapidly dissociate from a level (1) which contains three negative charges, and how it may get by a positive charge on level 6 seems difficult to understand.

Greenblatt et al. (1985) have studied several aspects of protein folding and compared the various published models. Therefore, the present considerations shall address only those findings that were not discussed. For instance, prediction of α-helical configuration shows a major peak in the center of the protein sequence, between the homology regions, that coincide exactly with that segment containing the polyglutamyl track assumed to be critical in the present context. Similarly, the power spectra screening for periodicities exhibit a major peak with a period of three residues. Yet, 3_{10}-helix is not accepted because it appears energetically disfavoured in soluble proteins. Guy and Seetharamulu (1986) have, in addition to amphipathicity, postulated the insertion of polypeptides in extended configuration into the membrane domain. Such a proposal would be compatible with complete saturation of hydrogen bondings only if bonded to specified side chain residues. Without a high resolution structure, such a postulate would appear rather difficult to accept.

CONCLUSION

Functions of membrane proteins include the transport of charge, such as electrons along a redox chain, and of ions across the hydrophobic membrane barrier. While the former task has been shown to be performed by a series of redox pigments, with a completely hydrophobic membrane domain insulating it from the environment, ion translocation may require completely different designs. Thus, although the polar peptide bonds provide a channel in the ionophoretic antibiotic gramicidine, such a structure is difficult to envisage if high speed of reaction to an electrical signal, and delicately poised conformational states are significant. Gating in any event implies movement of charges across the membrane, as has already been proposed by Hodgkin & Huxley (1952). The model proposed currently is mostly based on these functional aspects. It therefore views the protein as a cable, with a sodium "wire" spanning the membrane. An ionic core may indeed allow for both gating and high ion flux rates. Insulation from the surrounding hydrophobic membrane is, based on the known protein sequence, conceivable to exist by an ensemble of hydrophobic and amphipathic peptide segments. These concepts represent an application of the knowledge obtained with a polar bacterial transport protein, porin, as well as considerations of ion transport in bacteriorhodopsin (Rosenbusch, 1985; 1986). Consistent with the tacitly assumed neutralization of polarity in the polypeptide backbone by hydrogen bonding, the structure of this protein is suggested to contain also polar and ionizable residues within the membrane domain, provided that local and global electron neutrality is afforded and all potential hydrogen bonding donors and acceptors are saturated. Such a concept provides the possibility to consider the sodium transducing protein as being built according to a significantly different design, although the physical principles governing membrane protein structures are obviously the same as those applicable ubiquitously. The currently accepted notions on predicting the structures of membrane proteins will have to be reexamined in any event.

Another aspect that may deserve attention is the conscious refraining from any attempt to predict all presumed transmembrane segments. The structure of the anionic bands that are a constituent part of the model presented may well be assumed to be in a periodic array, but the type(s) would be determined by their respective environment (and *vice versa*). All predictions, however, must withstand experimental tests. Such experiments are currently in progress or in preparation, and include site-directed mutagenesis (Mishina et al, 1985), protein design (Mutter et al, 1988), molecular dynamics (with M. Kokkinidis), energy calculations and, eventually, high resolution structure analysis.

ACKNOWLEDGEMENTS

I wish to thank Drs. K. Heckmann, P. Läuger and S. Numa for illuminating discussions. This work was supported by the Swiss National Science Foundation, the Roche Research Foundation and Sandoz Research and Development.

REFERENCES

Allen JP, Feher G, Yester TO, Komiya H, Rees DC (1987). Structure of the reaction center from Rhodobacter sphaeroides R-26: The cofactor. Proc Natl Acad Sci USA 84:5730-5734.

Catterall WA (1986). Molecular properties of voltage-sensitive sodium channels. Ann Rev Biochem 55:953-985.

Chothia C (1974). Hydrophobic bonding and accessible surface area in proteins. Nature 248:338-339.

Deisenhofer J, Epp O, Mikki K, Huber R, Michel H (1985). Structure of the protein subunits in the photosynthetic reaction center of Rhodopseudomonas viridis. Nature 318: 618-624.

Engelman DM, Henderson R, McLachlan AD, Wallace BA (1980). Path of the polypeptide in bacteriorhodopsin. Proc Natl Acad Sci USA 77:2023-2027.

Engelman DM, Goldman A, Steitz TA (1982). The identification of helical segments in the polypeptide chain of bacteriorhodopsin. Meth Enzymol 88:81-88.

Finer-Moore J, Stroud R (1984). Amphipathic analysis and possible formation of the ion channel in an acetylcholine receptor. Proc Natl Acad Sci USA 81:155-159.

Geller S (1967). Crystal structure of the solid electrolyte $R6Ab_4I_5$. Science 157:310-312.

Greenblatt RE, Blatt Y, Montal M (1985). The structure of the voltage-sensitive sodium channel. FEBS Lett 193:125-134.

Guy HR (1984). A structural model of the acetylcholine receptor channel based on partition energy and helix packing calculations. Biophys J 45:249-261.

Guy HR, Hucho F (1987). The ion channel of the nicotinic acetylcholine receptor. Trends Neuro Sci 8:318-321.

Guy HR, Seetharamulu P (1986). Molecular model of the action potential sodium channel. Proc Natl Acad Sci USA 83:508-512.

Hodgkin AL, Huxley AF (1952). A quantitative description of membrane current and its application to conduction and excitation in nerve. J Physiol 117:500-544.

Holzwarth G, Doty P (1965). The ultraviolet circular dichroism in polypeptides. J Am Chem Soc 87:218-225.

Khorana HG, Gerber GE, Herlinhy WC, Gray CP, Anderegg RJ, Nihei H, Biemann H (1979). Amino acid sequence of bacteriorhodopsin. Proc Natl Acad Sci USA 76:5046-5050.

Kleffel B, Garavito RM, Baumeister W, Rosenbusch JP (1985). Secondary structure of a channel-forming protein: porin from E. coli outer membranes. EMBO J 4:1589-1592.

Kosower EM (1985). A structural and dynamic molecular model for the sodium channel of Electrophorus electricus. FEBS Lett 182:234-242.

Kyte J, Doolittle RF (1982). A simple method for displaying the hydropathic character of a protein. J Mol Biol 157:105-132.

Mishina M, Tokimatsu T, Imoto K, Tanaka K, Fujita Y, Fukuda K, Kurasaki M, Takahashi H, Morimoto Y, Hirose T, Inayama S, Takahashi T, Kuno M, Numa S (1985). Location of functional regions of acetylcholine receptor α-subunit by site-directed mutagenesis. Nature 313:364-369.

Mutter M, Altmann E, Altmann KH, Hersperger R, Müller K, Nebel K, Tuchscherer G, Vuilleumier S (1988). Artificial proteins: tertiary structures by assembly of amphipathic helices and β-sheets on template molecules. Nature: in press.

Noda M, Shimizu S, Tanake T, Takai T, Kayano T, Ikeda T, Takahashi H, Nakayama H, Kanaoka Y, Minamino N, Kangawa K, Miyata T, Numa S (1984). Primary structure of Electrophorus electricus sodium channel deduced from cDNA sequence. Nature 312:121-127.

Noda M, Ikeda T, Kayano T, Suzuki H, Takeshima H, Kurasaki M, Takahashi H, Numa S (1986). Existence of distinct sodium channel messenger RNAs in rat brain. Nature 320:188-192.

Ovchinnikov YuA, Abdulaev NG, Feigina MYu, Kiselev AV, Lobanov NA (1979). The structural basis of the functioning of bacteriorhodopsin. FEBS Lett 10:219-224.

Owens BB, Argue GP (1967). High conductivity solid electrolytes: MAg_4I_5. Science 157:308-309.

Page MGP, Rosenbusch JP (1986). Topographical labelling of pore-forming protein from the outer membrane of E. coli. Biochem J 235:651-661.

Paul C, Rosenbusch JP (1985). Folding patterns of porin and bacteriorhodopsin. EMBO J 4:1593-1597.

Rosenbusch JP (1985). Hydrogen bonding networks in bacteriorhodopsin may afford a proton pathway and structural stability. Bull Inst Pasteur 83:207-220.
Rosenbusch JP (1987). Three-dimensional structure of membrane proteins. In Inouye M, (ed.) "Bacterial outer membranes as model systems", New York, John Wiley, pp. 141-162.
Rosenbusch JP (1987). Is axonal sodium flux catalysed and regulated by solid state ion conduction? Chem Script, in press.
Rosenbusch JP (1987). Secondary and tertiary structure of membrane proteins. In Fehrenbach FJ (ed) "Bacterial protein toxins", New York, G. Fischer, pp 259-266.
Rosenbusch JP (1988). Ion transport across biological membranes: diffusion in water or conduction in the solid state? Biophys Chem, in press.
Schulz GE, Schirmer RH (1979). "Principles of protein structure". New York, Springer.
Tanabe T, Takeshima H, Mikami A, Flockeszi V, Takahashi H, Kangawa K, Kojima M, Matsuo H, Hirose T, Numa S (1987). Primary structure of the receptor for calcium channel blockers from skeletal muscle. Nature 328:313-318.
Tanford C (1972). Hydrophobic free energy, micelle formation and the association of proteins with amphiphiles. J Mol Biol 67:59-74.
Wallace BA (1986). Structure of gramicidine A. Biophys J 49:295-306.

The Ion Pumps: Structure, Function,
and Regulation, pages 133–138

MEMBRANE PROTEIN FOLDING: MOTIFS AND PREDICTIONS

B. A. Wallace
Department of Chemistry and Center for Biophysics
Rensselaer Polytechnic Institute
Troy, New York 12180, USA

The manners in which proteins fold are influenced not only by their chemical structure (Anfinsen et al, 1961), but also by the environment in which they are found (Wallace, 1983). The carbonyl and amide groups of peptide backbones may, in aqueous solutions, form hydrogen-bonds either intermolecularly with the surrounding solvent, or intermolecularly with other carbonyl and amide groups within the polypeptide. In membranes, however, where lipid fatty acid chains are the solvent, intermolecular hydrogen-bonding is not possible. Thus, in this environment, a polypeptide is more likely to form intramolecular bonds; the geometry of these bonds will dictate the secondary structure of the molecule.

Helical secondary structures can satisfy all their backbone hydrogen-bonds intramolecularly. Beta sheets satisfy those peptide groups that lie within the interior of the sheet, although carbonyl and amide groups at the edges of the sheet will be unpaired unless the molecule rolls up to form a beta barrel. On the other hand, beta-turns have at least two unsatisfied amide and carbonyl groups, and random coils may have many unpaired groups. The latter two types of structures should, therefore, be less likely to form in hydrocarbon solvents and membranes. Such considerations are important when we think about possible folding motifs for ion pump proteins in membranes: those portions of the pumps that are embedded in the membrane and which are sequestered away from aqueous solvent may not be well-modeled by empirical secondary structure prediction methods which are based on aqueous-soluble protein folding motifs (Chou and Fasman, 1974; Garnier et al, 1978; Burgess et al, 1974; Lim, 1974). Unfortunately, because of interest in deriving secondary structural information from sequence data, and because of their ease of use, such algorithms have often been applied to membrane proteins despite their potential inappropriateness. To ascertain if solvent-excluded regions do fold differently than solvent-accessible regions, we first compared secondary structure predictions by several of these algorithms with experimental results obtained for both soluble and membrane proteins (Wallace et al, 1986). Secondly, we compared the relative probabilities of amino acids in buried and exposed regions of soluble proteins being in each of the types of secondary structures (Mielke and Wallace, 1988).

With secondary structural information now available on membrane proteins from a variety of different physical methods, including X-ray crystallography and diffraction, and electronic and vibrational spectroscopy, it has been possible to test the empirical algorithms for their accuracy in predicting net membrane protein structures. The methods used for comparisons of predicted and experimental values are as described in Wallace et al (1986) and Mielke et al (1987). For the most popular of these methods, that designed by Chou and Fasman (1974), we found a poor correlation between the predicted and actual secondary structures of the membrane proteins (Figures 1 and 2), although the correspondence for the test set of soluble proteins was quite good. The correlation coefficient for 15 membrane proteins was 0.009, suggesting that this method is not appropriate for membrane proteins in general, and that the amino acids fold differently in hydrophobic and hydrophilic environments. Clearly, this is not meant to reflect negatively on the methodology in general, but rather to point to the inappropriateness of data basis sets derived from soluble proteins for predicting hydrophobic membrane protein structures. Indeed, if an equivalent data basis set could be made from data of known membrane protein structures, this method would likely be of use in some membrane protein predictions. Unfortunately, to date not enough crystal structures of membrane proteins have been determined to do this. Similarly poor correlations are found for two other empirical prediction algorithms (Burgess et al, 1974; Garnier et al, 1978), even when one of them is specifically biased (decision constants set to the correct values).

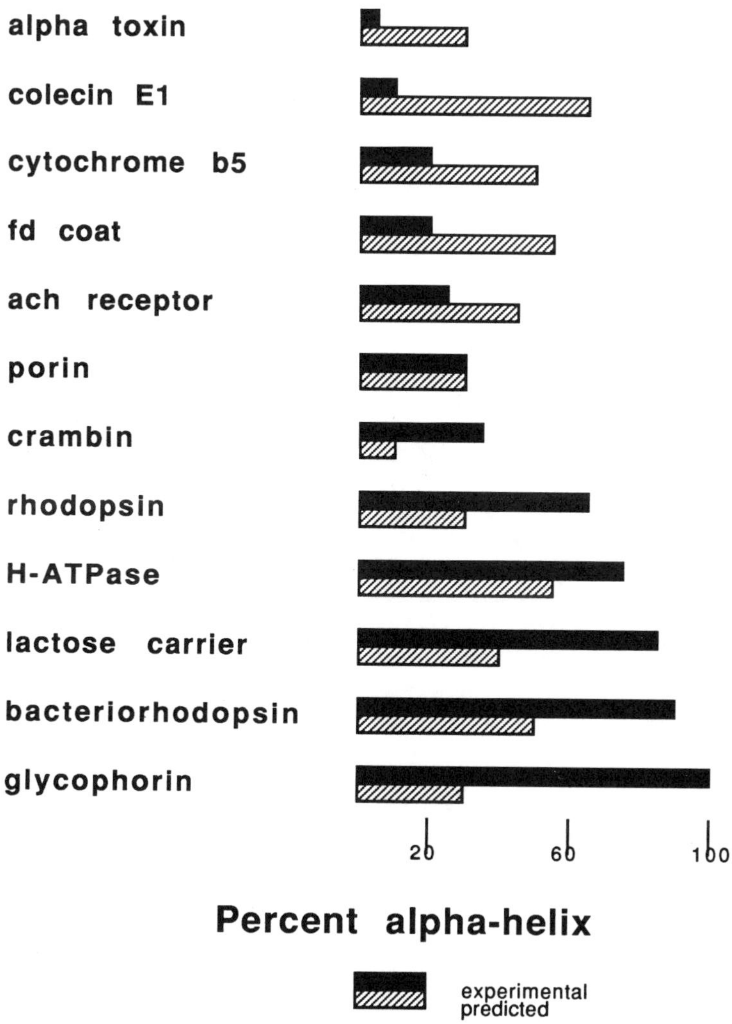

Figure 1: Bar graph showing lack of correspondence between calculated and experimental net helical secondary structures for a number of membrane proteins.

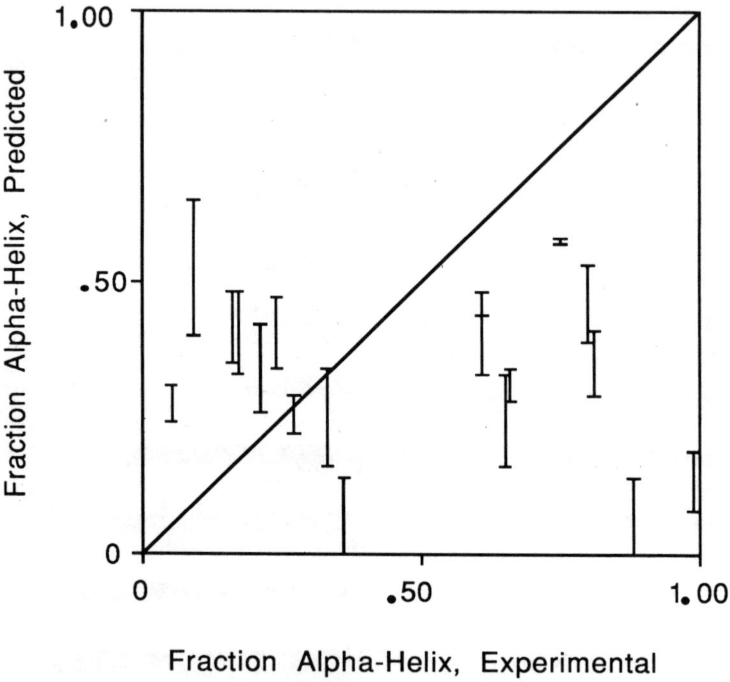

Figure 2: "Scatter" plots showing the lack of correspondence between the predicted and measured net helical contents of membrane proteins.

Another test for the different folding propensities of amino acid residues in hydrophilic and hydrophobic environments can be made by examining the structures of soluble proteins, for which there is a large data base of crystallographic information. The interiors of large globular proteins generally exclude water, and may therefore be more similar to the interiors of membranes. We have compared the relative probabilities for finding each amino acid in helical or sheet-type secondary structures, depending on whether they were in interior or exposed portions of the structure.

We used a Lee and Richards-type (1971) accessibility algorithm to identify ∿4000 residues buried from surface solvent in 51 large globular proteins, and found (Mielke and Wallace, 1988) that the probabilities for each type of amino acid in buried and exposed regions in the different types of secondary structures were, indeed, quite different. Eleven out of twenty amino acids have reversed helix-forming versus sheet-forming preferences in the buried residues. These results are another demonstration of environmental effects on secondary structure. It is also possible that this data from soluble protein interiors may be useful in formulating data bases for prediction of the secondary structures of the transmembrane regions for some membrane proteins.

In conclusion, using two different tests, it was found that amino acids in polypeptide chains which are located in hydrophobic milieux fold in very different ways than those in hydrophilic environments. This suggests that simple application of empirical methods based on soluble proteins motifs is not appropriate for prediction of membrane protein secondary structures, but that more environment-dependent considerations must be taken into account.

ACKNOWLEDGEMENTS:

I thank D. L. Mielke and M. Cascio for helpful discussions. This work was supported by NIH Grants AM31089 and GM27292.

REFERENCES:

Anfinsen CB, Haber E, Sela M, White FH Jr (1961). The kinetics of formation of native ribonuclease during oxidation of the reduced polypeptide chain. Proc Nat Acad Sci 47:1309-1314.

Burgess AW, Ponnuswamy PK, Scheraga HA (1974). Analysis of conformations of amino acid residues and prediction of backbone topography in proteins. Isr J Chem 12:239-286.

Chou PY, Fasman GD (1974). Prediction of protein conformation. Biochem 13:222-245.

Garnier J, Osguthorpe DJ, Robson B (1978). Analysis of the accuracy and implication of simple methods for predicting the secondary structure of globular proteins. J Mol Biol 120:97-120.

Lee B, Richards FM (1971). The interpretation of protein structures: Estimation of static accessibility. J Mol Biol 55:379-400.

Lim, VI (1974). Algorithms for prediction of α-helical and β-structural regions in proteins. J Mol Biol 88: 873-894.

Mielke DL, Cascio M, Wallace BA (1987). Membrane protein folding motifs: An examination of empirical predictions of secondary structure. In Burnett R, Vogel H (eds): "Biological Organization: Macromolecular Interactions at High Resolution," p. 319-333.

Mielke DL, Wallace BA (1988). Predictions of membrane protein secondary structures based on buried residues in globular proteins. Biophys J in press.

Wallace BA (1983). Gramicidin A adopts distinctly different conformations in membranes and in organic solvents. Biopolymers 22:397-402.

Wallace BA, Cascio M, Mielke DL (1986). Evaluation of methods for the prediction of membrane protein secondary structures. Proc Nat Acad Sci 83:9423-9427.

FUNCTION

The Ion Pumps: Structure, Function, and Regulation, pages 141–145

EDITOR'S INTRODUCTION: FUNCTION OF THE ION PUMPS

There is a dialectical interplay between studies of structure and of function. Function defines the systems that are important to study structurally, while structure is the foundation on which studies of function can usefully proceed. We see this clearly in the present section with the impressive studies of the function of the F_1F_0 ATPases. Penefsky shows us how analysis of the kinetics of ATP hydrolysis by this enzyme, studied under conditions where only a single molecule of substrate binds, reveals an exceedingly high affinity of the F_1 portion of the enzyme for its substrate. The association constant is of the order of 10^{12} M^{-1}! Yet this high affinity is reduced greatly when a carboxyl group on the F_0 portion of the enzyme is blocked by dicyclohexylcarbodiimide (DCCD). The distance between these two interacting regions of the F_1F_0 whole molecule is some 20 A^o. DCCD, of course, inhibits proton translocation through the F_1F_0 ATPase and, therefore, we are clearly here at the heart of the problem of energy transduction in this system: how the movement of the proton brings about the release of ATP, newly-synthesised by a chemical reaction close to equilibrium, but tightly bound at the site of synthesis on the F_1 portion.

Such studies of function show us that this sequence of events is the basis of energy-coupling but it is structural analysis that must give us the detailed picture of the molecular events that are occurring. The papers by Giraud and his colleagues and by Vredenberg and his, by Gromet-Elhanan, and by Carmeli take up the F_1F_0 system of the chloroplast and discuss the relation of ATP synthesis and proton movements in that system, where it is light absorption that drives the proton gradient (and gives us a remarkably convenient tool for switching on proton pumping). The problems of the binding of the nucleotide substrates and metals to the various subunits of this pump-enzyme are yielding to persistent analysis, aided by newer methods for the reconstitution of the enzyme from its subunits. Gerwert's paper on the light-driven pump, bacteriorhodopsin, describes how close we are to the hoped-for detailed understanding of this pump. Time-resolved Fourier transform infra-red studies show that the light-induced changes in the conformation of the protein's retinal chromophore bring about a switch of a proton between two internal carboxyl groups, located on aspartyl residues. Which aspartyl residues are involved in this pathway for proton movement through the protein is not yet known, but when this information is forthcoming we will have, for the first time, a detailed description of the route that an ion takes as it is being pumped through the protein. Studies on ion movement through the gramicidin channel (here discussed by Hinton) are bench-marks for our analysis of these routes. (Chirieri-Kovacs shows us that we may still have things to learn about bacteriorhodopsin which, she suggests, may have a direct role in the maintenance of the calcium gradient across the discs of the visual rods).

In the years ahead, we will surely be discussing ion pumping at a much more fundamental level. An indication of things to come appears in the papers by Gutman and his group who discuss the movements of protons on the surfaces of membranes and proteins. Studies with model compounds show that proton movements close to the water-protein interface are slow, so that dissociation of proton from the site on the pump to which it is bound will most probably be into the bulk solution rather than along the surface. Local proton gradients on a membrane's surface will not be maintained. Yam, from this group, analyses a system which provides a model for antennae for protons, groups which can bind protons and transfer these to neighbouring sites on a protein. The third paper shows how the dynamics of proton binding and debinding from phospholipid bilayers can be measured and simulated by dynamic calculations.

We have also to consider the details of movement of the other substrate, ATP. As Aflalo shows, using the ATP-sensitive enzyme luciferase which enables the concentration of ATP in the microenvironment of immobilised enzymes to be monitored, it is the diffusion of ATP towards or away from the enyzme that can be the rate-limiting step in certain reactions. Thus, with the details of ATP and proton movements elucidated we shall be in a position to reach a highly refined analysis of the coupling between these two substrates of the ATPases!

We are still far from this level in the analysis of the E_1E_2 ATPases, as Karlish shows us in his review of the sodium/potassium pump-enzyme. Most interesting is the concept that the enzyme can distinguish between nine different configurations of ions bound to it, making different choices as to whether to split ATP or release phosphate, or to translocate its bound ions, depending on which ions are bound. There is a suggestion that cation-binding sites have a non-aqueous environment. Karlish sets the framework for the discussion of the electrogenic aspects of cation pumping which the papers of Lauger and of Bamberg and his colleagues take up in detail. It seems that it is the form of the enyzme bound to three sodium ions that carries a (positive) charge, translocation of this form being the step that interacts with the prevailing trans-membrane potential. Grampp finds that the sodium pump of the lobster's stretch receptor in vivo is also voltage-regulated, pointing to a possible physiological role for this phenomenon.

In the related proton/potassium pump (reviewed by Sachs), there is no current flow in an overall cycle of ATP splitting, two potassium ions being exchanged for two protons during a complete cycle. But the individual steps of ion translocation are electrogenic and the gastric proton ATPase is subject to control of its activity by the potential across the epithelium. The rate of proton pumping is also controlled by the potassium permeability of the membrane in which the pump finds itself. This fact may be of physiological (and even pharmacological?) significance, when considered together with Keeling's demonstration that the compound SCH 28080 inhibits this pump by mimicing potassium ions at their binding-site on the pump-enzyme.

Your editor tries to elucidate the role of ATP-binding and -splitting in the changes of affinity of the pump for its cation substrates as it interconverts between inward- and outward-facing conformations. For the potassium half of the pump cycle, this change of affinity is merely a reflection of a "thermodynamic box" where the energy of binding (but not splitting) of ATP is traded off against the binding of potassium ions. For the sodium half of the cycle, the energy of ATP splitting is used, both in driving the overall cycle of pumping and in being directly harnessed to the affinity changes for sodium.

The calcium ATPase from sarcoplasmic reticulum provides interesting parallels with the sodium and proton systems. For example, Vinsen and Andersen study the occlusion of calcium ions by this enzyme. Using chromium ATP as a substrate, which slows down the rate of the occlusion and de-occlusion, they were able to show that the solubilised form of the enzyme can occlude calcium with much the same rates and capacity as does the membranous form. Important is their finding that blocking a carboxyl group in the enzyme reduces calcium's ability to bind, but only its affinity for the enzyme not the maximum capacity of binding. This suggests that the carboxyl group in question is near to the active site for calcium binding but does not comprise it. Interestingly, the size of the portion of the enzyme that occludes calcium (from radiation inactivation studies) is only one-third of the whole pump protein. An ingenious protocol of ATP addition, simultaneous with the addition of EGTA to the calcium-bound enzyme, enables Berman and Davidson to study single cycle reactions of the enzyme, and to define calcium binding sites. A first calcium binds with lower affinity, the second with high affinity at levels comparable with those in the muscle cytoplasm.

It had been thought that treatment of the calcium ATPase by, for instance, proteolytic digestion, uncoupled transport and ATP hydrolysis in that the latter was unaffected whereas calcium uptake was reduced. Two papers, by Shoshan-Barmatz and Chipman, and by Torok and Green, show clearly that true uncoupling does not occur but, rather, that calcium loss is stimulated by digestion of some other component of the vesicles. Two additional papers by Chipman and his colleagues deal with the hydrolysis of methyl phosphate in a partial reaction of the calcium-free ATPase and with the existence of a complex between the enzyme, calcium, and the inhibitor vanadate.

The parallels between the calcium ATPase and the sodium/potassium enzyme extend beyond the findings that each is an E_1E_2 enzyme, that each is phosphorylated during ATP splitting, and that each occludes its cation substrates - to the finding that ATP has both a catalytic and a regulatory action on the hydrolysis rate. For both enzymes, (Guillain reports for the calcium enzyme) the regulatory action is brought about by ATP in the absence of magnesium, whereas the catalytic role of ATP demands that magnesium also be bound. Davidson and Berman can show that the site of binding of the non-catalytic ATP molecule becomes more hydrophobic when ATP is bound, suggesting that the nucleotide-binding region becomes solvent inaccessible when the calcium ions are released within the membrane vesicles. Finally, Casadio links calcium ATPase activity and F_1F_0 ATPases in her study of calcium or magnesium-dependent ATPase activity of an oligomycin-sensitive enzyme present in chromatophores from photosynthetic bacteria. The magnesium-dependent activity is linked to the electrochemical gradient of protons; the calcium-dependent activity is not.

W.D. Stein
Editor

The Ion Pumps: Structure, Function,
and Regulation, pages 147–154

Ca^{2+} OCCLUSION IN MONOMERIC SARCOPLASMIC RETICULUM Ca-ATPase

Bente Vilsen and Jens P. Andersen

Institute of Physiology, University of Aarhus,
8000 Aarhus C, Denmark.

INTRODUCTION

The Ca^{2+} pump (Ca-ATPase) of sarcoplasmic reticulum is an integral membrane protein consisting of one or more identical 110,000 Da polypeptide chains. Previous work in this laboratory has demonstrated that a single Ca-ATPase peptide chain provides the full structural basis for coupling of ATP hydrolysis with E_1-E_2 conformational transitions (Andersen et al., 1985). It is, however, an open question whether an oligomeric unit is required for formation of the transport pathway through the membrane.

An important concept in relation to understanding the molecular nature of Ca^{2+} translocation is the "occlusion" of Ca^{2+} in the phosphorylated intermediate (Dupont, 1980). Rapid kinetic measurements have demonstrated that the Ca^{2+} ions occupying the transport sites on the protein are rendered unable to exchange with free Ca^{2+} at the cytoplasmic surface of the membrane as the enzyme becomes phosphorylated by the γ-phosphate of ATP. The Ca^{2+} bound in the phosphoenzyme will be translocated and released on the luminal side of the membrane, unless ADP is present to reverse the transphosphorylation reaction. The phosphate transfer from ATP to the enzyme is a rapid reaction having a rate constant of not less than 150 s^{-1} at room temperature. On the other hand, the release of bound Ca^{2+} at the luminal surface takes place rather slowly, with a rate constant in the range 10-20 s^{-1} under the same conditions. The kinetic measurements have generally been interpreted in two different ways:

1. Existence of two distinct conformational classes of phosphoenzyme with bound Ca^{2+}, E_1PCa_2 and E_2PCa_2. E_1PCa_2

represents an occluded transport intermediate in which the bound Ca^{2+} is unable to exchange with free Ca^{2+} on either side of the membrane. A relatively slow conformational transition to E_2PCa_2 leads to exposure of the bound Ca^{2+} at the luminal surface (de Meis, 1981; Andersen et al., 1985).

2. A model in which occlusion is relative rather than absolute. There would not necessarily be more than one major conformational state of the phosphoenzyme. The occluded Ca^{2+} would have access to the luminal surface, but only by slow diffusion through a channel-like structure containing ion-binding sites (cf. Brandl et al., 1986). In this model, the major conformational rearrangement leading to creation of the transport pathway would occur in relation to the transfer of phosphate to the protein.

If there are two distinct phosphoforms of the enzyme as implied in mechanism no. 1, and the transition between them can be blocked, it should be possible to isolate a stable occluded intermediate, from which the bound Ca^{2+} dissociates at a slower rate than 10-20 s^{-1}. The present work demonstrates the existence of such an occluded state formed by use of CrATP as phosphorylating substrate. We furthermore show that only one Ca-ATPase peptide chain is needed for Ca^{2+} occlusion. This result suggests that the transport pathway may be formed between segments belonging to the same peptide chain, rather than at the interface between subunits in an oligomeric structure. Another purpose of our work has been to locate Ca^{2+} binding sites in the protein structure and characterize the extent of the conformational changes associated with occlusion. We have therefore studied CrATP induced Ca^{2+} occlusion in enzyme specifically labelled with a fluorescent carbodiimide.

RESULTS AND DISCUSSION

Stabilization of the Ca^{2+} occluded form

We have found that the Ca^{2+} occluded form of the phosphoenzyme is very stable when a β,γ-bidentate chromium(III) complex of ATP (CrATP) has been used as substrate for phosphorylation of Ca-ATPase (Vilsen and Andersen, 1986,1987). Fig. 1 A and B show the respective time courses of occlusion and deocclusion of $^{45}Ca^{2+}$ in the presence of CrATP. For both processes the time scale is of the order of hours. This slow rate has made the measurement of Ca^{2+} occlusion very simple. The amount of occluded Ca^{2+} in membrane-bound enzyme could be determined after Millipore filtration in the presence of EGTA or an excess of non-radioactive Ca^{2+} in the washing buffer. In order to study occlusion of Ca^{2+} in Ca-ATPase which was in a defined monomeric state, the membrane was solubilized using the non-ionic detergent $C_{12}E_8$. For measurement of Ca^{2+} occlusion in the soluble enzyme, loosely bound Ca^{2+} and free Ca^{2+} was removed by rapid size-exclusion HPLC in a TSK G 3000 SW column. The peak of enzyme eluted from the HPLC column was monomeric as evidenced by sedimentation velocity and sedimentation equilibrium studies in the analytical ultracentrifuge (Andersen et al., 1986b; Vilsen and Andersen, 1986,1987).

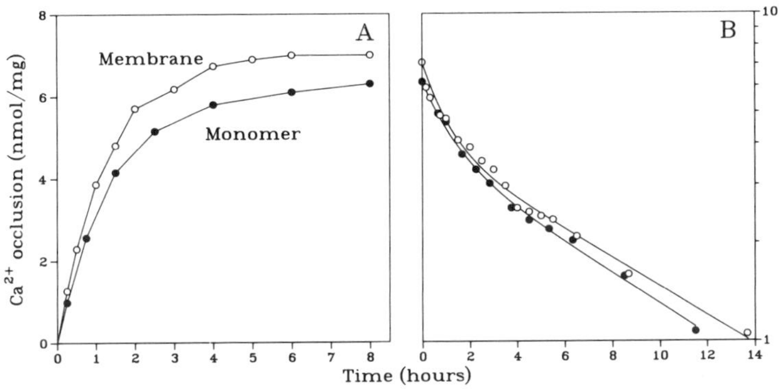

Fig. 1. Occlusion (A) and deocclusion (B) of Ca^{2+} in membranous and in soluble monomeric Ca-ATPase. The enzyme was incubated with CrATP and $^{45}Ca^{2+}$ as described in Vilsen and Andersen (1986). After 10 hours 5 mM EGTA was added to induce the deocclusion. Note the logarithmic ordinate scale in B.

As shown in Fig. 1, the rates of occlusion and deocclusion measured with soluble monomeric and with membrane-bound Ca-ATPase were very similar. For both enzyme preparations the deocclusion induced by removal of free Ca^{2+} with EGTA had a biphasic time course (Figure 1B). This may represent heterogeneity of two Ca^{2+} sites present on the same peptide chain or, possibly, the existence of more than one type of phospho-intermediate. We have found that addition of 10 mM ATP to saturate the low affinity nucleotide site on the enzyme increases the proportion of the rapid deocclusion phase relative to the slow phase (not shown).

The stoichiometry of Ca^{2+} occlusion measured in Fig. 1 in steady state was 6-7 nmol/mg of protein for membrane-bound as well as for soluble monomeric enzyme. The total number of high affinity Ca^{2+} binding sites in the membranous enzyme preparation was 7-8 nmol/mg of protein as determined under equilibrium binding conditions in the absence of phosphorylating substrate. The phosphorylation by CrATP did not create additional binding sites. The number of substrate sites determined by vanadate binding or ATP binding measurements was 4-5 nmol/mg of protein. Thus it can be concluded that there are two Ca^{2+} sites per substrate site and that most of the Ca^{2+} bound at the high affinity sites becomes occluded during phosphorylation with CrATP.

It has been shown that tryptic cleavage is useful for assignment of the conformational state of Ca-ATPase (Andersen et al., 1985,1986a). When the enzyme was phosphorylated with ATP at alkaline pH at a high Mg^{2+} concentration and low K^+ concentration the tryptic cleavage pattern was found to be similar to that of the unphosphorylated E_2 form. It is well established that the major fraction of phosphoenzyme formed from ATP under these conditions is ADP insensitive and has low Ca^{2+} affinity (cf. Andersen et al., 1985). After phosphorylation with CrATP in the same buffer medium, the tryptic cleavage pattern was indistinguishable from that of an unphosphorylated E_1 form (Vilsen and Andersen, 1987). This suggests that there are at least two phosphoenzyme conformations and that the presence of chromium at the substrate site blocks the interconversion of these states. The stabilization of a Ca^{2+} occluded intermediate in the presence of CrATP may therefore result from accumulation of E_1PCa_2 with bound Ca^{2+} that is prevented from access to the aqueous phase.

Ca^{2+} dependence of occlusion

The Ca^{2+} concentration dependence of the initial rate of CrATP induced Ca^{2+} occlusion is shown in Fig. 2. Activation of the occlusion reaction occurs in the micromolar Ca^{2+} concentration range and is characterized by positive cooperativity with a Hill number 1.8-2.0 at pH 7.0 for membrane bound Ca-ATPase as well as for soluble monomeric enzyme. This activation pattern is similar to that of Ca^{2+}-transport energized by ATP in intact vesicles, in accordance with the idea that the occlusion of Ca^{2+} in phosphoenzyme is an integral part of the transport reaction. The retention of the cooperative activation pattern after solubilization in monomeric form is consistent with existence of two interacting Ca^{2+} sites per peptide chain.

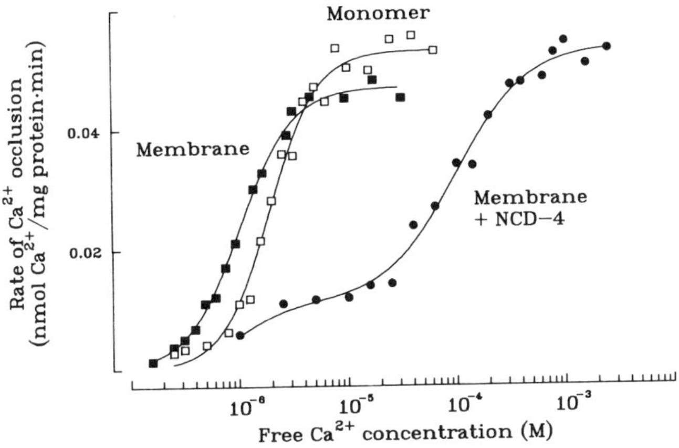

Fig. 2. Ca^{2+} concentration dependence of the rate of CrATP induced Ca^{2+} occlusion. Experiments were performed with the enzyme preparations in Fig. 1 as well as with a membrane preparation covalently labelled with NCD-4 (N-cyclohexyl-N'-(4-dimethylamino-naphtyl)carbodiimide). About 80% of the enzyme was labelled as estimated from remaining ATPase activity.

Ca^{2+} occlusion in carbodiimide labelled enzyme

The third Ca^{2+} activation curve shown in Fig. 2 indicates the effect of specific labelling of 1-2 carboxyl groups in the N-terminal part of the peptide with the fluorescent carbodiimide NCD-4. We have confirmed the observation by Chadwick and Thomas (1983) that in the absence of Ca^{2+} this carbodiimide labels sites in a relatively hydrophobic domain of the A_2 tryptic fragment. In the presence of micromolar Ca^{2+}, these sites are protected from carbodiimide. According to the structure model proposed on basis of the aminoacid sequence, the specific carbodiimide labelling sites would be located in the "stalk" domain (Brandl et al., 1986). It can be seen in Fig. 3 that the modification with NCD-4 in the absence of Ca^{2+} decreases the apparent affinity for the Ca^{2+} activation of occlusion about two orders of magnitude. On the other hand, the maximum rate of occlusion observed at around 1 mM Ca^{2+} is unaffected. Furthermore, the steady state level of occlusion reached after about 5 hours incubation with CrATP at 1 mM Ca^{2+} was found to be as high as 5.5 nmol/mg of protein, i.e. a value very close to that observed with unmodified enzyme (cf. Fig. 1).

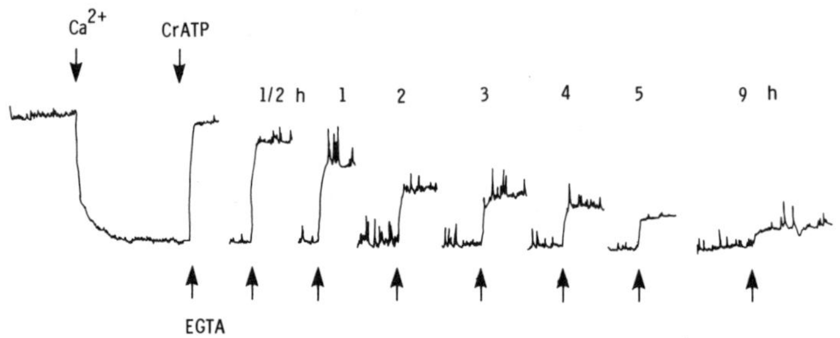

Fig. 3. Fluorescence measurement of Ca^{2+} binding and occlusion in NCD-4 modified enzyme. CrATP was added following 100 µM Ca^{2+} and after various times aliquots were tested for EGTA induced fluorescence increase. The wavelengths were 330 nm (ex.) and 415 nm (em.).

Fig. 3 shows the fluorescence responses of the covalently attached carbodiimide probe elicited by the binding and dissociation of Ca^{2+}. After addition of CrATP, the fluorescence response induced by EGTA addition disappears concurrently with progress of the occlusion reaction. The fact that occlusion **does** take place in the modified Ca-ATPase indicates that the labelled carboxyl groups do not participate directly in coordination of the bound Ca^{2+} ions. On the other hand, the fluorescent carbodiimide clearly senses the Ca^{2+} binding. From the data in Fig. 3, it may be surmised that the conformational rearrangement associated with phosphorylation and Ca^{2+} occlusion is much less extensive than that occurring in relation to the initial binding event. This is consistent with the observation of similar tryptic cleavage patterns of the phosphoenzyme containing occluded Ca^{2+} and the dephosphoenzyme with bound Ca^{2+}.

Radiation Inactivation Analysis

In order to measure the size of the Ca^{2+} occluding unit in the membrane-bound state, we have applied the radiation inactivation technique. Frozen samples of Ca-ATPase membrane and of standard proteins were irradiated with 10 MeV electrons at -20°C as described by Jensen and Ottolenghi (1985). Results are shown in Fig. 4. The target size for the CrATP induced Ca^{2+} occlusion was 94,000 Da as calculated by comparison of the decay curve with that of glucose-6-phosphate dehydrogenase activity, assuming a target size of 104,000 Da for this enzyme. Similar analysis of the fragmentation of albumin measured by SDS-polyacrylamide gelelectrophoresis provided a target size of 74,000 Da, in reasonable agreement with the molecular weight of 66,000. In accordance with Hymel et al. (1984), we found a relative target size for Ca-ATPase activity (154,000 Da), which was higher than corresponding to one peptide chain. Yet, it was too small to be consistent with a dimeric unit. It seems possible that the lipid annulus contributes to the target for ATP hydrolysis activity, since it is well established that the partial reactions following phosphorylation in the pump cycle have a strict requirement for hydrocarbon chains.

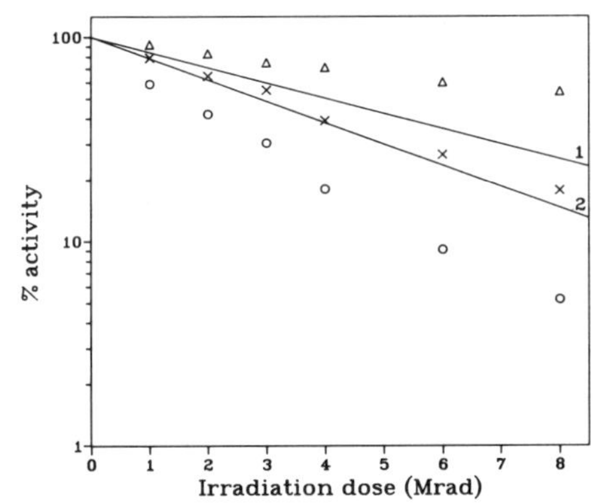

Fig. 4. Radiation inactivation of membranous Ca-ATPase. (○), Ca-ATPase activity; (X), occlusion of Ca^{2+} induced with CrATP after irradiation; (△), remaining occlusion of Ca^{2+} when the incubation with CrATP had been performed before irradiation. 1: irradiation induced fragmentation of albumin; 2: activity of glucose-6-phosphate dehydrogenase.

Interestingly, the target size for Ca^{2+} occlusion was found to be as small as 33,000 Da, when the occluded complex had been formed before irradiation (Fig. 4). This suggests that less than one third of the peptide chain is directly involved in the occlusion of the Ca^{2+} ions. This result is in agreement with the predicted domain structure (Brandl et al., 1986). The phosphorylation domain and the Ca^{2+} binding stalk constitute a total of about 300 residues.

REFERENCES

Andersen JP, Jørgensen PL, Møller JV (1985). Direct demonstration of structural changes in soluble, monomeric Ca^{2+} -ATPase associated with Ca^{2+} release during the transport cycle. Proc Natl Acad Sci USA 82:4573-4577.

Andersen JP, Vilsen B, Collins JH, Jørgensen PL (1986a). Localization of E_1-E_2 conformational transitions of sarcoplasmic reticulum Ca-ATPase by tryptic cleavage and hydrophobic labeling. J Membrane Biol 93:85-92.

Andersen JP, Vilsen B, Nielsen H, Møller JV (1986b). Characterization of detergent-solubilized sarcoplasmic reticulum Ca^{2+}-ATPase by high-performance liquid chromatography. Biochemistry 25:6439-6447.

Brandl CJ, Green NM, Korczak B, MacLennan DH (1986). Two Ca^{2+} ATPase Genes: Homologies and mechanistic implications of deduced amino acid sequences. Cell 44:597-607.

Chadwick CC, Thomas EW (1983). Inactivation of sarcoplasmic reticulum (Ca^{2+}+Mg^{2+})-ATPase by N-cyclohexyl-N'-(4-dimethylamino-naphtyl)carbodiimide. Biochim Biophys Acta 730:201-206.

De Meis L (1981). "The sarcoplasmic reticulum". New York: John Wiley & Sons.

Dupont, Y (1980). Occlusion of divalent cations in the phosphorylated calcium pump of sarcoplasmic reticulum. Eur J Biochem 109:231-238.

Hymel L, Maurer A, Berenski C, Jung CY, Fleischer S (1984). Target size of calcium pump protein from skeletal muscle sarcoplasmic reticulum. J Biol Chem 259:4890-4895.

Jensen J, Ottolenghi P (1985). Radiation inactivation at -20°C of pig kidney Na,K-ATPase. In Glynn I, Ellory C (eds): "The Sodium Pump", Cambridge: The Company of Biologists Ltd, pp 77-81.

Vilsen B, Andersen JP (1986). Occlusion of Ca^{2+} in soluble monomeric sarcoplasmic reticulum Ca-ATPase. Biochim Biophys Acta 855:429-431.

Vilsen B, Andersen JP (1987). Characterization of CrATP-induced calcium occlusion in membrane-bound and soluble monomeric sarcoplasmic reticulum Ca^{2+}-ATPase. Biochim Biophys Acta 898:313-322.

The Ion Pumps: Structure, Function,
and Regulation, pages 155–164

"UNCOUPLING" OF THE SARCOPLASMIC RETICULUM Ca^{2+}-ATPase ?

Varda Shoshan-Barmatz and D.M. Chipman
Department of Biology, Ben Gurion University
of the Negev, Beer Sheva, Israel

INTRODUCTION

The Ca^{2+}-ATPase of the sarcoplasmic reticulum (SR) is one of the best understood examples of a coupled "pump", in which ATP hydrolysis is coupled to Ca^{2+} transport. Several laboratories have reported a variety of treatments of SR vesicles that lead to apparent uncoupling of Ca^{2+} transport from ATP hydrolysis. Such treatments include low and high pH incubations (Berman et al., 1977, Duggan and Martonosi, 1970), chemical modification reagents (Hidalgo et al., 1982) and proteolysis by trypsin (Stewart and MacLennan, 1974, Scott and Shamoo, 1982). It has been suggested that the observed stimulation of ATPase and/or inhibition of Ca^{2+} accumulation are due to specific modifications of the Ca^{2+}-ATPase (Berman, 1982) or of boundary lipids (Hidalgo et al., 1982), which lead to changes in the operation of the enzyme itself.

According to our current understanding of the action of ion pumps, the coupling between ion transport and ATP hydrolysis is a result of a catalytic cycle obeying "coupling rules"; i.e., only certain reactions are allowed, while others are forbidden, for any state of the enzyme. We would expect, therefore, that any true uncoupling would be reflected in the catalytic cycle. An analysis of the changes in the catalytic cycle of the ATPase which lead to such uncoupling would thus contribute a great deal to the understanding of coupled ion transport processes.

In this communication we examine several treatments which lead to apparent uncoupling, in order to determine whether they indeed produce uncoupling by altering the ATPase molecular itself. To this end, we studied the catalytic activity and membrane permeability of the modified SR vesicles and proteoliposomes reconstituted from them.

RESULTS

Fig. 1 shows the effect on Ca^{2+} accumulation and ATP hydrolysis of pretreating, in the presence of ATP, of SR vesicles with acetic anhydride, methylbenzimidate (MBI), or trypsin. Upon these treatments the SR vesicles appear to be completely "uncoupled", i.e., the ATPase activity remains as in untreated vesicles, or is stimulated by over 2-fold, while Ca^{2+} accumulation is less than 10% of control vesicles. The loss of Ca^{2+} accumulation capacity upon the above treatments is accompanied by greatly accelerated passive efflux of Ca^{2+} from the SR vesicles. Fig. 2 compares the

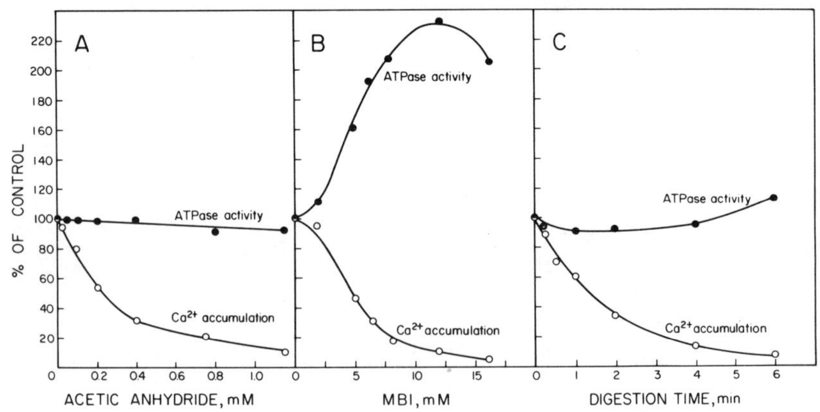

Fig. 1. Effect of chemical modifications or trypsin digestion of SR vesicles on their Ca^{2+} accumulation and ATPase activity.

SR vesicles (1 mg/ml) were preincubated with: (A) acetic anhydride (Shoshan-Barmatz, 1986); (B) methylbenzimidate (MBI) (Shoshan-Barmatz, 1987a); (C) trypsin (Shoshan-Barmatz et al., 1987) in the presence of 3 mM ATP. Ca^{2+} accumulation and ATPase activity were assayed as described previously (Shoshan-Barmatz, 1986).

Ca^{2+} efflux from control and treated membranes, passively loaded with Ca^{2+}. As shown, the modified membranes released about 80% of their trapped Ca^{2+} within 3 sec. In contrast, the unmodified membranes released about 3% of their Ca^{2+} content during the same period of time. The "leakiness" of the modified membranes is obviously a possible reason for the decreased Ca^{2+} accumulation capacity.

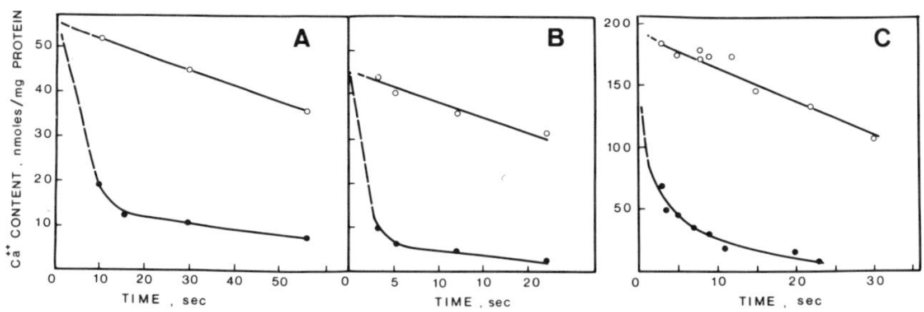

Fig. 2. Ca^{2+} efflux from unmodified and from membranes pretreated with acetic anhydride, MBI or trypsin.

Control or vesicles treated in the presence of ATP with acetic anhydride (1 mM), MBI (15 mM) or trypsin (3 min) as described in Fig. 1, were passively loaded with ^{45}Ca by 24 hr incubation with 5 mM $^{45}CaCl_2$ (A and B) or 20 mM (C); 20 mM MOPS (pH 6.8), and 100 mM KCl. Ca^{2+} efflux was measured by the filter washing method, in Ca^{2+}-free medium, pH 6.8 (Shoshan-Barmatz, 1987a).

In order to clarify whether the "uncoupling" produced by the various treatments was due to the modification of the Ca^{2+}-ATPase itself, several partial reactions of the transport cycle were tested. The pre-steady state phosphoenzyme (EP) formation from ATP by control and MBI-modified membranes is shown in Fig. 3. The same rate of EP formation and steady state level of EP were obtained with the unmodified or MBI-modified vesicles. Fig. 4 shows that a tryptic digestion which leads to a 90% decrease in Ca^{2+} accumulation capacity of the vesicles, has no effect on the Ca^{2+} dependence of the ATPase activity. $K_{0.5}$ for Ca^{2+}-activation remains close to 0.2 µM, and high Ca^{2+} levels inhibit the activity in treated as well as control vesicles ($K_{0.5}$ = 0.1 mM).

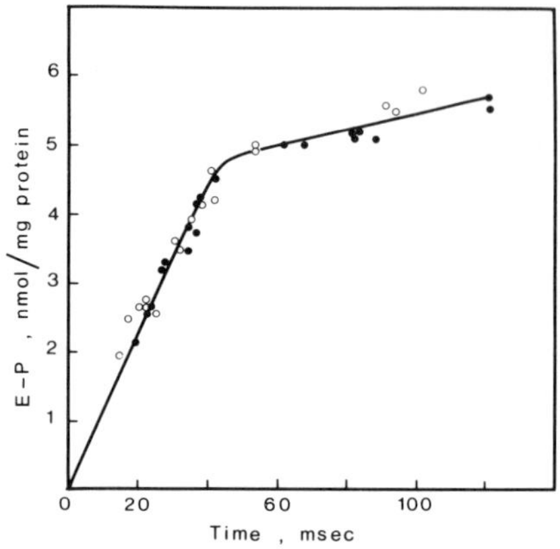

Fig. 3. Pre-Steady-State phosphoenzyme formation rate in control and MBI-modified SR vesicles.

Phosphorylation of control (o) and MBI-modified (●) membranes with [γ-^{32}P]ATP was carried out, using rapid mixing and quenching technique, as described (Shoshan-Barmatz, 1987a).

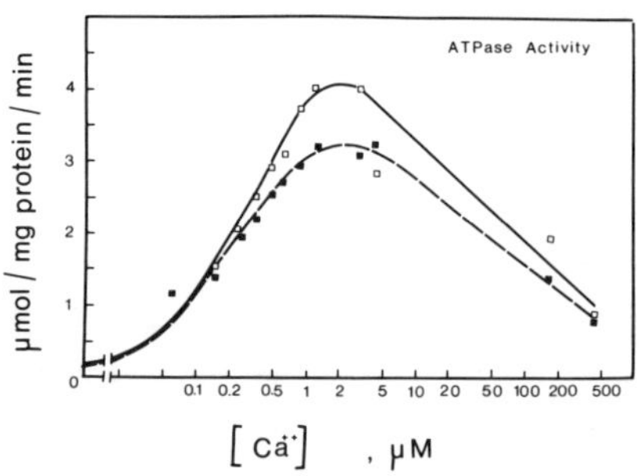

Fig. 4. Ca^{2+}-dependence of the ATPase activity of native (■) and trypsin cleaved (□) SR vesicles.

ATPase activity was assayed as in Fig. 1 except that free Ca^{2+} was controlled by different EGTA/Ca^{2+} concentrations.

The ATP - Pi exchange catalyzed by the ATPase at a low Ca^{2+} concentration is affected by tryptic digestion of the vesicles, but the activity measured at 5 mM is not affected (Fig. 5). The behaviour of the trypsin digested vesicles is as expected if the ATP-Pi exchange requires a high internal Ca^{2+} concentration (Carvalho et al., 1976). The following reactions also were not affected by the above modifications: the decay of EP formed from ATP, induced by ADP or EGTA; the ADP-ATP exchange and the phosphatase activity (the hydrolysis of dinitrophenyl phosphate). Thus, no partial reaction of the Ca^{2+} transport reveals changes in the propertiesof the Ca^{2+}-ATPase upon the different treatments.

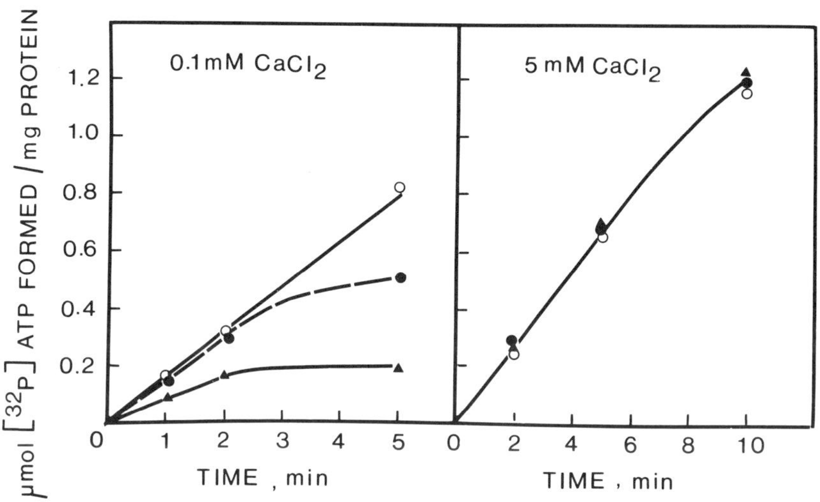

Fig. 5. Effect of trypsin treatment on ATP-Pi exchange.
SR vesicles were treated with trypsin (1/30, w/w) in the presence of 3 mM ATP, for 0 (o), 15 sec (●), or 5 min (▲). Aliquots were assayed for ATP-Pi exchange in the presence of 0.1 mM or 5 mM $CaCl_2$ as described (Shoshan-Barmatz et al., 1987).

To test whether the Ca^{2+}-ATPase molecule is involved in the increased Ca^{2+} efflux produced by the different modifications, experiments were conducted on the purified enzyme in a reconstituted system. Table 1 shows the effect of acetic anhydride modification of proteoliposomes on their Ca^{2+}-efflux properties. As shown, treatment with acetic anhydride stimulates Ca^{2+} efflux from SR vesicles and from proteoliposomes reconstituted with the partially purified ATPase (R_2) but not from proteoliposomes reconstituted with the more purified ATPase fraction (R_3).

Table 2 shows the Ca^{2+} accumulation by proteoliposomes reconstituted with R_2 or R_3 isolated from native or MBI-modified SR membranes. The ability of proteoliposomes reconstituted with R_2 isolated from MBI-modified membranes to accumulate Ca^{2+} is about 30% of that of proteoliposomes reconstituted with R_2 isolated from unmodified membranes.

TABLE 1. Effect of acetic anhydride modification on Ca^{2+} efflux from Ca^{2+}-loaded liposomes reconstituted with partially purified (R_2) or more purifed ATPase (R_3) fractions.

Protein preparations and proteoliposomes were prepared, loaded actively with Ca^{2+} phosphate, modified with acetic anhydride (1.5 mM) and assayed for Ca^{2+} efflux as described previously (Shoshan-Barmatz, 1986).

	Ca^{2+} release/min (% of control)	
Vesicle preparation	Control	Acetic anhydride modified
Sarcoplasmic reticulum	16	65
R_2-liposomes	16	52
R_3-liposomes	3	5

However, proteoliposomes reconstituted with R_3 derived from the unmodified or MBI-modified membranes shows no difference in their Ca^{2+} accumulation ability. These results suggest that some other component(s) lost during the purification of the Ca^{2+}-ATPase, may be responsible for the leakiness of the modified membranes. Another experiment, in which proteoliposomes prepared from the purified ATPase were treated with trypsin, is shown in Fig. 6. The results demonstrate that cleavage of the ATPase peptide(s) does <u>not</u> lead to increased Ca^{2+} permeability (Fig. 6B). Despite the fact that the cleavage of the ATPase (at the second tryptic site) proceeds similarly in the proteoliposomes and in the native vesicles (Fig. 6A), Ca^{2+} efflux from passively loaded proteoliposomes

TABLE 2. Ca^{2+} accumulation by proteoliposomes reconstituted with R_2 or R_3 isolated from native or MBI-modified SR membranes.

R_2 and R_3 were purified from control and MBI-modified membranes and reconstituted into liposomes as described previously (Shoshan-Barmatz, 1987a). Ca^{2+} accumulation was assayed as in Fig. 1.

Vesicle preparations	Ca^{2+} accumulation (μmol/mg of proteins/min)	Inhibition (%)
Sarcoplasmic reticulum (SR)	0.525	-
MBI-modified SR	0.131	75
R_2	0.301	-
MBI-modified R_2	0.109	64
R_3	0.175	-
MBI-modified R_3	0.173	1

is completely unaffected by tryptic digestion while efflux from passively loaded SR vesicles is greatly accelerated (Fig. 6B). Cleavage of some other protein component(s) of the native SR vesicles must be responsible for their increased Ca^{2+} permeability after trypsin treatment.

An experiment similar to that of Fig. 6A was used for demonstrating the lack of satisfactory correlation between the loss of Ca^{2+} accumulation and the cleavage of the ATPase (see Shoshan-Barmatz et al., 1987).

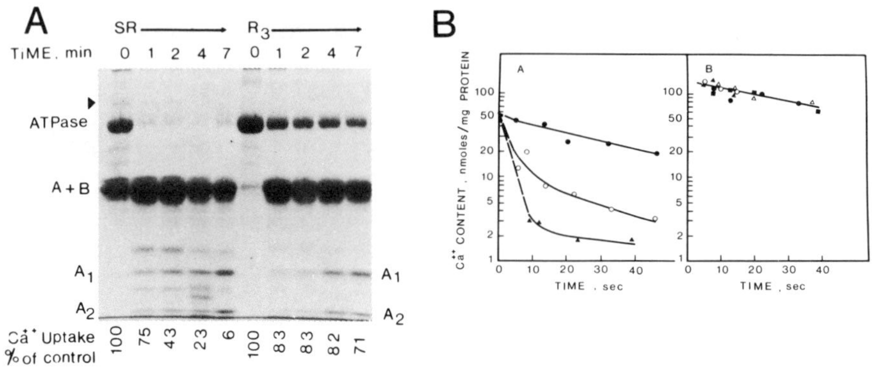

Fig. 6. SDS-PAGE analysis (A) and Ca^{2+} efflux (B) of trypsin treated SR vesicles (left) or proteoliposomes (right).

Proteoliposome were prepared from R_3, in 10 mM tricine, pH 8.0, containing 0.2 M KCl and 3 mM ATP. The liposomes (right) and SR vesicles (left) were treated at 1 mg/ml SR protein and 30 µg trypsin/ml for various times. Aliquots were analyzed by the Laemmli (1970) SDS-PAGE technique, and for Ca^{2+} uptake. Note 160 kDa band (arrow). In experiment B vesicles or proteoliposomes were passively loaded with ^{45}Ca for 3 hr at 22°C (see Fig. 2), and Ca^{2+} efflux initiated by filter washing in Ca^{2+}-free medium. Trypsin treatment times were: 0 (●), 1 (o), 2 (▲) 4 (△) or 7 (■) min.

DISCUSSION

The results presented here indicate that the decreased net Ca^{2+} accumulation accompanied by increased or unchanged ATPase activity, produced by the treatments studied here, is due in each case to increased membrane permeability, rather than to modification of the catalytic cycle of the ATPase. Furthermore, we suggest that these treatments increase Ca^{2+} permeability by action on membrane component(s) other than the Ca^{2+}-ATPase, possibly the Ca^{2+} release system of the SR (see Shoshan-Barmatz, 1986, 1987a, 1987b). These conclusions are based on the following observations: A. the treatments with amine modifying reagents or trypsin digestion lead to marked increases in the Ca^{2+} efflux from the modified vesicles.

It should be noted that in other studies of "uncoupling", Ca^{2+} permeability has sometimes been measured by loading vesicles with calcium oxalate, and that under such conditions dissolution of calcium oxalate crystals may be partly rate limiting for efflux. In other cases, high Ca^{2+} concentrations used in loading and/or efflux may lead to reclosing of efflux pathways (Shoshan et al., 1981; Nagasaki and Kasai 1983). B. The kinetics of the ATPase reaction and its partial reactions, such as the rate of EP formation and its decay, the [Ca^{2+}]-dependence of the ATPase, ATP-Pi and ATP-ADP exchanges, are essentially unchanged by the treatments. C. The uncoupling phenomenon disappears when the ATPase isolated from treated vesicles is reconstituted into liposomes, and the treatments do not affect Ca^{2+} efflux or accumulation by proteoliposomes reconstituted from unmodified ATPase. D. Furthermore, in contrast to a previous suggestion (Scott and Shamoo, 1982) our results show that in the case of trypsin treatment the "uncoupling" is poorly correlated with the specific cleavage of the ATPase polypeptide, (Andersen and Jorgensen, 1985).

The treatments with MBI or acetic anhydride (Shoshan-Barmatz, 1986, 1987a) require the presence of ATP for producing "uncoupled" membranes, which may well be related to the suggestion that a Ca^{2+} release channel of SR membranes has an ATP binding site (Meissner 1984, Martonosi, 1984).

In conclusion, our results suggest that the hydrolytic activity and coupled Ca^{2+} transport functions of the Ca^{2+}-ATPase may well be unaffected by the various treatments, but the apparent uncoupling is due to stimulation of Ca^{2+} efflux in these vesicles.

REFERENCES

Andersen P, Jorgensen PL (1985) Conformational states of sarcoplasmic reticulum Ca^{2+}-ATPase as studied by proteolytic cleavage. J Membr Biol 88: 187-198.

Berman MC (1982) Energy coupling and uncoupling of active calcium transport by sarcoplasmic reticulum membranes. Biochem Biophys Acta 694: 45-121.

Berman MS, McIntosh DB, Kench JE (1977) Protein inactivation of Ca^{2+} transport by sarcoplasmic reticulum. J Biol Chem 252: 994-1001.

Duggan PF, Martonosi A (1970) Sarcoplasmic reticulum. IX. The permeability of sarcoplasmic reticulum membranes. J Gen Physiol 56: 147-167.

Hidalgo C, Petrucci DA, Vergara C (1982) Uncoupling of Ca^{2+} transport in sarcoplasmic reticulum as a result of labelling lipid amino groups and inhibition of Ca^{2+}-ATPase activity by modification of lysine residues of the Ca^{2+}-ATPase polypeptide. J Biol Chem 252: 8446-8454.

Laemmli UK (1970) Cleavage of structural proteins during the assembly of the head of bacteriophage T_4. Nature 227: 680-685.

Martonosi AN (1984) Mechanisms of Ca^{2+} release from sarcoplasmic reticulum of skeletal muscle. Physiol Rev 64: 1240-1319.

Meissner G (1984) Adenine nucleotide stimulation of Ca^{2+} induced Ca^{2+} release in sarcoplasmic reticulum. J Biol Chem 259: 2365-2374.

Nagasaki, K, Kasai M (1981) Calcium induced calcium release from sarcoplasmic reticulum. J Biochem 90: 749-755.

Scott TL, Shamoo AE (1982) Disruption of energy transduction in sarcoplasmic reticulum by trypsin cleavage of (Ca^{2+}-Mg^{2+})-ATPase. J Membrane Biol 64: 147-167.

Shoshan-Barmatz V (1986) Chemical modification of sarcoplasmic reticulum; stimulation of Ca^{2+} release. Biochem J 240: 509-517.

Shoshan-Barmatz V (1987a) Chemical modification of sarcoplasmic reticulum by methylbenzimidate; stimulation of Ca^{2+} efflux. Biochem J 243: 165-173.

Shoshan-Barmatz V (1987b) Stimulation of Ca^{2+} efflux from sarcoplasmic reticulum by preincubation with ATP and inorganic phosphate. Biochem J 247: 497-504.

Shoshan-Barmatz V, Ouziel N, Chipman DM (1987) Tryptic digestion of sarcoplasmic reticulum inhibits Ca^{2+}-accumulation by action on a membrane component other than the Ca^{2+}-ATPase. J Biol Chem 262: 11559-11564.

Shoshan V, MacLennan DH, Wood DS (1981) A proton gradient controls a calcium release channel in sarcoplasmic reticulum. Proc Natl Acad Sci USA 78: 4828-4832.

Stewart PS, MacLennan DH (1974) Surface particles of sarcoplasmic reticulum membranes, structural features of the adenosine triphosphatase. J Biol Chem 249: 985-993.

The Ion Pumps: Structure, Function, and Regulation, pages 165–169

INHIBITION OF Ca^{++}-ATPase BY LIMITED PROTEOLYSIS

Katalin Török and N. Michael Green

Laboratory of Protein Structure,
National Institute for Medical Research,
Mill Hill, London NW7 1AA, England.

INTRODUCTION

The sarcoplasmic reticulum (SR) Ca^{++}-ATPase is cleaved by trypsin at two sites, T1 (Arg 505) and T2 (Arg 198). Cleavage at T1 is complete, but at T2 it is incomplete. In this report a number of conditions were tested in an attempt to increase the degree of specific digestion at T2. The effects of the cleavages on the function of the Ca^{++}-ATPase were also studied. It has been reported that while the activities of the singly cleaved enzyme are not impaired, the Ca^{++} transport activity is lost in the doubly cleaved enzyme (Scott and Shamoo, 1982). This loss was interpreted as uncoupling, since the ATPase activity was found unchanged.

The significance of this observation was not clear because the ATPase activity was measured in sealed vesicles. In these conditions inhibitory concentrations of Ca^{++} are accumulated and the apparent activity is increased by any leakage of Ca^{++} from the vesicles. If ionophor is used, however, accumulation of Ca^{++} in the vesicles is prevented and expression of the maximal or intrinsic ATPase activity is allowed. We have therefore measured activity both in the presence and in the absence of Ca^{++} ionophor so that the effects of trypsin digestion on the rate of Ca^{++} leakage can be distinguished from effects on the intrinsic ATPase activity. We found that the intrinsic activity is decreased in parallel with the T2 cleavage, but that the apparent Ca^{++} uptake decreases even more rapidly.

RESULTS

Conditions of Trypsin Digestion

A variety of conditions were used to try to increase the degree of specific cleavage at T2. The variations included pH 7-9, $0^{o}C$ and room temperature, repeated addition of trypsin, inclusion of nucleotide analogs (ADP, AMP-PNP) and DTT in the digestion medium, digestion of detergent solubilised Ca^{++}-ATPase and addition or omission of sucrose and glycerol (some of these conditions had been previously used to stabilise the A1 fragment (Saito et al., 1984; Imamura et al., 1984). The results showed that it was necessary to include sucrose or glycerol, or a nucleotide and mM Ca^{++} to prevent extensive degradation, and that while the rate of the cleavages could be substantially varied by changing the other parameters, the plateau of digestion at T2 remained invariably 50-60%. A typical digestion medium included 50 mM MES pH 7.0, 100 mM KCl, 5 mM $MgCl_2$, 2 mM $CaCl_2$, 20% glycerol.

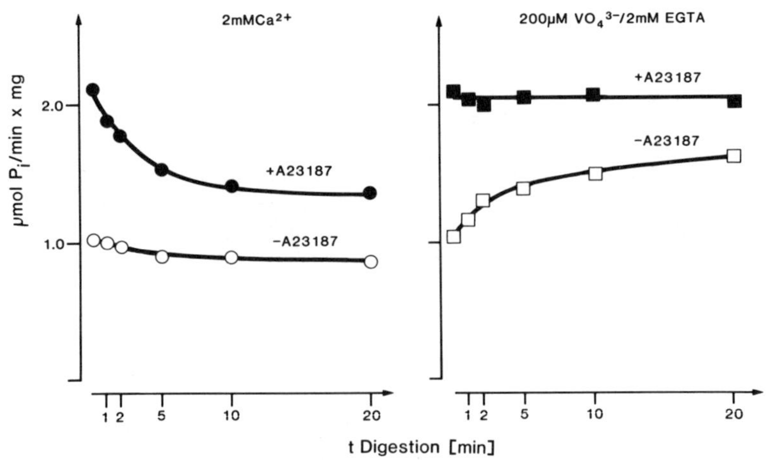

Figure 1. ATPase activity of SR vesicles containing singly and doubly cleaved Ca^{++}-ATPase.

Effects of Tryptic Digestion on Ca^{++}-ATPase Activity

Ca^{++}-sensitive ATPase activity of the digested samples was determined in the absence and presence of the Ca^{++} ionophor A 23187. In Fig. 1a, ATPase activities of the Ca^{++}-ATPase digested in the presence of Ca^{++} are shown. In the absence of ionophor virtually no change in the ATPase activity could be observed. In the presence of the Ca^{++} ionophor, however, inactivation of the ATPase up to about 30% was found. If the cleavage was restricted to T1 only using vanadate/EGTA, the intrinsic activity was not affected (Fig. 1b), but there was an increase in the activity measured without Ca^{++} ionophor consistent with a Ca^{++} leak caused by the tryptic digestion.

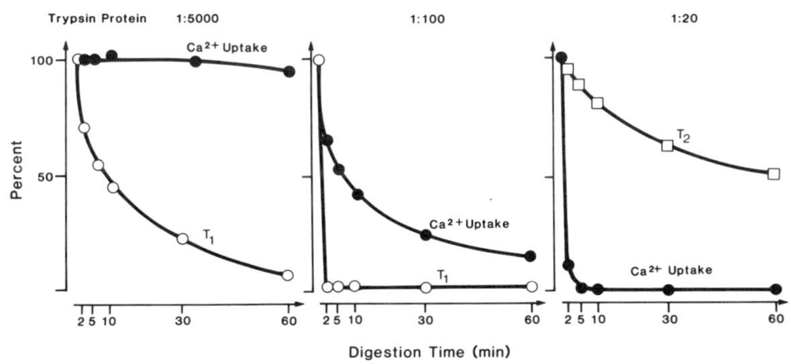

Figure 2. Lack of correlation between Ca^{++} uptake and cleavage of Ca^{++}-ATPase at the T1 and T2.

Is Ca^{++} Uptake Affected by Cleavage at T1 or T2?

The Ca^{++} leakage inferred from the ATPase assays was further characterised by direct Ca^{++} uptake measurements. In order to test for any correlation between tryptic cleavage of the Ca^{++}-ATPase and leakiness of the SR vesicles, three different levels of trypsin were used. Ca^{++} uptake of the digested SR vesicles was measured and the cleavage rates followed by densitometry of SDS-gels (Fig. 2). It can be seen that the very low level tryptic digestion (1/5000) completed cleavage at T1 without affecting Ca^{++} uptake. At the intermediate trypsin level (1/100) cleavage at T1 was very rapid, while loss of Ca^{++} uptake was slower. Finally, at high trypsin level (1/20) Ca^{++} uptake was rapidly lost, while T2 cleavage was slow. These results show that loss of Ca^{++} transport activity of the digested SR vesicles is not correlated with either the T1 or the T2 cleavage of the Ca^{++}-ATPase. The loss of Ca^{++} uptake therefore appears to be caused by digestion of some other protein(s) of the SR vesicles.

Specific Activity of the Doubly Cleaved Ca^{++}-ATPase

It was possible to calculate the Ca^{++}-ATPase activity of the doubly cleaved enzyme on the basis of a simple model containing the following constraints:

1. digestion at T1 is complete and the activity of the A and B fragments is identical to that of the undigested ATPase,
2. the total activity of the digested Ca^{++}-ATPase is composed of the activity of the fraction of the ATPase digested only at T1 and the activity of the fraction digested at both T1 and T2.

The observed activity would therefore be given by the formula a = bX + cY, where
"a" is the total ATPase activity measured in the presence of Ca^{++} ionophor,
"b" is the specific activity of the fraction of the Ca^{++}-ATPase digested at T1 only (A+B fragments),
"c" is the activity of the fraction of the Ca^{++}-ATPase digested at both T1 and T2 sites (A1+A2+B fragments),
X is the fraction of the ATPase digested at T1,
Y is the fraction of the ATPase digested at both T1 and T2.
Quantitation of the tryptic fragments was based on densitometry of SDS-gels of the digested Ca^{++}-ATPase samples. Table 1 shows some typical data on the basis of which the activity attributable to the Ca^{++}-ATPase cleaved at both T1 and T2 (A1+A2+B) was found to be 60% ± 8% of the undigested Ca^{++}-ATPase.

In summary, the limitation of cleavage at the T2 site to about half of the ATPase molecules suggests that there may be two different forms of Ca^{++}-ATPase in the presence

TABLE 1. Calculation of ATPase Activity of Doubly Cleaved Ca^{++}-ATPase

Time of digestion	a (μmol P_i/ min x mg)	%X	%Y	c/b
0 min	2.06	100	0	
2 min	1.93	83.6	16.4	0.61
30 min	1.74	52.2	47.8	0.68
2 hr	1.51	42.7	57.3	0.53

b = 2.06 μmol P_i/min x mg

of Ca^{++} which are not in a conformational or oligomeric equilibrium. Cleavage at the T2 site causes an about 40% inhibition of the Ca^{++}-dependent ATPase activity. This inhibition is characterised in detail in our full paper (Török et al., 1987). The Ca^{++} permeability change observed in the course of the tryptic digestion of SR vesicles is shown here not to correlate with either the T1 or the T2 cleavage of the Ca^{++}-ATPase and is therefore likely to be the consequence of digestion of some other membrane components of the SR vesicles.

REFERENCES

Imamura Y, Saito K, Kawakita M (1984). Conformational change of Ca^{++}+Mg^{++}+ATPase of sarcoplasmic reticulum upon binding of Ca^{++} and AMP-PNP as detected by trypsin sensitivity analysis. J Biochem (Tokyo) 95:1305-1313.

Saito K, Imamura Y, Kawakita M (1984). Effect of temperature and added ligands on the susceptibility of Ca^{++}+Mg^{++}+ATPase of the sarcoplasmic reticulum. J Biochem (Tokyo) 95:1297-1304.

Scott TL, Shamoo AE (1982). Disruption of energy transduction in sarcoplasmic reticulum by trypsin cleavage of Ca^{++}+Mg^{++}+ATPase. J Membrane Biol 64:137-144.

Török K, Trinnaman BJ, Green NM (1987). Tryptic cleavage inhibits but does not uncouple Ca^{++}-ATPase of sarcoplasmic reticulum. Eur J Biochem (submitted).

The Ion Pumps: Structure, Function, and Regulation, pages 171–176

BINDING OF CALCIUM TO THE VANADATE COMPLEX OF FITC-LABELLED Ca-ATPase

Stephania Markus, Zvi Priel and David M. Chipman

Departments of Biology (S.M. and D.M.C.) and Chemistry (Z.P. and S.M.), Ben Gurion University, Beer Sheva 84105, Israel

INTRODUCTION

We have recently reported studies of the interaction of vanadate with the Ca-ATPase from sarcoplasmic reticulum (SR), in which the initial velocity of the ATPase reaction was used as a measure of enzyme not complexed to vanadate (Markus et al., 1986). Contrary to previous reports (e.g., O'Neal et al., 1979; Pick, 1982; Medda and Hasselbach, 1983; Andersen and Moller, 1985) we found compelling evidence for the existence of a ternary complex Ca_2EVan (Markus et al., 1986). The initial velocities show cooperative dependence on $[Ca^{2+}]$ and non-competitive inhibition by vanadate. Equilibrium constants for dissociation of vanadate from EVan and Ca_2EVan were $K_V = 4x10^{-7}M$ and $K_V' = 7x10^{-6}M$, respectively, at pH 7.0 and 25°.

We report here studies of fluorescence intensity changes in FITC-labelled Ca-ATPase (Pick, 1982; Pick and Karlish, 1982), which support our previous conclusion. Furthermore, the kinetics of the changes provide additional information about the nature of the ternary complex(es).

RESULTS

FITC-labelled SR vesicles were prepared by the method of Pick and Karlish (1982). As reported, the fluorescence emission at 516 nm (excitation 496) of the bound fluorescein is quenched by Ca^{2+} addition, and enhanced by vanadate addition. Fig. 1A shows isotherms for the effect of Ca^{2+} on labelled vesicles preincubated with vanadate. The

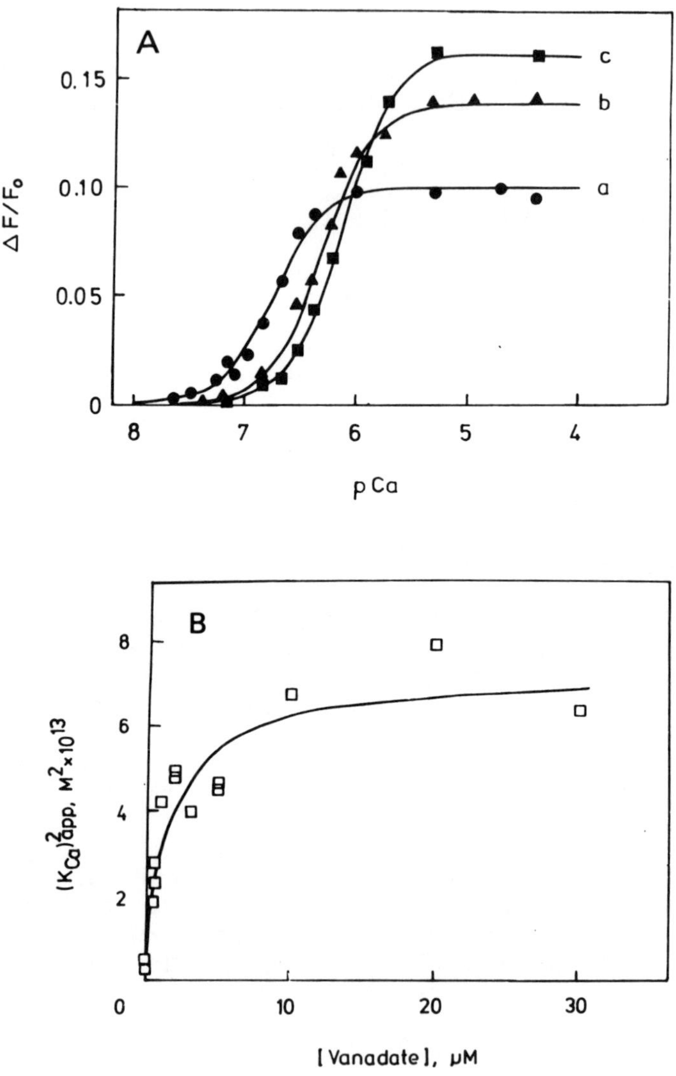

Figure 1. (A) Isotherms for Ca^{2+} interaction with FITC-Ca-ATPase in the absence (a) and presence of 0.2 uM (b) or 2 uM (c) vanadate. F_o is the fluorescence before Ca^{2+} addition. (B) The dependence on vanadate of $(K_{Ca})_{app}$ calculated from isotherms as in A.

results at any vanadate concentration fit cooperative binding (n = 2) of Ca^{2+}, with an apparent dissociation constant, $(K_{Ca})_{app}$, which shows hyperbolic dependence on [vanadate] (Fig. 1B). The limiting values of $(K_{Ca})_{app}$ are 6.8 x10^{-14} M^2 and 7.4 x10^{-13} M^2, at zero and saturating vanadate, respectively. The hyperbolic dependence requires simultaneous vanadate and Ca^{2+} binding, with $(K_{Ca})_{app}$ at saturating vanadate representing dissociation of Ca^{2+} from Ca_2EVan. The ratio of the constants determined at limiting [vanadate] is in good agreement with K_V/K_V' determined previously.

The observed final fluorescence intensity (relative to unliganded FITC-labelled vesicles) at saturating Ca^{2+} is vanadate independent. Observations similar to ours have previously been interpreted as evidence that Ca^{2+} and vanadate bind exclusively to labelled enzyme (e.g., Pick, 1982). We suggest here, rather, that this implies that the environment of the fluorophor in Ca_2EVan is like that in Ca_2E, e.g., "E_1-like".

The kinetics of the fluorescence changes further support the existence of Ca_2EVan. Addition of Ca^{2+} to

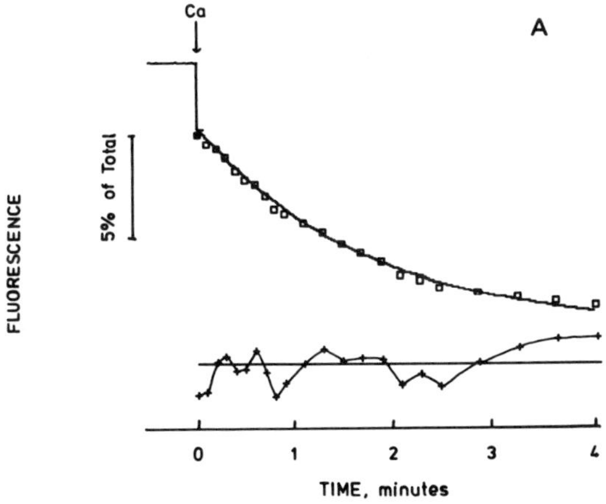

Figure 2A. Kinetics of fluorescence changes upon addition of Ca^{2+} to FITC-Ca-ATPase preincubated with vanadate. Example of time course, for [vanadate] = 24 uM, [Ca^{2+}] = 5 uM, with slow phase fit by exponential decay with k = 0.58 min^{-1}.

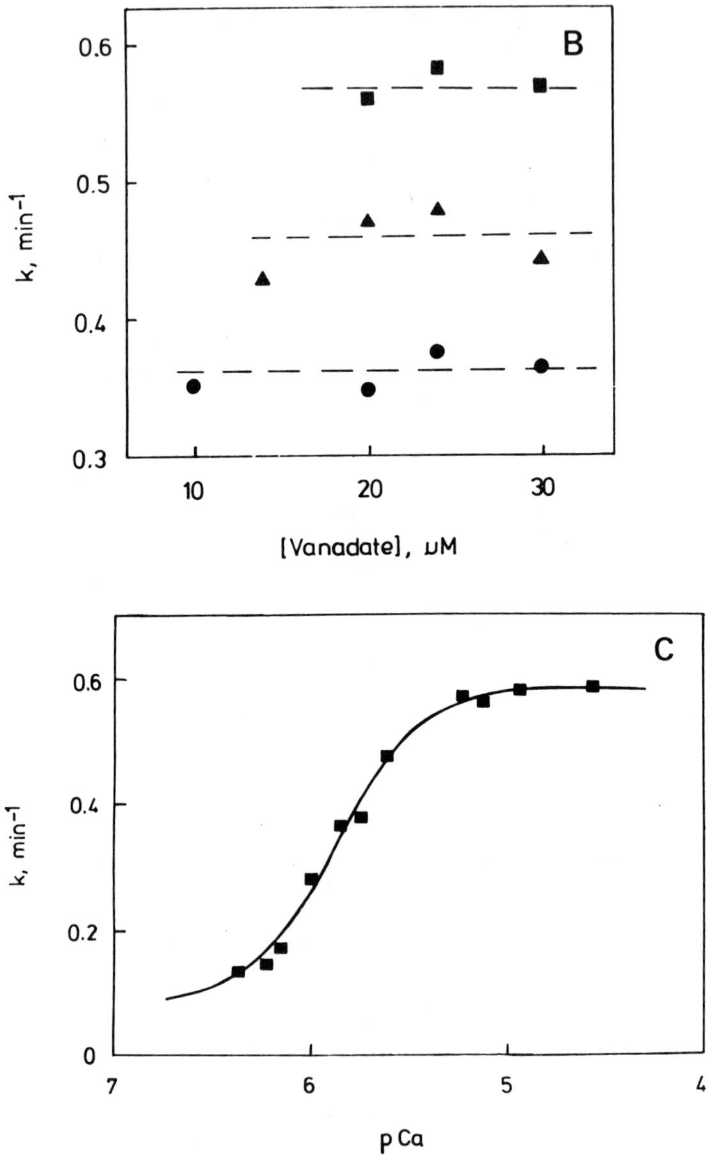

Figure 2B and C. Kinetics of fluorescence changes upon addition of Ca^{2+} to FITC-Ca-ATPase preincubated with vanadate. (B) Vanadate dependence of slow decay rate, for $[Ca^{2+}]$ = 1.8 uM (●), 5 uM (▲) or 19 uM (■). (C) Calcium dependence at constant vanadate (24 uM).

labelled enzyme preincubated with vanadate leads to biphasic quenching of fluorescence (Fig. 2A). The rate of the slow phase is independent of [vanadate] at saturating concentrations (Fig. 2B), but is cooperatively dependent on [Ca^{2+}] ($n = 2.1$, $K^* = 1.4 \times 10^{-6}$ M) (Fig. 2C).

Since the decay rate is independent of [vanadate] at high concentrations, the rate determining step for the observed process cannot be the association of Ca^{2+} to the residual vanadate-free enzyme existing in equilibrium with EVan. Since the process shows saturating dependence on [Ca^{2+}], it also cannot be due to rate-determining initial Ca^{2+} binding to EVan. The data are in accord, however, with the scheme below, in which a slow conformational change ($k_{+3} + k_{-3} = 0.6$ min^{-1}) follows the initial calcium binding to EVan.

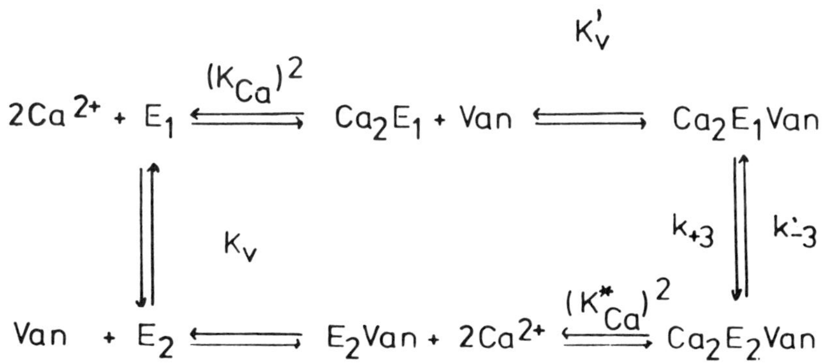

CONCLUSIONS

Equilibrium measurements by two techniques - initial activity of the native enzyme (Markus <u>et al.</u>, 1986) and fluorescence intensity of FITC-labelled enzyme - indicate the existence of a ternary Ca_2EVan complex, with K_V' about an order of magnitude greater than K_V. The kinetics of the formation of Ca_2EVan from EVan and free calcium suggest rapid formation of one Ca_2EVan complex, which slowly isomerizes to a stable "E_1-like" conformation, similar to Ca_2E in the environment of the fluorophor site. If the scheme shown is correct, we can estimate from $(K_{Ca})_{app}$ at vanadate saturation (Fig. 1B), and the calcium dependence of the slow decay phase (Fig. 2C), that the final Ca_2EVan complex is more stable by a factor of 5-10 than the initial complex.

It is tempting to speculate that the difference in conformation, between the stable complex with an "E_1" like environment at the nucleoside site and the Ca_2EVan complex initially formed on addition of Ca^{2+} to EVan, is related to the difference between the form of Ca_2E-P with occluded calcium, and that which is presumed to release Ca^{2+} into the lumen of the SR vesicle. The structure of the Ca_2EVan complex is, in any case, worth further study.

REFERENCES

Andersen JP, Moller JV (1985) The role of Mg^{2+} and Ca^{2+} in in the simultaneous binding of vanadate and ATP at the phosphorylation site of sarcoplasmic reticulum Ca^{2+}-ATPase. Biochim Biophys Acta 815: 9-15.

Markus S, Priel Z, Chipman DM (1986) Simultaneous binding of calcium and vanadate to the Ca^{2+}-ATPase of sarcoplasmic reticulum. Biochim Biophys Acta 874: 128-135.

Medda P, Hasselbach W (1983) The vanadate complex of the calcium transport ATPase of the sarcoplasmic reticulum, its formation and dissociation. Eur J Biochem 139: 7-14.

O'Neal SC, Rhoads RB and Racker E (1979) Vanadate inhibition of sarcoplasmic reticulum Ca^{2+} ATPase and other ATPases. Biochem Biophys Res Commun 89: 845-850.

Pick U (1982) The interaction of vanadate ions with the Ca-ATPase from sarcoplasmic reticulum. J Biol Chem 257: 6111-6119.

Pick U, Karlish SJD (1982) Regulation of the conformational transition in the Ca-ATPase from sarcoplasmic reticulum by pH, temperature and calcium ions. J Biol Chem 257: 6120-6126.

The Ion Pumps: Structure, Function, and Regulation, pages 177–182

SPECIFICITY OF THE E-P STATE OF THE SR Ca-ATPase: IMPLICATIONS FOR THE COUPLING MECHANISM

David M. Chipman and William P. Jencks

Dept of Biology, Ben Gurion University, Beersheva, Israel (DMC) and Graduate Dept of Biochemistry, Brandeis University, Waltham, Massachusetts, U.S.A. (WPJ)

INTRODUCTION

The coupling of ion transport and ATP hydrolysis by the Ca-ATPase is made possible by a catalytic cycle, at each stage of which the enzyme obeys certain "coupling rules;" for each enzyme intermediate certain reactions are allowed while others are forbidden (Jencks, 1983). This chemical and vectorial specificity of the enzyme is different at different stages of the cycle. The calcium-free form of the enzyme readily catalyzes the rather remarkable reversible phosphorylation of an active site asparate carboxyl group from inorganic phosphate, with removal of water from the phosphate (eq. 1, R=H) (Kanazawa

$$E + ROPO_3^{=} \rightleftharpoons E\text{-}COO\text{-}PO_3^{=} + ROH \quad (1$$

and Boyer, 1973). We report here an examination of the specificity of this phosphate transfer, aimed at understanding the mechanism responsible for the variable specificity of the enzyme during its catalytic cycle.

RESULTS

We have followed the rate of transfer of the phosphoryl group of EP to methanol, and compared it to the rate of transfer to water, under the conditions used by McIntosh and Boyer (1983) to study the Ca-ATPase catalyzed phosphate-water oxygen exchange (pH 6.5, 20 mM Mg^{2+}). The enzyme is active and stable for hours under these conditions in the presence of up to 15% methanol. However, despite the fact that methanol is an excellent nucleophile towards phosphoryl groups (e.g., Kirby and Varvoglis, 1967), the observed formation of methyl phosphate in 10% methanol is very slow (table 1). In addition, it appears that a significant part of the observed reaction is not even due to the transfer of phosphoryl groups from the phosphorylated Ca-ATPase, since the observed activity in the presence of Ca^{2+} (which inhibits phosphorylation almost completely; see e.g., table 1) is higher than, although within

TABLE 1. Incorporation of ^{32}P into Methyl Phosphate[a]

	A	B	"Ca-inhibitable" Reaction [b]
$[Ca^{2+}]$,mM	$<10^{-5}$	0.05	
Incorporation of ^{32}P, nmol(mg E)$^{-1}$min^{-1}	0.064+.026	0.125+.041	<0.029
Steady State EP, nmol(mg E)$^{-1}$	0.37	<0.004	
Calc'd transfer to MeOH mol(mol E)$^{-1}$min^{-1}			<0.078
k_{MeOH}, $M^{-1}min^{-1}$			<0.032

a) SR vesicles were incubated at 25° in 10% methanol (2.5 M in MeOH) in 50 mM Mes buffer, pH6.5, 20 mM $MgSO_4$, 50 mM KCl and 0.05 mM EGTA. At 45 min, aliquots were taken and analyzed by an ammonium molybdate extraction protocol which removed protein, acid-labile phosphates and 99.95% of P_i, and allowed recovery of >97% of authentic methyl phosphate. EP was determined separately. Total site concentration was 2.0 nmol(mg E)$^{-1}$.
b) Upper limit for relevant rates, assuming reaction of interest is the difference between that observed in absence and presence of Ca^{2+} (A-B, within 85% confidence limits).

experimental error of, that in its absence. The observed reaction may thus be chiefly due to a minor component of the SR preparation, and the real rate of the reaction of interest may be much lower. By a conservative estimate, the upper limit for the reaction due to the Ca-ATPase (see table 1) is less than half of the observed reaction.

From the estimate for the rate of formation of methyl phosphate by the Ca-ATPase, the level of E-P under the experimental conditions, and the methanol concentration (table 1) we conclude that the second order rate constant for transfer of the phosphoryl group from E-P to methanol, k_{MeOH}, is at most 0.032 $M^{-1}min^{-1}$, or $5.3x10^{-4}$ $M^{-1}sec^{-1}$. This is at least 1600 times slower than the second order rate constant for transfer to water, k_{H2O}=0.85 $M^{-1}sec^{-1}$, which can be calculated from the data of McIntosh and Boyer (1983).

If the E-P form of the Ca-ATPase is in fact capable of transfering its phosphoryl group to methanol, the enzyme should be able to catalyze the hydrolysis of methyl phosphate. We prepared ^{32}P-labelled methyl phosphate (Wolfenden and Williams, 1983) and followed its hydrolysis to P_i in the absence of Ca^{2+}. In a typical experiment, Ca-ATPase (3.8 uM in phosphorylation sites) was added to reaction mixtures in the buffer used to study methyl phosphate formation (table 1) containing 0.13 mM methyl [^{32}P]-phosphate. Aliquots taken over four hours were analyzed for P_i formation by ammonium molybdate extraction. This reaction is also very slow, with a pseudo-first order rate of $7x10^{-4}$ min^{-1}. Since K_M for methyl phosphate appears to be >0.2 mM, we can calculate k_{cat}/K_M for the enzymic hydrolysis: $(k_{cat}/K_M)_{MeOP}$ is apparently 3.0 $M^{-1}sec^{-1}$. However, since the reaction is nearly unaffected by 0.5 mM Ca^{2+} or 2 mM Tb^{3+}, and is much more resistant to denaturation at 65° than the Ca-ATPase activity, we believe that it, too, is due in large part to contaminating enzymes. No more than a third of the hydrolysis rate is likely to be due to the Ca-ATPase; that is, $(k_{cat}/K_M)_{MeOP}$ <1 $M^{-1}sec^{-1}$. From the data of McIntosh and Boyer (1983), k_{cat}/K_M for phosphate-water oxygen exchange under the same conditions is 2200 $M^{-1}sec^{-1}$; since this rate is for any of four phosphate oxygens the rate for hydrolysis of a given O-P bond of P_i would be 550 $M^{-1}sec^{-1}$.$(k_{cat}/K_M)_{MeOP}$ is thus at least 500-fold slower than the comparable reaction with water as leaving group.

The SR Ca-ATPase can hydrolyze a number of acyl phosphates or activated phosphate esters in the absence of Ca^{2+}

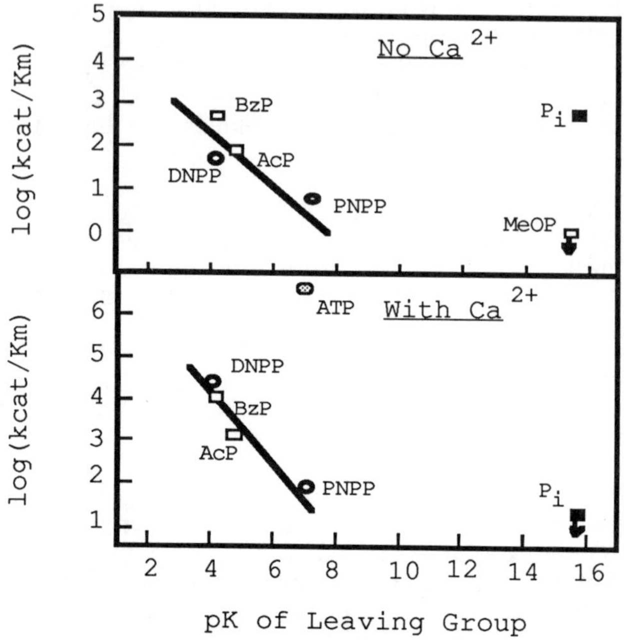

Figure 1. Rate constants (k_{cat}/K_M) for hydrolysis of phosphate derivatives $ROPO_3^=$ by Sr Ca-ATPase in the absence (top) and presence (bottom) of Ca^{2+}, as a function of the pK of the leaving group ROH. P_i, inorganic phosphate (data of McIntosh and Boyer, 1983, discussed in text. Rate in presence of Ca is upper limit). MeOP, methyl phosphate (data presented here; upper limit). AcP, acetyl phosphate (data of Bodley and Jencks, 1987). BzP, benzoyl phosphate (from pseudo-1st order disappearance corrected for spontaneous hydrolysis; data not shown). DNPP, 2,4-dinitrophenyl phosphate (from substrate dependence of initial rates measured spectroscopically; data not shown. Rate in absence of Ca shown by inhibition to be substantially due to ATPase). PNPP, *p*-nitrophenyl phosphate (as for DNPP, data not shown).

(Pucell and Martonosi, 1971; Rossi et al., 1979; Bodley and Jencks, 1987). In fig. 1 we compare values of k_{cat}/K_M for hydrolysis of a number of such compounds, in linear free energy plots, in the presence and absence of calcium. The second order constant k_{cat}/K_M reflects the free energy difference between the transition state for the observed reaction and free enzyme plus free substrate. The constants for several of these phosphates are approximately correlated with the acid dissociation constants of their leaving groups (fig. 1), as expected if negative charge is developed on the leaving oxygen at the transition state. While methyl phosphate may fall on the same line (depending on how much slower than our "upper limits" its Ca-ATPase-catalyzed hydrolysis may actually be), P_i clearly reacts orders of magnitude more rapidly than expected from the acidity of the leaving group, water.

DISCUSSION

In the absence of Ca^{2+}, the SR Ca-ATPase catalyzed reaction of eq. 1 in which R is the methyl group is far slower, in either direction, than the reaction in which R is a proton; the transition state with methanol as leaving group or nucleophile is at least 2 to 3 kcal/mol higher in energy relative to its ground state than that with water as leaving group/nucleophile. This difference can not be due to simple steric exclusion of the methyl group; as we see from fig. 1 a mechanism is available for hydrolysis of even more hindered phosphates. Several explainations are possible for the special reactivity of water in this state of the enzyme: the hydrogen bonding of both protons of the attacking/leaving water might be crucial to the stabilization of the transition state; the attacking/leaving group might be an H-bonded $MgOH^+$ (Herschlag and Jencks, 1987); or the fit of the phosphoryl group and water in the active site might be so snug (Pickart and Jencks, 1984) that any additional group distorts the site so as to prevent correct alignment of catalytic groups.

Whatever the detailed explaination for the unique specificity of the Ca^{2+}-free state of the enzyme might be, the arrangement of the active site and mechanism of phosphoryl transfer in this state must be profoundly different from those in the presence of Ca^{2+}.

REFERENCES

Bodley AL, Jencks WP (1987). Acetyl phosphate as a substrate for the calcium ATPase of sarcoplasmic reticulum. J Biol Chem (in press).

Herschlag D, Jencks WP (1987). The effect of divalent metal ions on the rate and transition-state structure of phosphoryl transfer reactions. J Am Chem Soc 109:4665-4674.

Jencks WP (1983). What is a coupled vectorial process. Curr Top Membr Trans 19:1-19.

Kanazawa T, Boyer PD (1973). Occurence and characteristics of a rapid exchange of phosphate oxygens catalyzed by sarcoplasmic reticulum vesicles.J Biol Chem 248:3163-3172.

Kirby AJ, Varvoglis AG (1967). The reactivity of phosphate esters. Monoester hydrolysis. J Amer Chem Soc 89:415-423.

McIntosh DB, Boyer, PD (1983). ATP modulation of catalytic intermediates of calcium ion activated ATPase of sarcoplasmic reticulum subsequent to enzyme phosphorylation. Biochemistry 22:2867-2875.

Pickart CM, Jencks WP (1984). Energetics of the calcium transport ATPase. J Biol Chem 259:1629-1643.

Pucell A, Martonosi A (1971). Sarcoplasmic reticulum. XIV. Acetylphosphate and carbamylphosphate as energy source for Ca^{2+} transport. J Biol Chem 246:3389-3397.

Rossi B, Leone FdeA, Gache C, Lazdunski M (1979). Pseudosubstrates of the sarcoplasmic Ca^{2+}-ATPase as tools to study the coupling between substrate hydrolysis and Ca^{2+}-transport. J Biol Chem 254:2302-2307.

Wolfenden R, Williams R (1983). Affinities of phosphoric acids, esters and amides for solvent water. J Amer Chem Soc 105:1028-1031.

The Ion Pumps: Structure, Function,
and Regulation, pages 183–188

MODULATION BY FREE ATP OF SARCOPLASMIC RETICULUM ATPase DEPHOSPHORYLATION.

Florent Guillain*, Philippe Champeil* and
David McIntosh+

* Service de Biophysique. CEN-Saclay
91191-Gif sur Yvette. France
+ Dpt of Chemical pathology, University of Cape
Town Observatory 7925, Cape Town, South Africa

INTRODUCTION

Mg.ATP binds to Sarcoplasmic Reticulum ATPase with micromolar affinity and allows calcium transport by the Ca-pump to proceed at the expense of the hydrolysis of one ATP molecule for two Ca^{++} ions transported (fig.1). Millimolar concentrations of substrate however, further stimulate turnover resulting in a non michaelian behavior of the enzyme (Inesi et al, 1967). Under standard conditions, the predominant species during steady state ATP hydrolysis is a phosphorylated form of the enzyme. Therefore, phosphoenzyme must be a privileged target for the stimulating effect of high concentrations of ATP.

Since the ATPase cycle is fully reversible (fig.1) it is possible first to phosphorylate the enzyme from P_i yielding the E_2-P species with approximately full phosphorylation (Guillain et al, 1984; Champeil et al, 1985) and then to measure the rate of dephosphorylation of the E_2-P species and characterize any ATP acceleration of the dephosphorylation reaction.

Fig. 1

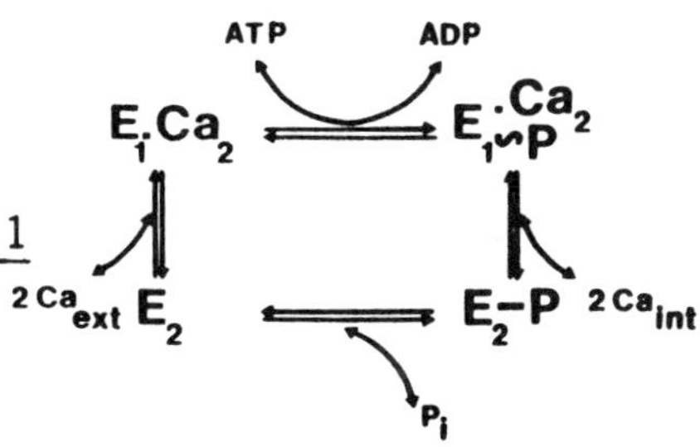

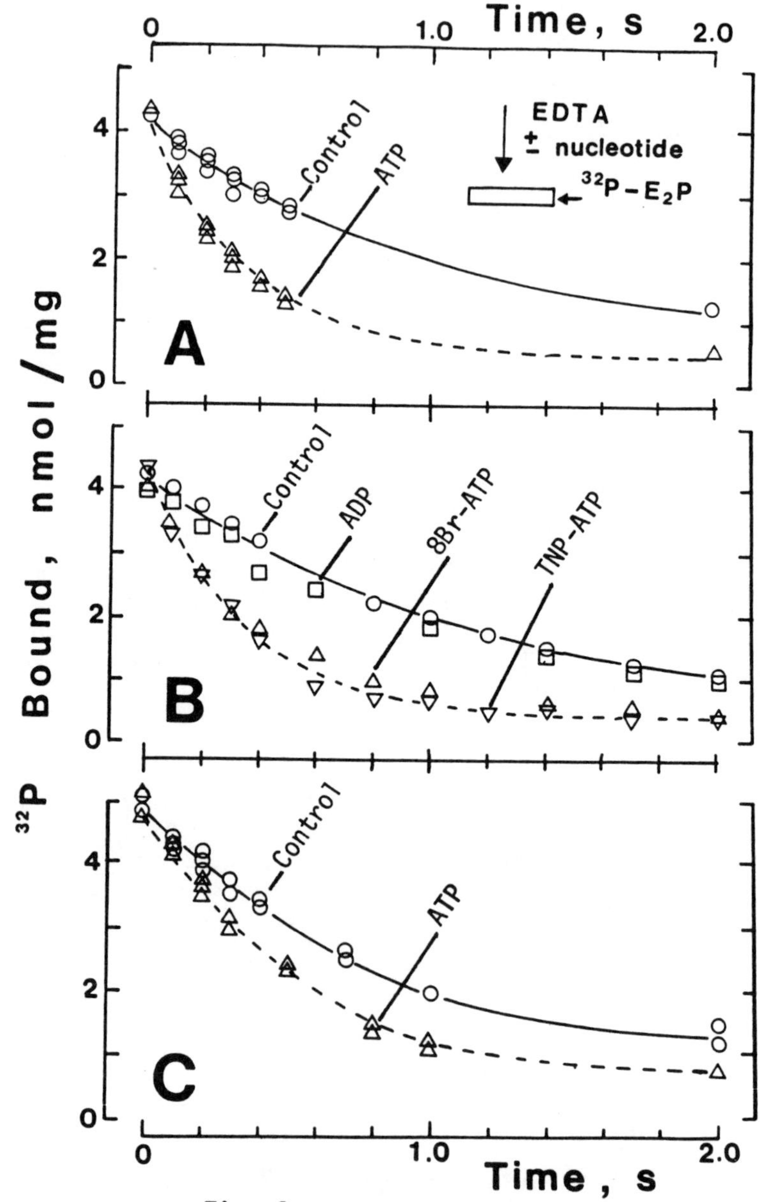
Time, s
0
1.0
2.0
EDTA
± nucleotide
^{32}P – E$_2$P
Control
ATP
A
ADP
8Br-ATP
TNP-ATP
B
C
^{32}P Bound, nmol/mg
0
2
4

Fig. 2

ATP STIMULATION OF E_2-P HYDROLYSIS

Acceleration of dephosphorylation of the E_2-P species has been tested by the rapid filtration technique (Champeil and Guillain, 1986). Experiments were mostly conducted at 20°C and pH 6, in the absence of potassium following the procedure described in fig. 2 legend. Fig.2 shows that when ATP or analogs (panels A and B) were included in the EDTA-containing dephosphorylation medium, dephosphorylation was accelerated approximately 2-fold, whereas inclusion of ADP (1 mM, panel B) had no effect.

When the dephosphorylation medium contained 20 mM $MgCl_2$ instead of 5 mM EDTA, the dephosphorylation rates in the presence or absence of ATP were the same (data not shown), suggesting that the metal-free ATP, and not the Mg.ATP complex, stimulates dephosphorylation.

The filtration technique only allows rough estimation of the affinity for the nucleotide-induced stimulation but experiments similar to the one in fig.2 panel A, carried out at different ATP concentrations clearly show that this affinity is high:

$$10\ \mu M \leqslant K_d \leqslant 100\ \mu M$$

Figure 2. Effect of nucleotides on the rate of dephosphorylation of native ATPase and FITC-ATPase measured by rapid filtration.

SR vesicles (0.3 mg/ml) were incubated in a phosphorylation medium containing 15% v/v DMSO, 20 mM $MgCl_2$, 2mM EGTA, 0.2 mM ^{32}P-P_i and 130 mM Mes-Tris (pH 6.0, 20°C). 1 ml of this suspension was layered onto an HA Millipore filter in the Biologic fast filtration device, and then flushed for various periods with a dephosphorylation solution containing 5 mM EDTA, 150 mM Mes-Tris (pH 6.0, 20°C), and various nucleotides.

Panel A : Rate of dephosphorylation of native ATPase, in the absence (○) or presence (△) of 1 mM ATP.

Panel B : Experiments were performed under control conditions (○) or in the presence of 1 mM ADP, (□) 0.005 mM TNP-ATP (▽), or 1 mM 8-Br-ATP (△).

Panel C : Rate of dephosphorylation of FITC-ATPase, in the absence (○) or presence (△) of 1 mM ATP. Modification with FITC: 20 min incubation at pH 8 and 20°C in the following medium: 10 mM Tris-Tricine, 0.3 M sucrose, 1 mM $MgCl_2$, 0.01 mM $CaCl_2$, 0.016 mM FITC, 2mg/ml SR-ATPase. The modification was slowed down by lowering the temperature to 0°C.

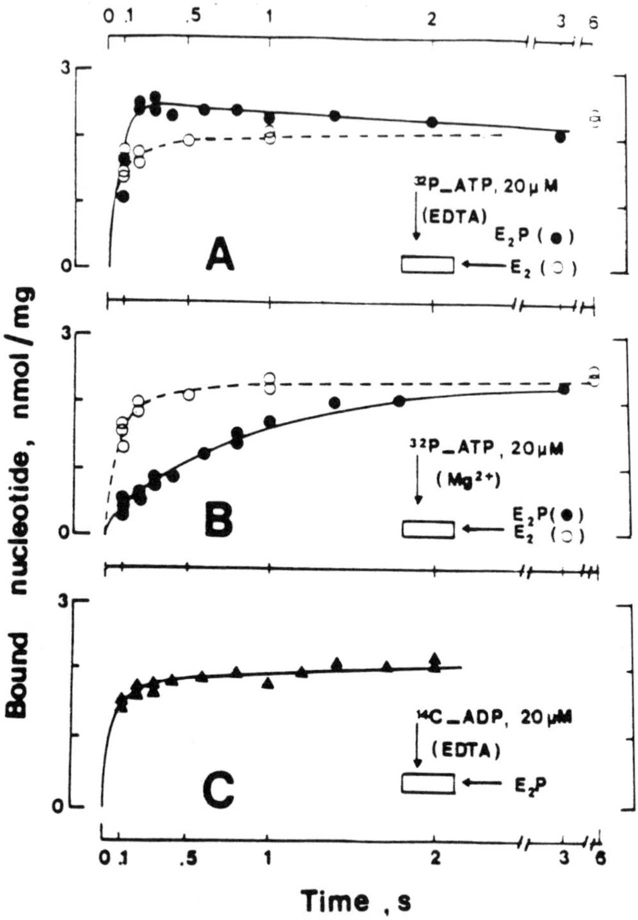

Figure 3. Time course of ATP and ADP binding to phosphorylated and unphosphorylated native ATPase.

The procedure used was similar to that described in the legend of fig. 2 except that the ATPase on the filter had been preincubated either in the absence of P_i (○, unphosphorylated ATPase, designated "E_2") or with 2mM non-radioactive P_i instead of 0.2 mM ^{32}P-P_i (●, ▲, fully phosphorylated ATPase, "E_2-P"), and that the perfusion medium contained 0.02 mM radioactive nucleotide (^{32}P-ATP for panels A and B, ^{14}C-ADP for panel C), 150 mM Mes-Tris (pH 6.0, 20°C) and either 5 mM EDTA (panels A and C) or 20 mM $MgCl_2$ and 2 mM EGTA (panel B).

LOCUS OF THE REGULATORY NUCLEOTIDE SITE

Two types of models need to be considered for the locus of the catalytic and regulatory nucleotide sites
1) Two different sites: one catalytic site + one regulatory site (for instance either two sites per ATPase chain or one site per ATPase chain with a putative polymeric form for the Ca-pump and a putative allosteric activation)
2) One single catalytic site per ATPase chain, being transformed into regulatory site during turnover (for instance after ADP departure, see fig. 1).

One way to discriminate between the two possible models is to modify the enzyme with FITC (Pick and Karlish, 1980). Modifying the SR-ATPase with FITC blocks access of the adenine moiety of the catalytic site but does not inhibit basic catalytic properties of the enzyme as shown by the unaltered acetylphosphate-supported Ca^{++} uptake. Therefore, a dephosphorylation experiment similar to the one shown in fig. 2, panel A may be repeated with FITC-ATPase. The experiment (fig.2, panel C) shows that FITC modification renders dephosphorylation insensitive to ATP. Thus, binding of one FITC molecule per SR-ATPase chain blocks, at the same time, both ATP binding at the catalytic site and ATP regulation of the dephosphorylation reaction. This strongly favors model 2 in which there is only a single ATP site, alternatively catalytic or regulatory.

FREE ATP BUT NOT Mg.ATP BINDS TO E_2-P

Inclusion of radioactive ATP in dephosphorylation medium allows measurement of ATP binding during dephosphorylation of non radioactive E_2-P. In fig.3, panel A shows that free ATP binds rapidly to E_2-P, with a rate faster than the dephosphorylation process. In contrast, panel B shows that in the presence of 20 mM $MgCl_2$, Mg.ATP binds at a rate comparable to dephosphorylation. This confirms that the Mg.ATP complex only binds to E_2 after dephosphorylation of the E_2-P species.

In fig.3, panel C, ^{14}C-ADP was included in the dephosphorylation medium. A fast binding of ADP to E_2-P was observed although ADP did not stimulate the dephosphorylation rate (fig. 2, panel B).

DISCUSSION

From the data presented above we conclude that free ATP and ADP bind to E_2-P but that only ATP (or analog) accelerates dephosphorylation. The complex Mg.ATP does not bind to E_2-P. Modification of the active site with FITC blocks the accelerating effect of ATP.

This is compatible with the locus of the regulatory binding site being the catalytic site itself. This single ATP binding site per ATPase molecule (110,000 dalton) is successively a catalytic site during the early steps of the ATPase cycle and a regulatory site after phosphoenzyme formation.

These results open the question of the possibility that, after the first phosphorylation, ATP enters the calcium transport cycle via its binding to phosphoenzyme.

REFERENCES

Champeil P, Guillain F, Venien C, Gingold M P (1985). Interaction of Magnesium and Inorganic Phosphate with Calcium-Deprived Sarcoplasmic Reticulum Adenosinetri phosphatase as Reflected by Organic Solvent Induced Perturbation. Biochemistry 24 : 69-81

Champeil P, Guillain F, (1986). Rapid Filtration Study of the Phosphorylation-Dependent Dissociation of Calcium from Transport Sites of Purified Sarcoplasmic Reticulum ATPase and ATP Modulation of the Catalytic Cycle. Biochemistry 25 : 7623-7633

Guillain F, Champeil P, Boyer P (1984). Sarcoplamic Reticulum ATPase Phosphorylation from Inorganic Phosphate. Theoretical and Experimental Reinvestigation. Biochemistry 23 : 4754-4761

Inesi G, Goodman JJ, Watanabe S (1967). Effect of Diethyl Ether on the ATPase Activity and the Calcium Uptake of Fragmented Sarcoplasmic Reticulum of Rabbit Skeletal Muscle. J Biol Chem 242 : 4637-4643

Pick U, Karlish SJD (1980). Indications for an Oligomeric Structure and for Conformational Changes in Sarcoplasmic Reticulum Ca-ATPase Labelled Selectively with Fluorescein. Biochim Biophys Acta 626 : 255-261

The Ion Pumps: Structure, Function, and Regulation, pages 189–194

Ca^{2+} UPTAKE AND AFFINITY OF THE Ca^{2+} PUMP DURING SINGLE CYCLE CATALYSIS OF THE SARCOPLASMIC RETICULUM ATPase.

George A. Davidson and Mervyn C. Berman

Medical Research Council Biomembrane Research Unit, Department of Chemical Pathology, University of Cape Town Medical School, Observatory 7925, Cape Town, Republic of South Africa.

INTRODUCTION

Energy transduction of the Ca^{2+}-ATPase of skeletal muscle sarcoplasmic reticulum incorporates two Ca^{2+} sites that alternate from high affinity ($K_{0.5}$= 1 uM) to low affinity (Kd=1 mM) during the catalytic cycle in synchrony with orientation to the cytosolic and lumenal surfaces of the membrane respectively. The externally oriented high affinity Ca^{2+} sites are cooperatively linked and appear to be non-identical (Inesi et al., 1980; Champiel et al., 1983; Dupont 1982). Initial binding to a relatively low affinity site on E_2 is followed by an isomerization step that increases affinity of the second Ca^{2+} site on E_1 as a sequential process (Inesi et al., 1980; Dupont 1982).

We describe here an alternative approach to characterization of intermediate events of Ca^{2+} activation and its dynamic interrelationship with catalytic events. Simultaneous initiation of catalytic events is followed to completion, such that each enzyme subunit undergoes a single transport cycle. The dynamic formation and decay of individual subspecies may be related kinetically and their synchronicity established by means of phosphoenzyme levels and fluorescent probes.

RESULTS

The basis of the single cycle experiment is the simultaneous addition of substrate and EGTA to the Ca^{2+} saturated enzyme, ($2Ca.E_1$). Total phosphoenzyme (E-P) was measured at various times after addition of $2Ca.E_1$ to [γ-^{32}P]ATP and EGTA, and reached a transient of intermediate species of 2.3 nmol/mg at 100 msec, followed by a monoexponential decay with k_{obs} = 3.6 s-1 (Fig. 1). The kinetics are compatible with rapid formation of E-P and its decay according to the scheme, $2Ca.E_1 \xrightarrow{fast} E\langle 2Ca\rangle\text{-}P \xrightarrow{slow} E + P_i + Ca_{in}$. (Occluded $Ca^{2+} = \langle Ca\rangle$).

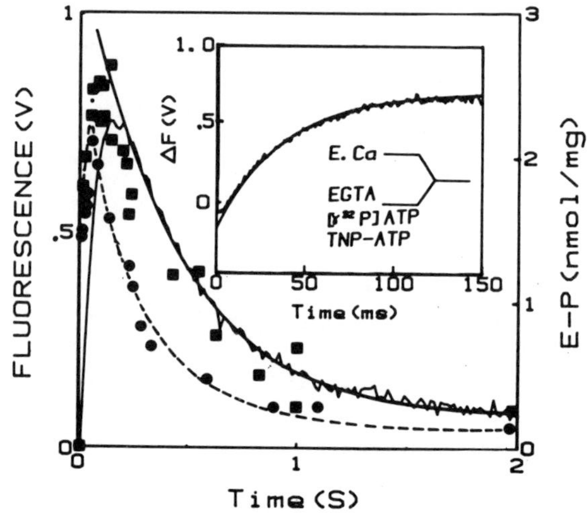

Figure. 1 Single cycle E^{32}-P and TNP-ATP fluorescence. Conditions for rapid mixing experiments (Durrum 133) at 10°C were syringe A, 0.2 mg/ml SR, 50 uM Ca^{2+}, 20 mM Mops/Tris, pH 7.0, 5 mM $MgCl_2$ mixed with syringe B, 100 uM [γ-^{32}P]ATP, 0.5 mM EGTA in the absence (●) or presence (■) of 4 uM TNP-ATP, aged for 17 ms up to 2 s, and stopped with 7% TCA and 30 mM H_3PO_4 in syringe C. TNP stopped flow fluorescence measurements (line) were performed as above on a Dionex D110 at λ_{ex} = 418 and λ_{em} = 530 nm.

Inclusion of the ATP analog, 2',3'-O-(2,4,6-trinitrocyclohexyldienyldine)- adenosine-5'-triphosphate (TNP-ATP) delayed the E-P transient (Fig. 1) by competition with ATP for the catalytic site (Bishop et al., 1985). Enhanced fluorescence of bound TNP-ATP is a sensitive index of the ADP-insensitive phosphoenzyme species, E_2-P (Davidson and Berman, 1987). Stopped flow analysis under the above conditions showed a transient of TNP-ATP fluorescence that closely follows the kinetics of the E-P transient. The inset to Fig. 1 shows details of the first 150 ms of the cycle, showing a rate of E_2-P formation of 25 s-1. The transients indicate that no further turnover occurs after Ca^{2+} is chelated by EGTA or occluded by the enzyme.

$^{45}Ca^{2+}$ uptake and ^{32}Pi release have been followed using varying [Ca^{2+}] preincubations prior to initiation of single transport cycles as shown in Fig 2. The maximal values for Ca^{2+} uptake and Pi release were 8.3 and 4.4 nmol/mg respectively, showing a stoichiometric ratio of Ca^{2+} uptake to Pi release of 1.88. The apparent Kd of the Ca^{2+} site(s) that initiate a Ca^{2+} pump cycle were 0.058 and 0.065 uM for uptake and Pi release respectively. The Ca^{2+} dependence was non-cooperative with $n_{H(Ca^{2+})}$ values in the range 0.8 to 0.95. The [Ca^{2+}] dependence was not altered at decreased rates of E-P formation at low temperatures. Equilibrium Ca^{2+} binding by filtration in the absence of ATP, showed $K_{0.5}$ = 0.88 uM with Hill coefficient, n(H) = 1.89, consistent with previously obtained values (Dupont, 1982; McIntosh and Davidson, 1984). Thus the apparent Ca^{2+} affinity under turnover conditions, using the identical Ca-EGTA concentrations as for equilibrium binding, is 10-fold higher and does not show positive cooperativity.

The effects of addition of increasing concentrations of EGTA as a quench reagent during a single transport cycle would be expected to modify Ca^{2+} uptake. Quench with excess EGTA (> 1mM) decreased Ca^{2+}uptake from 10 nmol/mg at 0.5 mM EGTA to 1 nmol/mg at 10 mM EGTA, indicating that the "off" rate of $Ca^{2+}$from high affinity sites exceeds 150 s-1. This would account for the low values of Ca^{2+} uptake measured in previous studies using high [EGTA quench] (Chiesi and Inesi, 1980; Ikemoto et al, 1981). These studies also assumed that chelation of Ca^{2+} by EGTA is relatively instantaneous and that significant Ca^{2+} transport occuring a few milliseconds after EGTA addition is due to the slow release of Ca^{2+} from high affinity sites. However, chelation of free Ca^{2+} by EGTA has been shown more recently to be a relatively slow process with an observed second-order rate constant of 1.4 to 1.5 X 10^6 M^{-1} s^{-1} at pH 7.0, 25°C and ionic strength of 0.1 M (Smith et al. 1984; Harafuji and Ogawa, 1980), under similar conditions to those reported here. Such considerations imply that the rate of removal of Ca^{2+} from the medium is influenced by competition between the chelator and binding sites on the enzyme. Rapid mixing of 2Ca.E and ATP, with addition of excess EGTA (10 mM) after 33 ms, also showed an apparent [Ca^{2+}] dependence of 0.070 uM. This indicates the independence of the apparent Ca^{2+} affinity from the mixing and chelation times of EGTA.

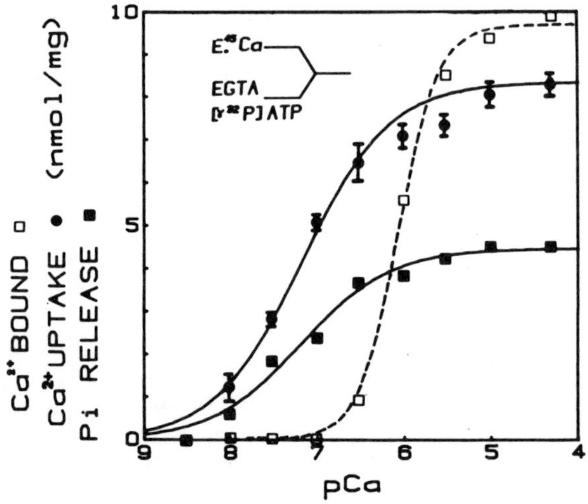

Figure. 2. [Ca^{2+}] dependence for Ca^{2+} binding (□), Ca^{2+} uptake (●) and Pi release (■) at 20°C. Equilibrium Ca^{2+} binding was performed by filtration of 0.2 mg/ml SR in standard buffer, with 50 uM $^{45}Ca^{2+}$ buffered with various [EGTA]. Under the identical conditions, Ca^{2+} uptake and Pi release were performed by mixing buffered 2Ca.E into equal volumes of 0.5 mM quench EGTA and 100 uM [γ-^{32}P]ATP and trapped on GF/F filters with a 5ml $LaCl_3$ wash for Ca^{2+} uptake or centrifuged in charcoal to determine free ^{32}Pi in the supernatent.

The ATP concentration dependence of Ca^{2+} uptake was measured under single cycle conditions (app.$K_{0.5(ATP)}$ = 0.85 uM), indicating a similar apparent affinity to that of equilibrium binding in the presence of EGTA (Kd = 1.7 uM) (Ross and McIntosh,1987). In addition, the [Ca^{2+}] dependence for single cycle turnovers initiated at micromolar or millimolar [ATP] remained unchanged ($K_{0.5(Ca^{2+})}$ = 0.065 uM).

DISCUSSION

Single cycle catalysis is based on the addition of the Ca^{2+}-ATPase to both ATP and EGTA. The results are interpreted in terms of Ca^{2+} sequestration either by occlusion as E_1-<2Ca>-P or chelation by EGTA. It is suggested that under the present conditions, rapid enzyme phosphorylation k_{phos} = 100 s-1 is limited by binding of the second Ca^{2+} to a high affinity externally oriented site, whose intrinsic affinity K_2 is expressed as the $[Ca^{2+}]$ dependence of single cycle turnovers.

Inesi and Hill (1983) have analysed equilibrium binding of H^+ and Ca^{2+} in a two-site model and have inferred the existence and the affinities of two Ca^{2+} binding sites that have affinity constants that vary by two orders of magnitude following isomerization of E_2 to E_1. The first Ca^{2+} ion binds rapidly but with relatively low affinity, Kd = 10-100 uM (Champeil et. al.,1983) or Kd = 25 uM (Dupont, 1982).

The intrinsic binding constants, K_1 and K_2, and isomerization equilibrium, E_2/E_1, were analysed in terms of the Monod-Wyman-Changeaux formulation for cooperative site interaction (Monod et al., 1965). Known values for Ca^{2+} binding were substituted as:- Apparent Ca^{2+} binding, $K_{0.5}$ = 1.0 uM; n_H = 1.8, K_1 = 10-100 uM, and E_2/E_1 = 100, resulting in K_2 =0.07 uM. It follows that the second Ca^{2+} binding sites are of significantly higher affinity, consistent with the theoretical values of 60 nM (Inesi et al., 1980) and of 160 nM (Dupont, 1982; Fernandez-Belda et al., 1984) employed in simulations of the catalytic cycle and are consistent with the value of 65 nM obtained in the present study.

Cooperativity is accounted for by a predominance of the low affinity state (E_2) over the high affinity state (E_1). The lack of cooperativity observed during single cycle catalysis is explained by acceleration of E_2 --> E_1 by micromolar ATP (Champeil et al., 1986) such that E_2/E_1 is less than 30.

The stoichiometry of Ca^{2+} uptake to Pi release during single cycle catalysis of 1.88 (Fig 2) is consistent with a set of rules formulated by Jencks (1980), that two Ca^{2+} ions are required to bind before phosphorylation can proceed. Binding of the second Ca^{2+} ion may therefore act as regulator of phosphorylation.

Events described and discussed in the study of single cycle catalysis by the Ca^{2+}-ATPase may, in fact, be relevant to Ca^{2+} uptake into the sarcoplasmic reticulum of skeletal muscle in vivo. The effective $K_{0.5}$ of 60 nM would favour rapid Ca^{2+} reaccumulation to achieve submicromolar cytoplasmic Ca^{2+} levels and relaxation of the contractile proteins. Analysis of Ca^{2+} binding under dynamic conditions in the presence of ATP therefore appears to be physiologicaly relevant.

REFERENCES

Bishop JE, Al-Shawi MK, Inesi G (1987) J Biol Chem 262: 4658-4663.

Bishop JE, Johnson JD, Berman, MC (1984) J Biol Chem 259: 15163-15171.

Champeil P, Guillain F (1986) Biochemistry 25: 7623 7633.

Chiesi M, Inesi G (1979) J Biol Chem 254: 10370-10377.

Dupont Y (1982) Biochem Biophys Acta 688: 75-87.

Fernandez-Belda F, Kurzmack M, Inesi G (1984) J Biol Chem 259: 9687-96198.

Harafuji H, Ogawa Y (1980) J Biochem (Tokyo) 87: 1305-1312.

Ikemoto N, Garcia AM, Kurobe Y, Scott TH (1981) J Biol Chem 256: 8593-8601.

Inesi G, Hill TL (1983) J Biophys Soc 44: 271-280.

Inesi G, Kurzmack M, Coan C, Lewis DE (1980) J Biol Chem 255: 3025-3031.

Jencks WP (1980) Adv Enzymol 51: 75-106.

Monod J, Wyman J, Changeux JP (1965) J Mol Biol 12: 88-118.

Ross DC, McIntosh DB (1987) J Biol Chem 262: 2042-2049.

Smith PD, Liesgang GW, Berger RL, Czerlinski G, Podolsky RJ (1984) Analytic Biochemistry 143: 188-195.

Watanabe T, Lewis D, Nakamoto R, Kurzmack M, Fronticelli C, Inesi G (1981) Biochemistry 20: 6617.

The Ion Pumps: Structure, Function, and Regulation, pages 195–200

PROPERTIES OF THE CATALYTIC SITE FOLLOWING THIOL MODIFICATION OF THE Ca^{2+}-ATPase OF SARCOPLASMIC RETICULUM FROM SKELETAL MUSCLE

George A. Davidson and Mervyn C. Berman

Medical Research Council Biomembrane Research Unit, Department of Chemical Pathology, University of Cape Town Medical School, Observatory 7925, Cape Town, Republic of South Africa

INTRODUCTION

Energy transfer between the scalar reaction of ATP hydrolysis and vectorial transfer of Ca^{2+} across sarcoplasmic reticulum membranes is believed to occur while Ca^{2+} ions are bound to phosphorylated intermediates of the catalytic cycle. The mechanism of the energy transducing step is, as yet, unknown. However, it is believed that at least two conformational states, E_1 and E_2, are involved. The former reacts with ATP, but not with Pi and H_2O, while the latter reacts with Pi and H_2O, but not with ATP (Tanford, 1983).

Enhanced fluorescence of the ATP analogue, 2',3'-0-(2,4,6-trinitrocyclohexyldienyldine)-adenosine 5'-tri phosphate (TNP-ATP), bound to the Ca^{2+}-ATPase is closely related to phosphoenzyme (E-P) levels and has an emission maximum consistent with decreased polarity of the TNP-ATP binding site (Dupont and Pougeois, 1983; Davidson and Berman, 1985). Binding to non-phosphorylated SR vesicles at the catalytic site is associated with a moderate increase in fluorescence (Bishop et al., 1984). However, TNP-ATP fluorescence is increased ten-fold following E-P formation, caused by the addition of either ATP plus Ca^{2+}, or by Pi, in the absence of Ca^{2+}. The probe has been suggested to bind to the adenine nucleotide binding site of the phosphorylated species, after ADP has been released in the forward reaction (Bishop et al., 1987). It is uncertain which of the E-P conformational states are responsible for enhanced TNP-ATP fluorescence. Bishop et al. (1986) suggested that all phosphoenzyme intermediates are related to enhanced TNP-ATP fluorescence, based on TNP-ATP fluorescence lifetime studies, whereas Wakabayashi et al. (1987) have related fluorescence to the ADP-insensitive (E_2-P) species. We have addressed this question in the present study by redistribution of phosphoenzyme intermediates following specific thiol group modification.

Kawakita et al. (1980) have characterized discrete classes of thiol groups, related to formation (SH_F) and decomposition (SH_D) of E-P species. These groups may be specifically labelled by treatment of SRV with N-ethylmaleimide (NEM) or its fluorescent analogue, anilinonapthylmaleimide (ANM) (Yasuoka-Yabe et al.,1983). The presence of a non-hydrolysable ATP analogue, AMP-P(NH)P, is necessary for protection of the SH_F group. SH_N is then specifically labelled, at either Cys 344 or 364 in the phosphorylation domain with no functional effects (Kawakita et al., 1980; Kawakita and Yamashita, 1987). Labelling at both SH_N sites (SH_D), results in inhibition of the $E_1{\sim}P$ to E_2-P transition, with a net accumulation of $E_1{\sim}P$ in the forward reaction. Nakamura and Tonomura (1982) have recently shown that, following SH_D modification, Ca^{2+} is occluded in the $E_1{\sim}P$ conformation, and that E_2-P formation, in the forward reaction, is inhibited, suggesting that this is a key event of the translocation process.

The aim of the present experiments was to favour accumulation of $E_1{\sim}P$ by blocking those thiol groups (SH_D) that are specifically involved in E-P decomposition and to relate the effects of this blockade to the polarity of the catalytic site by TNP-ATP and ANM fluorescence. In this study, derivatization of SH_D thiol groups with NEM was found to inhibit both ATP- and Pi-induced TNP-ATP fluorescence. The significance of this phenomenon is explored with respect to energy transduction by the Ca^{2+}-ATPase.

RESULTS

Derivitization with 400 uM NEM, in the presence of 50 uM Ca^{2+}, 1 mM AMP-P(NH)P, pH 7.0 and 25 ^{o}C for 30 min, selectively modified the -SH groups (SH_D), essential for phosphoenzyme decomposition, resulted in decreased ATPase and uptake while E-P levels from ATP remained unaltered.

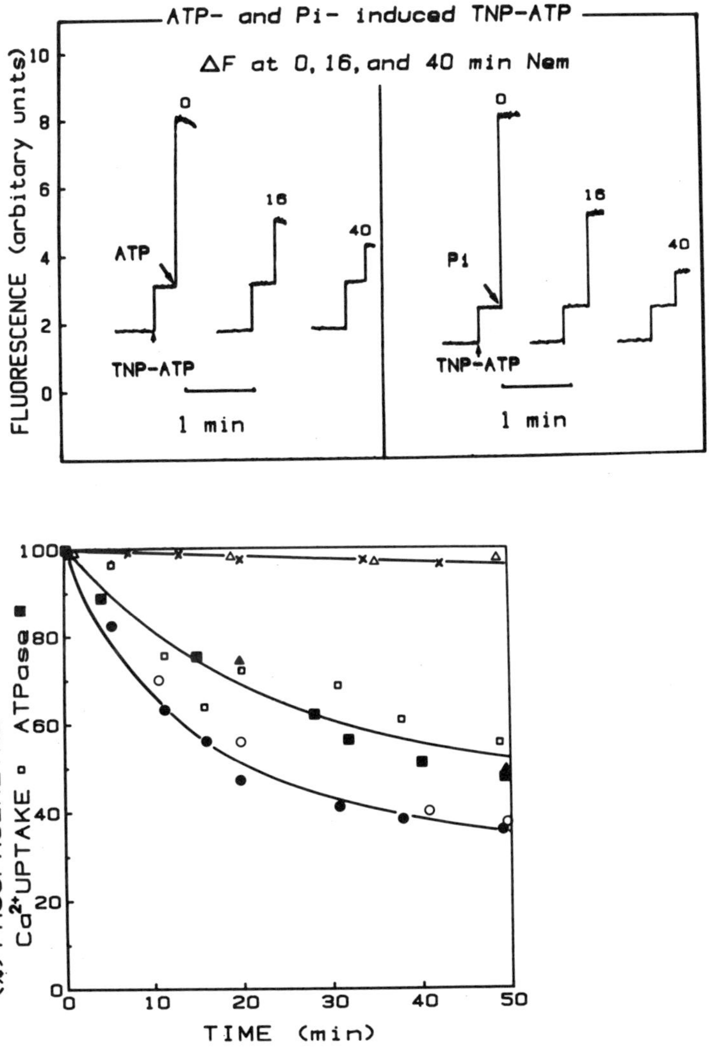

Figure 1. Time-dependence of TNP-ATP fluorescence in SH_D (400 uM NEM) treated SRV, from ATP and Ca^{2+} (●) or Pi (○). E-P levels (% of untreated SRV) from ATP (△) or Pi (▲). Fluorescence following SH_N modification (40 μM NEM) (X).

Phosphoenzyme levels from [^{32}P]Pi were diminished by approximately 50% following derivitization, while the E_1 to E_2 transition, as monitored by the change in intrinsic tryptophan fluorescence (Dupont and Leigh, 1978), was unaffected. This suggests that SH_D thiol groups are also involved in the reverse reaction of the formation of covalent E-P from Pi, with the exclusion of water.

NEM modification did not alter the relatively low TNP-ATP fluorescence from the nonphosphorylated enzyme prior to the addition of ATP or Pi, indicating no change in probe binding (Fig. 1A). However, ATP- and Pi-induced fluorescence enhancement by either ATP or Pi were diminished to approximately similar extents (Fig. 1A + B). Inner filter effect correction amounted to less than 5% of total fluorescence (Lakowitz, 1983). Thus, under these conditions, the decline in Pi-dependent TNP-ATP fluorescence is readily explained by decreased E-P formation. However, the decrease in TNP-ATP fluorescence from ATP plus Ca^{2+} occurs when total E-P levels are unaltered.

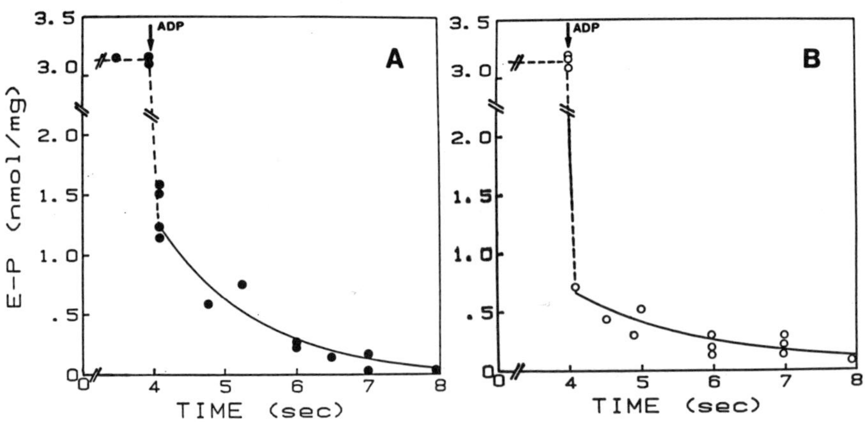

Figure 2. (A) Control SRV and (B), SH_D-treated (50 min) SRV phosphorylated with 100 uM [γ-^{32}P]ATP for 4 s followed with 5 mM ADP plus 0.5 mM EGTA. Rate constants for decay of slow phase in A and B were 0.71 and 0.65 s^{-1} respectively. ADP insensitive values were 1.17 and 0.65 nmol/mg, respectively.

The relative proportions of ATP-dependent E_1P and E_2P species have been determined following rapid quenching by ADP and EGTA (Fig. 2). This shows a rapid decay in E-P from 3.1 nmol/mg, complete within the mixing time, which was followed by a slower decay in the ensuing 4 s. Following SH_D modification, total E-P levels remained relatively unaffected, but E_2-P decreased from 56% to 15%.

The possibility that the presence of NEM prevents TNP-ATP fluorescence in E_2-P was considered. Labelling SH_N with 27 uM ANM has no functional effects, and resulted in enhanced ANM fluorescence with an emission maximum at 430 nm (Yasuoka-Yabe et al., 1983). The presence of AMP-(NH)P did not modify SH_N derivitization, indicating that access of the two SH_N groups is not inhibited. Fluorescent energy transfer from ANM (λ_{em} = 430 nm) to TNP-ATP (λ_{ex} = 420 nm) was minimal under turnover or non-turnover conditions, indicating that donor and acceptor fluorophore distances are either outside the Forster radius or that they have poor orientation with respect to electromagnetic coupling.

DISCUSSION

The aim of the present study was to determine the nature of the phosphorylated intermediate species that result in increased hydrophobicity of the non-catalytic binding site on the Ca^{2+}-ATPase, to which TNP-nucleotides bind and show enhanced fluorescence. Under conditions of NEM modification that did not alter total E-P levels during catalysis, E_2-P was decreased. This would indicate that $E_1{\sim}P$ is the predominant species after SH_D modification.

Two possible mechanisms for decreased fluorescence were considered. The first is that sulphydryl modification at Cys 346 and Cys 366 directly quenches TNP-ATP fluorescence. Results with only SH_N modification however, showed no alteration of enhanced TNP-ATP fluorescence. In addition substitution with ANM showed minimal fluorescence energy transfer from ANM to TNP-ATP, indicating poor electromagnetic interaction with the probe at the nucleotide binding domain under turnover conditions. The second is that E_2-P formation, and its level, determine enhanced TNP-ATP fluorescence. Decreased fluorescence during turnover, of approximately 50% following SH_D modification, correlates well with the decrease of E_2-P from 1.3 to 0.7 nmol/mg.

The significance of the increased hydrophobicity at the locus of the non-catalytic, or regulatory, site in the E_2-P conformation may be to prevent access of water to the acyl phosphate. Dupont and Pougeois (1983) have suggested that the increase in fluorescence indicates that H_2O molecules are expelled from the nucleotide binding site as part of the catalytic cycle. However, the observation that ANM fluorescence does not change quantum yield during turnover would indicate that Cys 344 and 364 may be distant to the phosphorylation site, Asp 351. TNP-ATP fluorescence may then signify that only the nucleotide binding domain becomes solvent inaccessible during Ca^{2+} release to the lumen of the vesicle.

REFERENCES

Bishop JE, Al-Shawi KM, Inesi G (1987) Biophys J 49: 237

Bishop JE, Johnson JD, Berman MC (1984) J Biol Chem 259: 15163-15171.

Bishop JE, Nakamoto RK, Inesi G (1986) Biochemistry 25: 696-703.

Champeil P, Gingold MP, Guillain F (1983) J Biol Chem 258: 4453-4458.

Davidson GA, Berman MC (1985) J Biol Chem 260: 7329.

Dupont Y, Chapron Y, Pougeios R (1982) Biochem Biophys Res Commun 106: 1272-1279.

Dupont Y, Leigh JB (1978) Nature 273: 396-398.

Dupont Y, Pougeois R (1983) FEBS Lett 156: 93-98.

Kawakita M, Yamashita T (1987) J Biochem (Tokyo) 102: 103-109.

Kawakita M, Yasuoka K, Kaziro Y (1980) J Biochem (Tokyo) 87: 609-617.

Lakowicz JR (1983) Principles of Fluorescence Spectroscopy, pp 43-47, Plenum Press, New York.

Nakamura Y, Tonomura Y (1982) J Biochem (Tokyo) 91: 449-461.

Tanford C (1983) Annu Rev Biochem 52: 379-409.

Wakabayashi S, Ogurusu T, Shigekawa W (1986) J Biol Chem 261: 9762-9767.

Wakabayashi S, Shigekawa Y (1987) J Biol Chem 262: 11524-11531.

Watanabe T, Inesi G (1982) J Biol Chem 257: 11510-11516.

Yamada S, Ikemoto N (1978) J Biol Chem 253: 6801-6807.

Yamaguchi M, Kanazawa T (1984) J Biol Chem 259: 9526-9531.

Yasuoka-Yabe K, Tsuji A, Kawakita M (1983) J Biochem (Tokyo) 94: 677-688.

The Ion Pumps: Structure, Function, and Regulation, pages 201–206

THE OLIGOMYCIN-SENSITIVE CA-ATPASE OF CHROMATOPHORES FROM PHOTOSYNTHETIC BACTERIA IS NOT COUPLED TO $\Delta\tilde{\mu}_{H^+}$ GENERATION.

Rita Casadio

Department of Biology, University of Bologna, V.Irnerio,42. I-40126 Bologna, Italy.

INTRODUCTION

Chromatophores from photosynthetic bacteria catalyze both Ca- and Mg-dependent ATP hydrolysis, which according to inhibitor and resolution/reconstitution studies, represent activities of the same membrane bound F_1 complex. Differently than in chloroplasts, these activities are manifest in the dark and are not enhanced or stabilized by treatment of chromatophores with thiol reagents (Baccarini Melandri and Melandri, 1978).

In illuminated chromatophores from few bacterial species, such as Rhodobacter capsulatus, the rate of ATP hydrolysis is about two fold stimulated as compared to that measured in the dark: a phenomenon which is referred to as "light-activation" of the ATPase activity. This activation has been related to a dynamic conformational change of the membrane-bound coupling factor, as detected with the triplet probe eosin-isothiocyanate, utilized as a spectroscopic tool covalently reacted with the enzyme complex (Casadio and Wagner, 1985). The light induced conformational change is independent of the extent of the free energy change for ATP hydrolysis or synthesis. This suggests that energization of the chromatophores promotes an active state of the membrane-bound enzyme which catalyzes either ATP hydrolysis or synthesis, depending on the "phosphate potential" values as compared to those of $\Delta\tilde{\mu}_{H^+}$. Consistently, ATPsynthase and ATPase activities in the light are regulated in parallel and to a similar degree by titrating chromatophores with increasing concentrations of ionophores and antibiotics (Casadio and Melandri, 1984). In the light, under the same steady state conditions of $\Delta\tilde{\mu}_{H^+}$, sustained by the activity of the redox pumps, the stimulation of the rate of ATP hydrolysis is detected only in the presence of Mg-ATP and not of Ca-ATP. Moreover, at the same $\Delta\tilde{\mu}_{H^+}$ value, illuminated chromatophores catalyze ATP synthesis when the substrates are Pi/Mg-ADP and not Pi/Ca-ADP.

In this paper, the relevance of Mg-ATP and Ca-ATP as substrates for the proton pumping activity of the ATPase is investigated. It is found that both substrates are hydrolyzed by the membrane-bound enzyme complex in an oligomycin-sensitive way. However, only the Mg-ATPase activity is coupled to $\Delta\tilde{\mu}_{H^+}$ generation.

MATERIALS AND METHODS

Procedures for the preparation of chromatophores and the evaluation of the ATPase activity in the dark and in the light are described elsewhere (Casadio and Wagner, 1985). ΔpH generation was detected from the fluorescence quenching of 9-amino-6-chloro-2-methoxy acridine (ACMA), properly calibrated in terms of ΔpH units (Casadio and Melandri, 1985). Concentrations of Ca and Mg ions in the chromatophore preparations were evaluated colorimetrically after complexation of Ca^{++} with cresolphthalein complexone and reaction of Mg^{++} with Titan Yellow. All experiments were carried out in the following assay medium: 50 mM Tricine pH 8.5, 50 mM KCl, 0.2 mM Na-succinate. Bacteriochlorophyll (BChl) concentration was 30 µM. Valinomycin, oligomycin and ACMA were added at concentrations of 2 µM, 0.7 ug nmole BChl and 5 µM, respectively.

RESULTS AND DISCUSSION

Mg- and Ca-ATPase activities of chromatophores. As for other enzymes, utilizing adenosin di- and triphosphate, the true substrate for the membrane-bound ATPase in chromatophores is the complex of ATP with divalent cations, particularly Ca^{++} and Mg^{++}. When the ATPase reaction is investigated with respect to the requirement for these cations, it is necessary to determine their concentration in the chromatophore preparations, and include it in the overall substrate concentration. Table 1 shows the typical cation contents of chroma-

TABLE 1. Evaluation of Mg^{++} and Ca^{+++} in chromatophores

	BChl (mM)	Mg^{++} (mM)	Ca^{++} (mM)
Mg-chromatophores*	1.60	10.0	1.7
Ca-chromatophores*	0.77	2.2	12.9

* Chromatophores were prepared from photosynthetic cells in a 50 mM glycylglycine buffer, at pH 7.5, containing 5 mM $MgCl_2$ or $CaCl_2$, respectively.

TABLE 2. Mg- and Ca-ATPase activities in chromatophores

	v^*_{Pi} (µmoles/µmole BChl hour) Dark	Light
Mg-ATP 2mM	27	42
" + oligomycin	8	10
Mg-ATP 0.25 mM + Ca-ATP 1.75 mM	31	38
" + oligomycin	9	7
Mg-ATP 0.15 mM + Ca-ATP 2.50 mM	33	32
" + oligomycin	9	8

* Rate of ATP hydrolysis colorimetrically detected from the release of inorganic phosphate (Pi).

tophores, when either Mg- or Ca-containing buffers are utilized during their preparation (to preserve F_1 interaction with its membrane counterpart). The Mg-ATPase activity is detected in the presence of EGTA at a concentration equimolar to that of endogenous Ca^{++}. This activity, as shown in Table 2, is sensitive to oligomycin and stimulated in the light. A selective chelating agent of Mg^{++} in the presence of Ca^{++} is lacking. A possible way of detecting the Ca-ATPase activity is to measure the rate of ATP hydrolysis when Ca-ATP is added to a fixed concentration of Mg-ATP, corresponding to that of endogenous Mg^{++}. A kinetic model of competitive inhibition between the two substrates, with similar K_m for both activities, is assumed. It can be therefrom predicted that at a relatively high concentration of Ca-ATP compared to Mg-ATP, the rate of Ca-ATP hydrolysis would be unaffected by Mg-ATP and very similar to that of the sole Mg-ATPase. The data shown in Table 2 agree with this prevision and indicate that, in the dark, both substrates are hydrolyzed in an oligomycin-sensitive way and with similar V_{max}. In the light, the stimulation of the ATPase activity is prevented by Ca-ATP.

Divalent cation requirement for ΔpH generation coupled to ATPase activity. In Fig.1 it is shown that, in the presence of Ca-EGTA (0.5 mM), upon addition of ATP to chromatophores at a concentration equimolar to that of Mg^{++} present in the assay (0.5 mM), a quenching of ACMA fluorescence is observed. When calibrated in terms of ΔpH units, the total percent of quenching corresponds to a ΔpH of 1.7-2.0, maintained under steady state conditions by the hydrolytic activity of the Mg-ATPase (in the dark and in the presence of valinomycin, to abolish the kinetic control of $\Delta\psi$ on the proton pumping activity). The original level of fluorescence is restored upon addition to the assay medium of oligomycin or of EDTA equimolar to Mg^{++}. This indicates that when Mg^{++} is subtracted from Mg-ATP by EDTA, the proton pumping activity of the enzyme complex is inhibited (free ATP is a competitive inhibitor of the Mg-ATPase with a K_i equal to 3.5+/-0.2mM, Casadio et al., unpublished). Consistently, when ATP is added in the presence of Mg-EDTA and of Ca-EGTA, no fluorescence quenching is observed and no oligomycin-sensitive ATPase activity is detected. In the experiment shown in Fig.2, ATP is added in the absence of chelating agents, at an equimolar concentration with Ca and Mg ions. Under steady state conditions for ion fluxes, ΔpH is not affected when EGTA or EDTA are added at concentrations equimolar to that of Ca^{++}. Upon a further addition of EDTA,

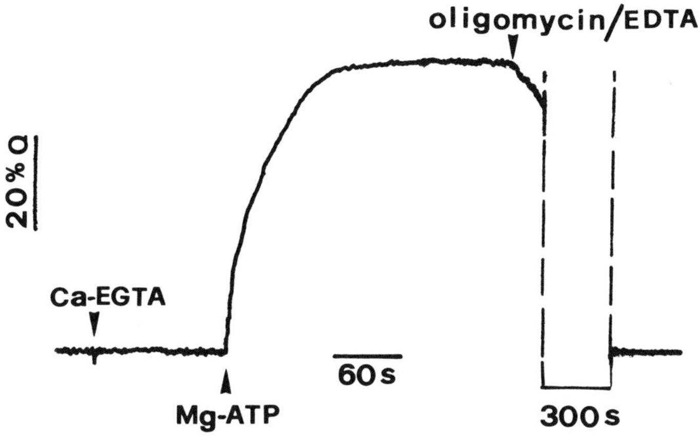

Figure 1. Quenching (Q) of ACMA fluorescence induced by addition of Mg-ATP to a chromatophore suspension in the dark and in the presence of valinomycin.

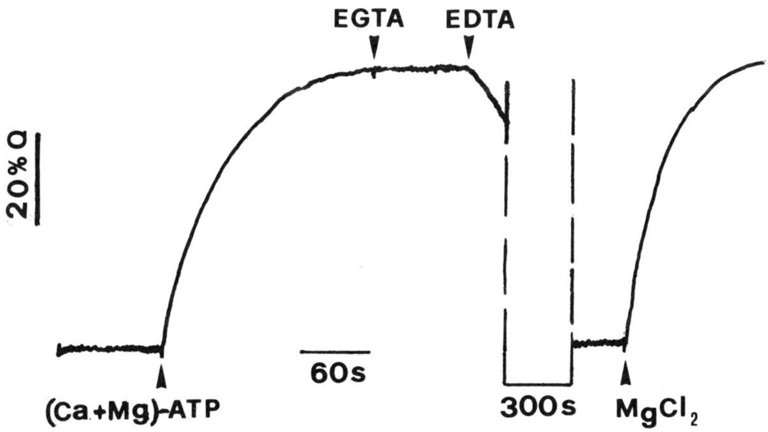

Figure 2. Differential effect of the chelation of Ca and Mg ions on the steady state quenching of ACMA fluorescence, induced by addition of (Ca+Mg)-ATP to chromatophores in the dark and in the presence of valinomycin. Ca- and Mg-ATP concentrations were 1.75 and 0.25 mM, respectively.

equimolar to Mg^{++}, or of EGTA equimolar to Ca^{++}, an oligomycin-like effect is observed. The fluorescence quenching is restored upon further addition of cations to the assay medium. To interpret these results, it is considered that the stability constant values of Ca-EGTA and Ca-EDTA are higher than those of Mg-EGTA and Mg-EDTA, and that those of Ca-ATP and Mg-ATP are mutually similar and lower than the former. Then, it results that ΔpH formation is induced when Mg-ATP is available as a substrate for the hydrolytic activity and that chelation of Ca^{++} does not impair the steady state ΔpH. These results indicate that the proton pumping activity of the enzyme complex is coupled only to the hydrolysis of Mg-ATP and suggest that binding of Ca-ATP at the active site/s causes intrinsic decoupling of the hydrolytic from the proton pumping activity. Both Ca- and Mg-ATPase are, however, sensitive to oligomycin. This may indicate that binding of the inhibitor at the membrane sector of the enzyme complex promotes conformational changes at the catalytic site, which result in the impairment of the hydrolytic activity. A similar mechanism was previously suggested to explain the decrease of the ATP affinity constants in the the catalytic sites upon binding of oligomycin to deenergized submitochondrial particles (Penefsky, 1985).

REFERENCES

Baccarini Melandri A, Melandri BA (1978). Coupling factors. In Clayton RK, Sistrom WR (eds): " The Photosynthetic Bacteria" New York: Plenum Press, pp 615-628.

Casadio R, Melandri BA (1984). On the regulation of the ATPase activity in chromatophores of Rhodopseudomonas sphaeroides. In Papa S et al. (eds): " H^+ ATPase: Structure, Function, Biogenesis" Bari: Adriatica Ed, pp 411-420

Casadio R, Melandri BA (1985). Calibration of the response of 9-amino acridine fluorescence to transmembrane pH difference in bacterial chromatophores. Arch Biochem Biophys 238: 219-225.

Casadio R, Wagner R (1985). Rotational mobility and domain flexibility of membrane-bound bacterial coupling factor detected with the triplet probe eosin-isothiocyanate. Biochim Biophys Acta 809: 215-227.

Penefsky HS (1985). Mechanism of inhibition of mitochondrial adenosine triphosphatase by dicyclohexylcarbodiimide and oligomycin: relationship to ATP synthesis. Proc Natl Acad Sci USA 82: 1589-1593.

The Ion Pumps: Structure, Function, and Regulation, pages 207–216

THE MECHANISM OF ACTIVE CATION TRANSPORT BY THE Na/K-PUMP

Steven J.D. Karlish

Biochemistry Department, Weizmann Institute of Science, Rehovot 76100, Israel

The central problem in study of Na/K pump function is identification of states of the protein which bind or occlude cations and definition of the detailed mechanism of active cation movements. As the protein's structure becomes better understood knowledge of functional sites will lead to increasingly realistic models of the trans-membrane path of the cations. This paper summarizes the currently accepted transport mechanism and some unexpected observations which provide insight into the cation binding sites and transport pathways.

THE BASIC TRANSPORT MECHANISM

A convenient starting point is the kinetic scheme proposed some years ago based on studies of phosphorylation-dephosphorylation, transport modes, nucleotide binding and conformational transitions using fluorescence probes and proteolytic digestion (Karlish et al.,1978). This is of course a variant of the classical Post-Albers mechanism (Post et al., 1965) Transfer of each ion involves a specific cation catalyzed bond breaking step and conformational transition. 3Na ions catalyze phosphorylation and are transported in the $E_1P \rightarrow E_2P$ transition. 2K catalyze dephosphorylation and are transported in the $E_2(K) \rightarrow E_1$ transition. ATP accelerates $E_2K \rightarrow E_1$ in addition to phosphorylating the protein. The crucial point as far as the cation transport is concerned is that this is consecutive or ping-pong, Na preceding the K. Na binds and K dissociates at the cytoplasmic face only on E_1 while K binds and Na dissociates at the extracellular face only on E_2P. As suggested by Post et al., (1972), the 2K were postulated to be occluded in $E_2(K)$, and for symmetry 3Na were suggested to be occluded in E_1P. There is now ample evidence for the tight binding or occlusion of K and Na (Glynn and Richards, 1982; Glynn et al., 1984). The maximal stoichiometries are 2K(Rb) and 3Na occluded per pump molecule (Glynn et al., 1984; Vilsen et al., 1987; Shani et al., 1987). This fits the classical pumping stoichiometry of 3Na/2K/ATP. A suggested role for occlusion is suggested below.

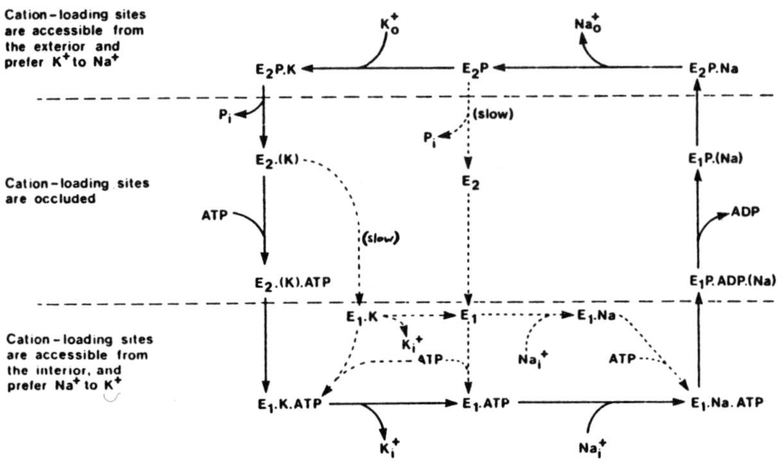

This mechanism accounted well for abnormal pump flux modes and their ligand requirements e.g. Na/Na, K/K exchange, 'uncoupled' Na flux etc. (see Glynn, 1984). Nevertheless several unexpected phenomena have been described recently which are not compatible with the simple concept and oblige its further elaboration. Some of these phenomena are discussed below, in particular experiments performed in this laboratory using renal (Na,K)ATPase reconstituted into phospholipid vesicles (Karlish and Pick, 1981). Important properties of the vesicle system for transport studies include (a) simplicity (b) control of sidedness of ligand effects (c) accessibility to the cytoplasmic surface of the pump and (d) a very low passive cation permeability or electrical conductance.

TIGHT COUPLING OF CATION FLOWS TO ATP HYDROLYSIS SLIPPAGE FLUXES. CATION SPECIFICITY

The classical stoichiometry of 3Na/2K/1ATP has been found in essentially all cells and conditions yet tested. Thus we were surprised some years ago to detect, in the vesicles, slow passive pump-mediated 'slippage' fluxes of K or Rb in the complete absence of other pump ligands. These included Rb/Rb exchange, net passive Rb uptake and Rb congener exchange such as with Na (Karlish and Stein, 1982). The striking point about these fluxes was their very low rate which was in no case greater than 1.5% of maximal Na/K exchange. These fluxes have now also been observed in red cells (Kenny and Kaplan, 1987). The characteristics of these fluxes suggested the following mechanism involving $E_2(Rb)$ the occluded Rb form. The important conclusion was that the occluded form $E_2(Rb)$ is a stable species in an energy well so that movement out of it in either direction is very slow, (except in the presence of ATP which forces K(Rb) ions in the forward direction and Pi which forces K(Rb) ions in the backward direction). The role of ion occlusion was therefore suggested to be minimization of the potentially gradient wasting slippage fluxes, such as K-Na exchange.

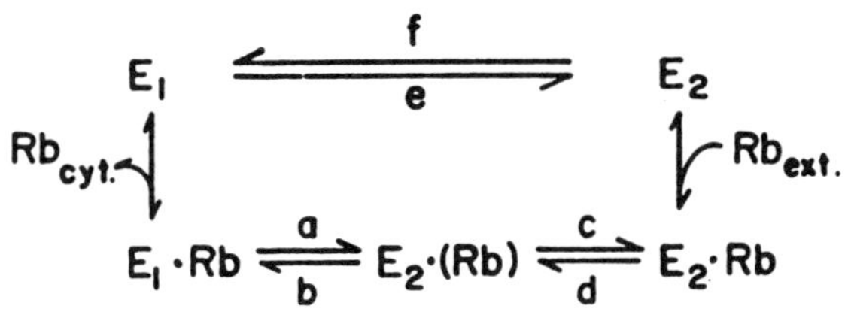

Slippage fluxes can be considered in the broader context of the conditions for tight coupling (see Jencks, 1983). A tightly coupled system must minimize energy wastage and work at an optimal rate. One requirement is an ordered chemical and transport reaction. Thus the high energy E_1P must not be hydrolyzed by water, nor E_2P formed by ATP, for in either case ATP hydrolysis would occur without cation transport. In addition, the cation should not dissociate from the form immediately involved in the transport, for again this would lead to uncoupling and provide a pathway for the passive movements of cations. This is the rationale for ion occlusion.

An equally critical requirement for tight coupling is specificity for cation binding and catalysis. Na binds tightly to E_1 and weakly to E_2P and the converse is true for K. This is important in allowing the ions to bind in the compartment of low concentration and dissociate into that of high concentration, i.e. it allows a rapid turnover of the pump. However, the more crucial element of specificity is catalytic. Na at the cytoplasmic side catalyzes phosphorylation and K the reverse reaction, while K at the extracellular face catalyzes dephosphorylation and Na the reverse reaction. Consider for example that on form E_1 3Na and 2K compete so there are 9 different forms possible, depending on the concentrations of Na and K: E_1 E_1Na $E_1$2Na $E_1$3Na E_1K E_1NaK $E_1$2NaK $E_1$2K E_1Na2K. Only when the full complement of 3Na are bound in $E_1$3Na is phosphorylation triggered and only with 2K bound in $E_1$2K does the reverse reaction occur. All other forms are kinetically dead-end species. The protein must be constructed so as to sense all 9 species and minimize reactions of forms other than $E_1$3Na and $E_1$2K. The sites must therefore be catalytically non-equivalent.

ORDERED RELEASE OF CATIONS

Recently Glynn et al. (1985) and Forbush (1985) have provided convincing evidence for ordered deocclusion of 2Rb ions from an E_2P form. Thus whereas the basic model proposes three locations of ion binding sites inwards, outwards or occluded, one must now invoke sub-conformations of E_2P with one K site occluded and the other exposed to the extracellular medium. In addition there is now evidence for three rather than only two kinetically distinguishable species of phosphoenzyme (Norby et al., 1983), the third form E*P being sensitive to both ADP and K (Yoda and Yoda, 1986) and binding ouabain with a high affinity (Fortes and Lee, 1984). The E*P appears to be able to occlude 2Na ions and participate in an ordered release of Na (Yoda and Yoda, 1987). Ordered release, of course, suggests non-equivalence of sites. It leads one to the notion that transport could occur via gradual opening of trans-membrane segments of the polypeptide chain *i.e.* as in a Zipper (see model below).

NON-EQUIVALENCE OF BINDING SITES. TRANS EFFECTS OF Na.

A long-standing dogma holds that 3Na bind to cytoplasmic sites and 2K bind to extracellular sites with equal intrinsic affinities. This is a surprising notion if only on electrostatic grounds. It arose out of transport experiments with red cells (Garay and Garrahan, 1973) showing that to a first approximation $V = V_{max}(S/S+K_s)^n$ where $S=Na_i$ or K_o, and n=3 for Na_i and n=2 for K_o. However, using the reconstituted vesicles, more accurate Na_{cyt} activation curves for Na/K exchange can be obtained than in red cells and these show a significant deviation from the above behaviour, as expected for a ping-pong mechanism (Karlish and Stein, 1985). Detailed curve fitting to the steady-state rate equation suggest that, intrinsically, the three cytoplasmic Na sites are very different. Making assumptions as to rate-constants, Na affinities of about 0.1 , 4 and 30 mM can be inferred (Karlish and Stein, unpublished).

Quite unexpectedly the shape of Na_{cyt} or Rb_{cyt} activation curves, sigmoid or hyperbolic, was found to be variable, depending on the absence or presence of Na_{exc} (Karlish and Stein, 1985). Na_{cyt} activation of Na/K exchange is sigmoid in the absence of Na_{exc} (n_H=1.9) but only moderately sigmoid in the presence of Na_{exc} (n_H=1.3). Conversely activation by Rb_{cyt} of Rb/Rb exchange is hyperbolic in the absence of Na_{exc} (n_H=1) but sigmoid in the presence of Na_{exc} (n_H=1.5). These trans-membrane effects of Na_{exc} are allosteric since the Na_{exc} is not itself transported.

The following explanations were proposed: (A) 3Na or 2Rb (K) must be bound to E_1 to catalyze the specific cation catalyzed transition. (B) The degree of sigmoidicity depends on the relative binding affinities of $3Na_{cyt}$ or $2Rb_{cyt}$. If affinities are relatively close, a sigmoid activation is observed. If affinities are very different, a more hyperbolic curve ensues. (C) Na_{exc} induces a trans-membrane effect which causes separation of binding affinities towards 3 Na_{cyt} and make them more equal to $2Rb_{cyt}$.

Sigmoid Na activation reflects the primary energetic requirement to pump out 3Na per ATP split. However, there is no fixed relation between the stoichiometry, which is constant, and degree of sigmoidicity which is variable. The trans-membrane effect of Na could be a mechanism evolved to increase kinetic efficiency by reducing the occupancy of dead-end forms with both Na and K bound (see Karlish and Stein, 1985).

CATION-BINDING RESIDUES

The structure of cation-binding sites *i.e.* the identity of residues binding cations is quite unknown. We have been looking at this question by a combination of approaches. (A) The first approach uses chemical modification. It has been observed that the hydrophobic carbodiimide, DCCD, inactivates Rb occlusion in a selective way. Incubating the protein with ^{14}C-DCCD shows that the α-chain is labelled, and Rb ions protect (M. Shani and S.J.D. Karlish, unpublished). These experiments suggest participation of at least one crucial carboxyl group in a non-aqueous environment. (B) Another approach to identifying the cation sites is to study the electrogenic aspects of Na/K pump function that is the voltage sensitivity of the pump and electrogenic potentials. Experiments summarized below provide information on the identity of charge-carrying steps and charged residues in the active site.

VOLTAGE-SENSITIVITY OF THE Na/K PUMP

The Na/K pump is electrogenic and its rate must therefore be sensitive to voltage. In principle, rates of all steps carrying charge through the electric field should be sensitive to membrane potential. These charges include the ions themselves, groups on the protein to which the ions bind, or movement of charges and displacement of dipoles unrelated directly to the transport. Theoretical analysis show that it should be possible to identify charge-carrying steps, deduce the number of charges moving and the dielectric distance through which the charge moves (Lauger and Apell, 1986). The most likely candidates for the charge-carrying steps are the transport linked conformational changes and/or ion binding or dissociation.

Effects of voltage have now been observed in several whole cells particularly heart cells (Gadsby et al., 1985; Nakao and Gadsby, 1986a), on the isolated renal (Na,K)ATPase reconstituted into phospholipid vesicles (Goldshleger et al., 1987a), and in addition, transient currents generated by Na/K pump have been recorded from membrane fragments bound to planar bilayers (Fendler et al., 1985; Borlinghaus et al., 1987; Apell et al., 1987). The outcome of these efforts is a unified and relatively simple picture of the charge carrying steps (Karlish et al., 1985; see De Weer et al., 1987).

It is now agreed that it is the Na-transporting steps of the cycle which carries charge, while the K-carrying steps are electrically silent. Charge movement accompanying Na transport steps have been deduced from (a) effects of diffusion potentials on steady-state Na/K exchange in the vesicles in conditions when the Na transporting steps are limiting (Karlish et al., 1985; Goldshleger, et al., 1987a) and (b) on the rates of conformational transition $E_1P \rightarrow E_2P$ measured directly on reconstituted FITC-labelled (Na,K)ATPase using acetyl-phosphate as substrate (Rephaeli et al., 1986a) (c) from effects of voltage on transient Na movements is heart cells (Nakao and Gadsby, 1986) and(d) from observation on transient currents accompanying the first turnover of Na transport on (Na,K)ATPase adsorbed to bilayers (Apell et al., 1987). The electrically neutral nature of K-movements has been deduced from lack of effects of diffusion potentials in the vesicles on (a) Na/K exchange when K-transporting steps are limiting (b) Rb/Rb exchange (Goldshleger et al., 1987a) and (c) rates of conformational transitions $E_2(K) \rightleftarrows E_1K$ with the reconstituted FITC-labelled protein (Rephaeli et al., 1986b).

The simplest model to explain voltage-insensitivity of $2K_-$ movements and voltage sensitivity of 3Na-movements would be that in the cation binding domain we have two negatively charged residues. When $2K^+$ bind, the protein carries no net charge and when $3Na^+$ bind, a net charge of +1 is carried. Two Na and two K at the same surface would presumably bind to the same two negative residues while the third Na binds to a neutral site. From electrostatics one would obviously expect non-equivalence of affinities at these sites. The negative residues could be glutamate or aspartate, but there is no necessity for these to be the same residues at the cytoplasmic and extracellular faces. Thus, models (below) involving breakage and formation of salt-bridges satisfy the requirements of net charge movements.

In the vesicle system, a second effect of voltage on an intrinsic Na binding constant in E_1 was deduced. This could be due to (a) a Na 'well' or (b) a direct effect of potential on protein conformation.

ELECTROGENIC POTENTIALS

Potentials generated by the Na/K pump have been detected in many cells. In red cells, Hoffman and colleagues (1979) haved detected Na/K exchange-mediated electrogenic potentials using a carbocyanine dye. Interestingly, it was found that so-called uncoupled Na efflux into a Na- or K-free medium is electroneutral and it was suggested that in this mode the pump performs a Na/anion co-transport, say 3Na and 3Cl (Dissing and Hoffman, 1983). The notion of an anion-binding site on a cation pump was sufficiently unorthodox to make this worth testing in a different system. Recently we have measured electrogenic potentials in the reconstituted vesicles using the dye Oxonol VI (Goldshleger et al., 1987b). Na/K exchange can generate very large potentials, up to 250 mV positive inside. This is due presumably to the high membrane resistance. In the 'uncoupled' Na flux mode electroneutrality has been confirmed for pH 7.0, but at pH 8.5 this flux does generate a potential. Electroneutrality at pH 7.0 could be due to 3Na/3Cl co-transport, but another possibility is 3Na/3H exchange. A Na/anion co-transport might be expected to show some anion selectivity. However, we have not been able to detect any such specificity. This fact together with the appearance of electrogenicity when the proton concentration was reduced is clearly more suggestive of 3Na/3H exchange at pH 7. If correct, this has an interesting implication. It means that the residues binding protons have pK's of about 7. Since as we have concluded above it is carboxyl groups which bind the cations, these must be in or transferred to a non-aqueous environment, for in water the pK's would be only 4-5.

A TRANSPORT MODEL WITH GATING BY SALT-BRIDGES

The speculative model below focusses on the possible role of intramembrane charged groups. Initially it was proposed (Karlish et al., 1985) by analogy to the Zipper concept for anion exchange (Wieth and Bierrum 1983). Based on experiments showing participation in anion exchange of arginines and carboxyls these authors suggested the existence of intramembrane salt-bridges acting as diffusion barriers. For the Na/K pump the transport could occur during the E_1-E_2 transitions by exchange of cations bound on one set of fixed carboxyls for a second set within the salt-bridge, and movement of the positive residues in the bridge towards the first set of carboxyls. Thus salt-bridges form and reform alternately at the opposite surfaces. An economic feature was that the TRANS effects of Na could be postulated to occur from sites forming part of the salt-bridge without having to invoke the existence of separate 'allosteric' sites.

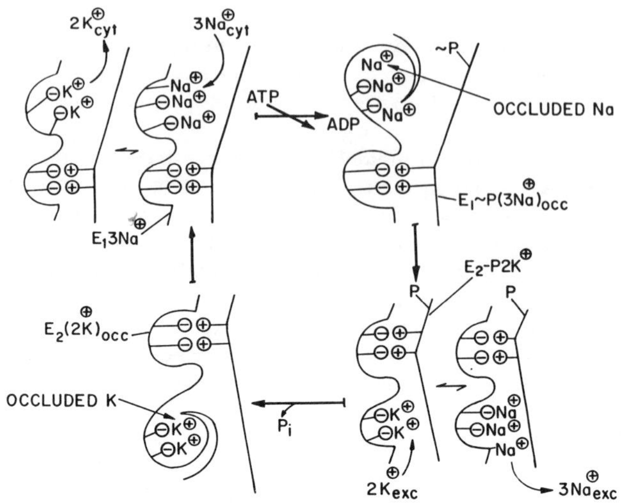

Subsequently, the notion of carboxyls (glu, asp) and positive residues (lys, arg) within trans-membrane segments has been shown correct from the primary sequence of the α-chain (Shull et al., 1985). Our own experiments suggest the involvement of membrane-embedded carboxyls. The value of models such as these is that they point to a possible role of intra-membrane charged groups. The challenge for the future is to obtain evidence concerning particular residues and define the nature of their role in the cation movements.

This work was supported by a grant from the US PHS GM32286.

REFERENCES

Apell HJ, Borlinghaus R, Lauger P (1987). Fast charge translocations associated with partial reactions of the Na/K pump. II. Microscopic analysis of transient currents. J Mem Biol. 97: 179-191.

Borlinghaus R, Apell HJ, Lauger P (1987). Fast charge translocations associated with partial reactions of the Na,K-pump. I. Current and voltage transients after photochemical release of ATP. J Mem Biol. 97: 161-178.

De Weer P, Gadsby DC, Rakowski RF (1987). Voltage dependence of the Na/K Pump. Ann Rev Physiol, In Press.

Dissing S, Hoffman JF (1983). Anion-coupled Na efflux mediated by the Na/K pump in human red blood cells. Current Topics in Membranes and Transport 19: 693-695.

Fendler K, Grell E, Haubs M, Bamberg E (1985) Pump currents generated by purified Na,K-ATPase from kidney on black lipid membranes. EMBO J 4: 3079-3085.

Forbush B (1985) Rapid ion movements in a single turnover of the Na pump in The Sodium Pump Eds. Glynn IM, Ellory JC The Company of Biologists pp 599-611.

Gadsby DC, Kimura JM, Numa A (1985). Voltage dependence of Na/K pump current in isolated heart cells. Nature 315: 63-65.

Garay RP, Garrahan PJ (1973) The interaction of sodium and potassium with the sodium pump in red cells. J Physiol 231: 297-325.

Glynn IM (1984) The Na^{+},K-transporting adenosine triphosphatase In the Enzymes of Biological Membranes. Ed Martinosi A Plenum Press N.Y. pp. 35-114.

Glynn IM, Richards DE (1982) Occlusion of rubidium ions by the sodium-potassium pump - its implications for the mechanism of potassium transport. J Physiol 330: 17-43.

Glynn IM, Howland JL, Richards DE (1985) Evidence for the ordered release of rubidium occluded within Na,K-ATPase of mammalian kidney. J Physiol 368: 453-469.

Goldshleger R, Karlish SJD, Rephaeli A, Stein WD (1987a). The effect of membrane potential on the mammalian sodium-potassium pump reconstituted into phospholipid vesicles. J Physiol 387: 331-355.

Goldshleger R, Karlish SJD, Shahak Y (1987b) Electrogenic and electroneutral transport modes of the mammalian renal Na/K pump. J Physiol 387: 98P.

Hoffman JF, Kaplan JH, Callahan TH (1979). The Na:K pump in red cells is electrogenic. Fed Proc 38: 2440-2441.

Jencks WP (1983) What is a coupled vectorial process? Current Topics in Membranes and Transport 19: 1-19.

Karlish SJD, Pick U (1981) Sidedness of the effects of sodium and potassium on the conformational state of the sodium-potassium pump J Physiol 312: 505-529.

Karlish SJD, Stein WD (1982) Passive rubidium fluxes mediated by Na,K-ATPase reconstituted into phospholipid vesicles when ATP- and phosphate-free. J Physiol 328: 295-316.

Karlish SJD, Rephaeli A, Stein WD (1985). Transmembrane modulation of cation transport by the Na,K-pump. In The Sodium Pump Eds. Glynn IM, Ellory JC. Company of Biologists pp 487-499.

Karlish SJD, Yates DW, Glynn IM (1978) Conformational changes between Na-bound and K-bound forms of (Na+K)ATPase studied with formycin nucleotides. Biochim Biophys Acta 525: 252-264.

Kenney LJ, Kaplan JH (1987) Cation and ligand effects on the K transport limb of the red cell sodium pump. Abstr. 5th Int. Conf. on Na,K-ATPase. Aarhus 14-19 June.

Lauger P, Apell HJ (1986). A microscopic model for the current-voltage behaviour of the Na/K pump. Eur Biophys J 13: 309-321.

Lee JA, Fortes PAG (1985) Anthroylouabain binding to different phosphoenzyme forms of Na,K-ATPase. In The Sodium Pump Eds. Glynn IM, Ellory JC The Company of Biologists. pp 277-282.

Nakao M, Gadsby DC (1986). Voltage dependence of Na translocation by the Na/K pump. Nature 323: 628-630.

Norby JG, Klodos I, Christiansen NO (1983) Kinetics of Na-ATPase activity by the Na,K Pump. J Gen Physiol 82: 727-757

Post RL, Sen AK, Rosenthal AS (1965) A phosphorylated intermediate in adenosine triphosphate dependent sodium and potassium transport across kidney membranes. J Biol Chem 240: 1437-1445.

Rephaeli A, Richards DE, Karlish SJD (1986a). Conformational transitions in fluorescein-labeled Na,K-ATPase reconstituted into phospholipid vesicles. J Biol Chem 261: 6248-6254.

Rephaeli A, Richards DE, Karlish SJD (1986b) Electrical potential accelerates the $E_1P(Na) \rightarrow E_2P$ conformational transition of Na,K-ATPase in reconstituted vesicles. J Biol Chem 261: 12437-12440.

Shani M, Goldshleger R, Karlish SJD (1987) Rb occlusion in renal (Na+K)-ATPase characterized with a simple manual assay. Biochim Biophys Acta 904: 13-21.

Shull GE, Schwarz A, Lingrel JB (1985) Amino-acid sequence of the catalytic subunit of the $(Na^+ + K^+)$-ATPase deduced from a complementary DNA. Nature 316: 691-695

Vilsen B, Andersen JP, Petersen J, Jorgensen PL (1987) Occlusion of ^{22}Na and ^{86}Rb in membrane-bound and soluble protomeric $\alpha\beta$-units of Na,K-ATPase. J Biol Chem 262: 10511-10517.

Wieth JO, Bjerrum PJ (1983) Transport and modifier sites in capnophorin, the anion transport protein of the erythrocyte membrane. Structure and Function of Membrane Proteins Eds. Quagliariello E, Palmieri F Elsevier pp 95-107.

Yoda S, Yoda A (1986) ADP- and K-sensitive phosphorylated intermediate of Na,K-ATPase. J Biol Chem 261: 1147-1162.

Yoda S, Yoda A (1987) Two different phosphorylation-dephosphorylation cycles of Na,K-ATPase proteoliposomes accompanying Na-transport in the absence of K. J Biol Chem In Press.

The Ion Pumps: Structure, Function, and Regulation, pages 217–224

ELECTROGENIC PROPERTIES OF THE NA,K-PUMP

P. Läuger

Department of Biology, University of Konstanz, D - 7750 Konstanz, F.R.G.

INTRODUCTION

The Na,K-pump under physiological conditions transports three Na^+ ions outward and two K^+ ions inward and thus generates a transmembrane current. As a consequence of the electrogenic nature of the Na,K-pump the transport rate becomes a function of voltage. According to the Albers-Post reaction scheme (Fig. 1) the pump carries out a cycle of conformational transitions and ion-binding and -release reactions. It is interesting to ask in which step(s) of the overall transport reaction charge is translocated. Information on the nature of the electrogenic steps may be obtained from two kinds of experiment, by measuring the current-voltage characteristic of the pump and by studying transient currents after a sudden change of an external parameter, such as voltage or ATP concentration.

VOLTAGE DEPENDENCE OF PUMP CURRENTS

Experimental studies of voltage effects on the activity of the Na,K-pump could be performed in the past only in a narrow voltage interval. Recently, by using the whole-cell recording technique with patch pipettes it has become possible measuring the current-voltage characteristic in a more extended voltage range (Gadsby et al., 1985). Furthermore, the current-voltage behaviour of the pump has been recently studied by electrophysiological techniques in *Xenopus* oocytes (Lafaire and Schwarz, 1986). In reconstituted lipid vesicles the electrogenic behaviour of the pump may be investigated

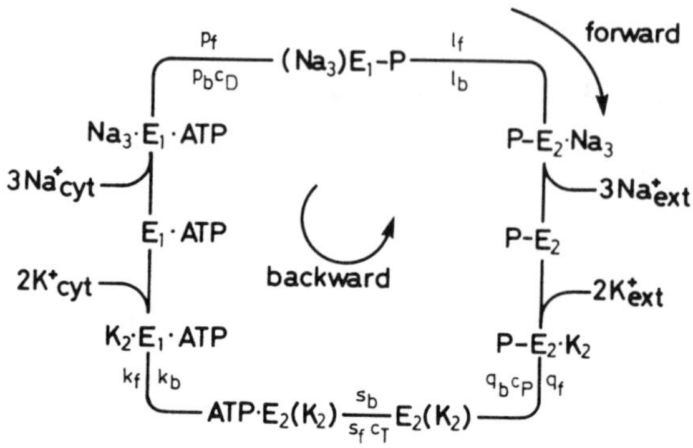

Figure 1. Albers-Post scheme for the pumping cycle of Na,K-ATPase. E_1 and E_2 are conformations of the enzyme with ion binding sites exposed to the cytoplasm and the extracellular medium, respectively. In the "occluded" states $(Na_3)E_1$ and $E_2(K_2)$ the bound ions are unable to exchange with the aqueous phase. Dashes indicate covalent bonds and dots indicate non-covalent bonds. p_f, l_f, q_f, $s_f c_T$, k_f and $p_b c_D$, k_b, s_b, $q_b c_p$, l_b are rate constants for transitions in forward and backward direction, respectively. c_T, c_D and c_p are the cytoplasmic concentrations of ATP, ADP and P_i (inorganic phosphate).

using potential-sensitive dyes (Apell et al., 1985; Goldschlegger et al., 1987).

The current-voltage characteristic of the pump is determined by the voltage-dependence of the rate constants of the single transport steps. Voltage effects on the kinetic parameters may be described introducing the energy profile of the ion along the transport pathway (Fig. 2). In Fig. 2 it is assumed that the ion-binding sites are connected with the aqueous medium by a series of low barriers, meaning that ion binding and release are equilibrium reactions. Since the potential energy of an ion in the binding site is modified by the presence of a transmembrane voltage V, the equilibrium dissociation constants become voltage dependent. If $\alpha' V$ is the fraction of voltage which drops between the sodium site and water in state E_1 (Fig. 2), the dissociation constant of Na^+ at the cytoplasmic side is given by:

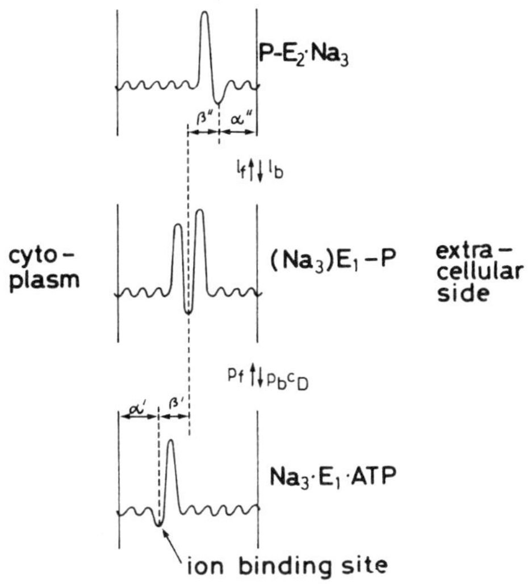

Figure 2. Hypothetical energy profile of a sodium ion along the transport pathway. The ion binding sites in state $Na_3 \cdot E_1 \cdot ATP$ are connected with the cytoplasmic side by a series of low barriers, but separated from the extracellular medium by a high barrier. In the "occluded" state $(Na_3)E_1$-P the energy barriers on either side are high. In state P-$E_2 \cdot Na_3$ the binding sites are easily accessible from the extracellular phase. α', α'' and β'' are dielectric distances, depending on the location of the ion binding site in the protein and on the dielectric properties of the protein and the surrounding medium.

$$K'_N = \tilde{K}'_N \exp(-\alpha' u) \tag{1}$$

$$u \equiv e_o V/kT \tag{2}$$

e_o is the elementary charge, k the Boltzmann constant, T the absolute temperature, and $\tilde{K}'_N$ the value of K'_N at zero voltage. If the potential of the cytoplasm is positive with respect to the external medium (u > o), the equilibrium constant $1/K'_N$ of sodium binding is increased by the Boltzmann factor $\exp(\alpha' u)$. The binding site acts as an "ion well", i.e., a change of electrical potential has a similar effect on the occupancy of the site as a change of external concentration.

In the process $Na_3 \cdot E_1 \cdot ATP \rightarrow (Na_3)E_1\text{-}P$ the protein is phosphorylated and the bound sodium ions become occluded. This transition is, in general, associated with a conformational change whereby the bound ions together with the ligand groups move over a certain distance. The effective dielectric distance over which the binding sites are translocated is described by the phenomenological parameter β' (Fig. 2). The electrostatic contribution of ions plus binding sites to the energy difference between states $Na_3 \cdot E_1 \cdot ATP$ and $(Na_3)E_1\text{-}P$ may be written as $(3+z_L)\beta' u$ where $z_L e_o$ is the charge of the empty ligand system; the energy is expressed in units of kT. If in the transition intrinsic charges of the protein (other than charged ligands) are translocated, an additional energy contribution of magnitude $\eta' u$ results.

According to the theory of absolute reaction rates, the forward and backward rate constants are given by

$$p_f = \tilde{p}_f \exp(\alpha_p u/2) \tag{3}$$

$$p_b = \tilde{p}_b \exp(-\alpha_p u/2) \tag{4}$$

$$\alpha_p \equiv (3+z_L)\beta' + \eta' \tag{5}$$

The parameter η' accounts for rotation of dipolar groups and translocation of intrinsic charges of the protein other than charged ligands (Läuger and Apell, 1986). The voltage dependence of the other rate constants is obtained in an analogous way.

Examples of current-voltage curves obtained by numerical simulation of the reaction cycle of Fig. 1 are represented in Fig. 3. If the ligand system is assumed to bear three negative charges ($z_L = -3$), the net charge of the alkali-ion binding sites is zero in state $Na_3 \cdot E_1 \cdot ATP$ and -1 in state $P\text{–}E_2 \cdot K_2$. Current-voltage curves are given for three different values of parameter β' representing the dielectric distance over which sodium ions move in the transition $Na_3 \cdot E_1 \cdot ATP \rightarrow (Na_3)E_1\text{–}P$. The current monotonically increases with voltage V and reaches asymptotic values at $V \rightarrow \infty$ and $V \rightarrow -\infty$. For $\beta' = 0.5$ the limiting currents (referred to a single pump molecule) are $I(\infty) = 13.5 \cdot 10^{-19}$ A and $I(-\infty) = -3.1 \cdot 10^{-19}$ A. The I(V) curves of Fig. 3 have been calculated assuming that the only voltage-dependent steps are the transitions $P\text{-}E_2 \cdot K_2 \rightarrow E_2(K_2)$ and $ATP \cdot E_2(K_2) \rightarrow K_2 \cdot E_1 \cdot ATP$ which involve a movement of the K^+-loaded binding sites (total charge $-e_0$) within the membrane dielectric. This results in a saturating behaviour. For increasing positive voltages ($V > 0$) the voltage-dependent transition rates in forward direction increase, until eventually the voltage-insensitive steps of the cycle become rate limiting. This means that for $V \rightarrow \infty$ (and also, by the same argument, for $V \rightarrow -\infty$) the current approaches an asymptotic value.

CURRENT TRANSIENTS AFTER SUDDEN CHANGE OF AN EXTERNAL PARAMETER

Further information on the nature of the electrogenic steps may be obtained by studying transient pump currents elicited by a sudden change of an external parameter. Using the whole-cell recording technique with patch pipettes, Nakao and Gadsby (1986) measured current transients after a voltage jump. Time-dependent pump currents after an ATP-concentration

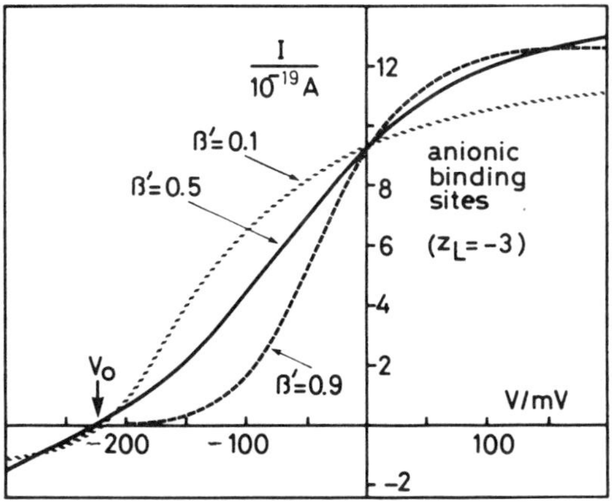

Figure 3. Current-voltage curve of a pump molecule with negatively charged ion-binding sites ($z_L = -3$). The net charge of the binding site is zero in state $Na_3 \cdot E_1 \cdot ATP$ and −1 in state $P\text{-}E_2 \cdot K_2$. I is the pump current (referred to a single pump molecule), V the transmembrane voltage and V_0 the reversal potential. The coefficient β' represents the dielectric distance over which sodium ions move in the transition $Na_3 \cdot E_1 \cdot ATP \rightarrow (Na_3)E_1\text{-}P$ (Fig. 2). From Läuger and Apell (1986).

jump can be observed in a compound membrane system consisting of a planar lipid bilayer with attached Na,K-ATPase membrane fragments (Fendler et al., 1985; Borlinghaus et al., 1987; Apell et al., 1987). In this system (Fig. 4) the planar bilayer acts as a capacitive electrode which couples charge translocations within the attached membrane fragment to the external measuring circuit. The membrane fragments contain a high density (10^3-10^4 μm^{-2}) of oriented Na,K-ATPase molecules. ATP is released by a light flash from an inactive, photolabile precursor ("caged ATP"). After an ATP-concentration jump in a solution containing Na^+ but no K^+, a transient current signal is observed (Fig. 5). The direction of the current corresponds to a translocation of positive charge from the solution toward the lipid bilayer (Fig. 4). Using estimated values of the circuit parameters of the compound membrane system, the intrinsic pump current $I_p(t)$ can be evaluated from the measured current signal I(t). The pump current $I_p(t)$ rises within about 30 ms to a peak and thereafter decays toward a much smaller quasistationary current I_p^∞ (Fig. 5).

A likely interpretation of the time course of I_p consists in the assumption that after activation by the ATP-concentration jump, the pump goes through the first steps of the cycle and then enters a long-lived state from which return to the initial state is slow (Apell et al., 1987). The rate-limiting step for the completion of the cycle is likely to be the dephosphorylation of the protein, which is extremely slow

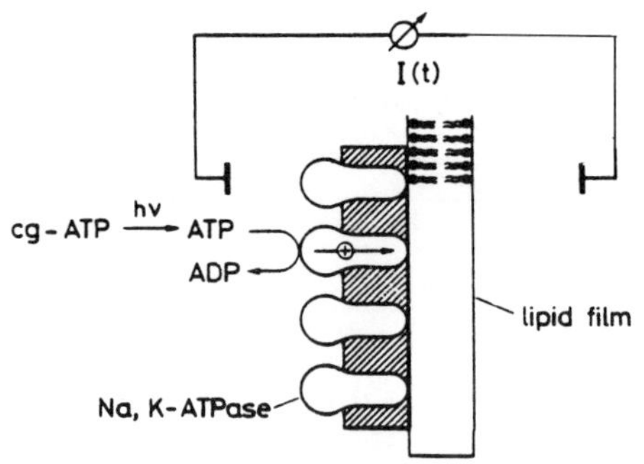

Figure 4. Compound membrane system consisting of a planar lipid bilayer with attached Na,K-ATPase membrane fragments. Photochemical release of ATP from "caged ATP" leads to transient charge translocation in the pump molecule.

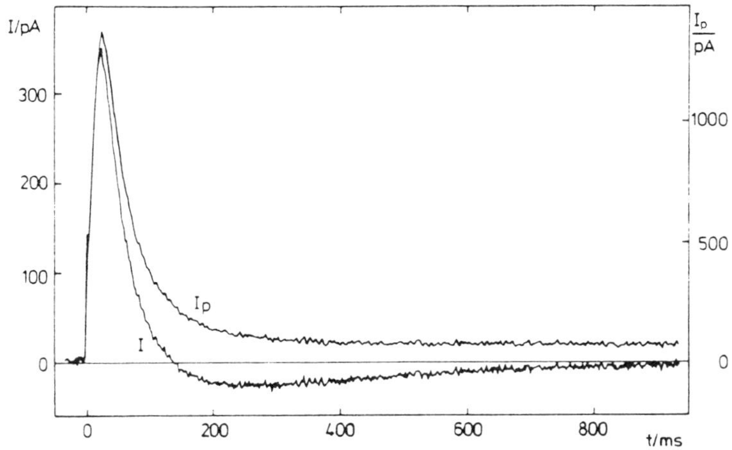

Figure 5. Intrinsic pump current $I_p(t)$ evaluated from the measured current signal I(t) by taking the circuit parameters of the compound membrane system into account (from Borlinghaus et al., 1987).

in the absence of potassium. As long as ATP is present in the solution, pump molecules re-entering the initial state may become phosphorylated again, which give rise to a small quasistationary current I_p^{∞}. Experiments with chymotrypsin-modified Na,K-ATPase (Borlinghaus et al., 1987) suggest that the main electrogenic event contributing to $I_p(t)$ is the transition $(Na_3)E_1\text{-}P \rightarrow P\text{-}E_2 \cdot Na_3$, followed by release of Na^+ to the extracellular side.

REFERENCES

Apell H-J, Borlinghaus R, Läuger P (1987). Fast charge translocations associated with partial reactions of the Na,K-pump. II. Microscopic analysis of transient currents. J Membrane Biol 97: 179-191.

Borlinghaus R, Apell H-J, Läuger P (1987). Fast charge translocations associated with partial reactions of the Na,K-pump. I. Current and voltage transients after photochemical release of ATP. J Membrane Biol 97: 161-178.

Chapman JB, Johnson EA, Kootsey JM (1983). Electrical and biochemical properties of an enzyme model of the sodium pump. J Membrane Biol 74: 139-153.

De Weer P (1984). Electrogenic pumps: theoretical and practical considerations. In Blaustein MP, Lieberman M (eds): "Electrogenic transport: Fundamental principles and physiological applications". New York: Raven Press, pp 1-15

Fendler K, Grell E, Haubs M, Bamberg E (1985). Pump currents generated by the purified Na^+,K^+-ATPase from kidney on black lipid membranes. EMBO J 4: 3079-3085.

Gadsby DC, Kimura J, Noma A (1985). Voltage dependence of Na/K pump current in isolated heart cells. Nature (London) 315: 63-65.

Goldschlegger R, Karlish SJF, Rephaeli A, Stein WD (1987). The effect of membrane potential on the mammalian sodium-potassium pump reconstituted into phospholipid vesicles. J Physiol (London) 387: 331-355.

Lafaire AV, Schwarz W 81986). Voltage dependence of the rheogenic Na^+/K^+-ATPase in the membrane of oocytes of _Xenopus laevis_. J Membrane Biol 91: 43-51.

Läuger P, Apell H-J (1986). A microscopic model for the current-voltage behaviour of the Na,K-pump. Eur Biophys J 13: 309-321.

Nakao M, Gadsby DC (1986). Voltage dependence of Na translocation by the Na/K pump. Nature (London) 323: 628-630.

The Ion Pumps: Structure, Function, and Regulation, pages 225–232

PUMP CURRENTS GENERATED BY RENAL NA^+K^+-ATPASE ON BLACK LIPID MEMBRANES

Klaus Fendler, Georg Nagel and Ernst Bamberg

Max-Planck-Institut für Biophysik, Heinrich-Hoffmann-Str. 7, 6000 Frankfurt am Main, W-Germany

INTRODUCTION

The Na^+K^+-ATPase from mammalian cells is an electrogenic pump, which exchanges 3 Na^+ for 2 K^+ ions across the cell membrane. The electrical properties of the Na^+K^+-ATPase were investigated in whole cells (for a review see De Weer, 1986), in reconstituted vesicles (see e.g. Dixon and Hokin, 1980, Goldshlegger et al., 1987) and on black lipid membranes (BLM) (Fendler et al., 1985, Nagel et al., 1987, Borlinghaus et al., 1987). According to a Na^+/K^+ stoichiometry of 3/2 it is electrogenic and generates stationary pump currents.

Adsorption of membrane fragments containing the ATPase to a BLM offers the possibility to monitor directly the charge translocation taking place during the ATPase pumping cycle. Pump currents were initiated by a light-induced concentration jump of ATP via the conversion of a photolabile inactive derivative, the caged ATP (Kaplan et al., 1978 and Forbush, 1985), into its active form. With this method the electrogenic transport activity of the Na^+K^+-ATPase could be detected with high sensitivity (0.5 pA) and high time resolution (1 ms).

MATERIALS AND METHODS

Preparation of caged ATP and Na K -ATPase. The triethylammonium salt of caged ATP (P^3-1-(2-nitro)-phenylethyl-adenosine 5'-triphosphate) was synthesized as reported previously (Fendler et al., 1985). Na^+K^+-ATPase was prepared from the red outer medulla of pig kidneys according to Jorgensen (1974). Membrane fragements in the form of cup shaped discs of 0.1 - 0.6 µm diameter were obtained. The suspension contained ca. 2.5 mg/ml protein with a typical specific activity of 30 µmol P_i mg^{-1} min^{-1} at 37° C.

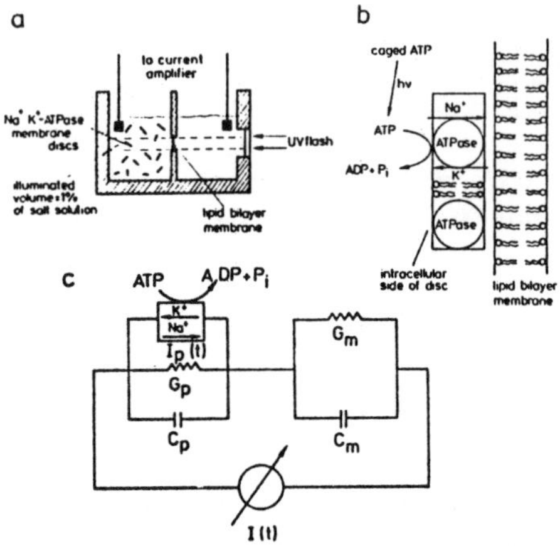

Figure 1: Schematic representation of the bilayer setup.
a: Teflon chamber with black lipid membrane (BLM) and adsorbed ATPase membrane fragments.
b: Proposed sandwich-like arrangement of discs and underlying lipid membrane. The two membranes are capacitively coupled.
c: Equivalent circuit diagram of the two membranes in series. G_m, G_p refer to the conductance of the planar bilayer (BLM) and the disc membrane, respectively, C_m and C_p to the capacitance of BLM and discs, respectively.

Lipid bilayer setup. Optically black lipid membranes with an area of 10^{-2} cm^2 were formed in a thermostated teflon cell with 1.5 ml of an appropriate electrolyte solution in each compartment. The temperature was kept at 22°C. The membrane forming solution contained 1.5 % w/v diphytanoyllecithin (Avanti Biochemicals, Birmingham Al. USA) and 0.025 % w/v octadecylamine (Riedel de Haen, Hannover, FRG) in n-decane. The membrane was connected to an external measuring circuit via platinum electrodes. To avoid photoeffects, the electrodes were separated from the aqueous compartments of the teflon cell by agar-agar salt bridges. The signals were filtered and recorded with a digital oscilloscope. For further details see Fendler et al., 1985.

15 µl of the Na^+K^+-ATPase or containing suspension together with different amounts of caged ATP were added under stirring to one compartment of the cuvette, containing an appropriate electrolyte. To photolyse the

caged ATP, light pulses of a Hg high pressure lamp of 125 ms duration or of an excimer laser (wavelength 308 nm) of 10 ns duration were focussed onto the membrane. One flash of the lamp or the laser converted 30 % and 10 %, respectively, of the caged ATP in the membrane plane into ATP.

PUMP CURRENTS OF THE NA^+K^+-ATPASE

15 min after addition of ATPase and caged ATP a current could be induced by a UV light pulse. In the absence of ionophores the current is transient, whereas the permeabilization of the underlying lipid bilayer yields a sustained current (figure 2) which decays slowly to zero due to the consumption and the dilution of the liberated ATP into the not irradiated volume. Because of their insensitivity to UV light a carrier system consisting of monensin as an alkali metal ion/H^+ exchanger together with the protonophore 1799 was used for these experiments.

The sign of the current was always same and corresponds to a positive charge movement toward the protein free side of the BLM. The discs are preferentially oriented with the extracellular side facing the planar bilayer.

Mg^{++} was necessary for the appearance of pump currents. If not otherwise indicated, in all experiments a concentration of free Mg^{++} of 3 mM was used. Together with additional control experiments (Fendler et al., 1985) these experiments show that the observed current reflects the pumping activity of the Na^+K^+-ATPase.

The appearance of stationary Na^+ pump currents can prove that the enzyme acts as an electrogenic Na^+ pump. They can be detected on the sandwich-like structure by increasing the conductivity of the supporting lipid bilayer with an appropriate carrier system. A stationary Na^+ pump current was obtained indeed after the addition of monensin and the protonophore 1799 to the membrane system (figure 4). It can be excluded that this effect is due to residual K^+ in the electrolyte (Nagel et al., 1987). The experiment therefore demonstrates the electrogenic activity of the Na^+-ATPase.

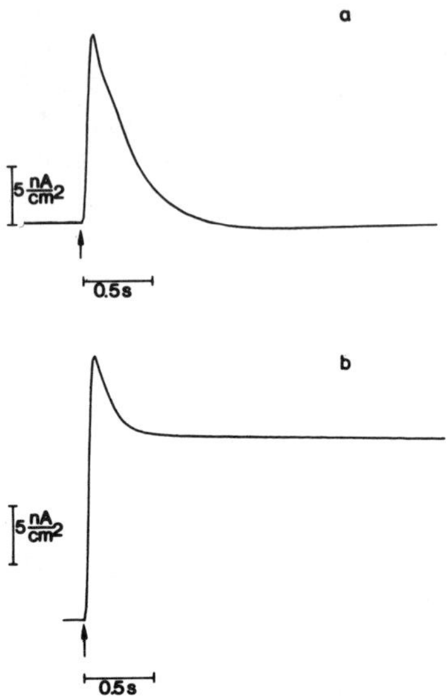

Figure 2: Short circuit currents in the absence and presence of ionophores. The membrane bathing solution contained 130 mM NaCl, 20 mM KCl, 3 mM $MgCl_2$, 25 mM imidazole-HCL pH 7.5. The arrows indicate opening of the shutter, which was opened for 125 ms.
a: 100 µM caged ATP, no addition of ionophores.
b: Same membrane system after addition of monensin and 1799 to a concentration of 10 µM and 0.6 µM, respectively.

TRANSIENT KINETICS OF THE NA^+K^+ATPASE

Using a 10 ns light pulse of an excimer laser together with fast recording instrumentation the current generated by the Na^+K^+-ATPase after an ATP concentration jump could be measured with a time resolution of 1 ms. This provides at the same time information about the electrogenicity as well as the rate constants of partial reaction steps of the Na^+K^+-ATPase.

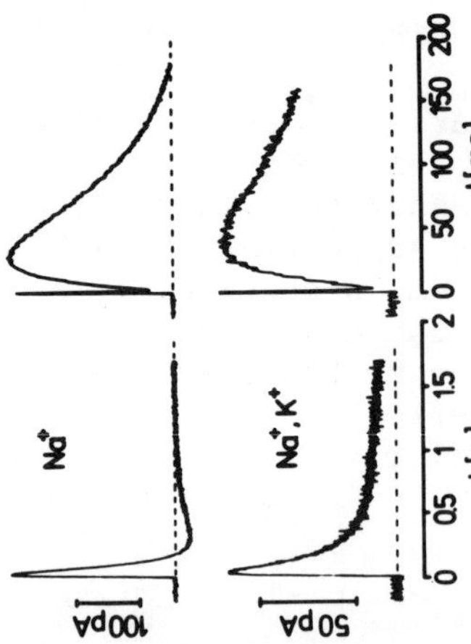

Figure 3: Short circuit currents without and with K^+. The signal was recorded simultaneously with two different sample rates (left hand side and right hand side). Conditions: 3 mM $MgCl_2$, 25 mM imidazole-HCl at pH 6.3, 7 µM ATP released per flash, 130 mM NaCl and 20 mM KCl (lower traces).

Inspection of the equivalent circuit of the disc-BLM compound membrane (see figure 1) shows that the current $I_p(t)$ of the pump is distorted by the network formed by the disc-BLM compound membrane. In particular the characteristic time constant $\tau_o = (C_m+C_p)/(G_m+G_p)$ of the equivalent circuit is introduced in the measured signal (Fahr et al. 1981).

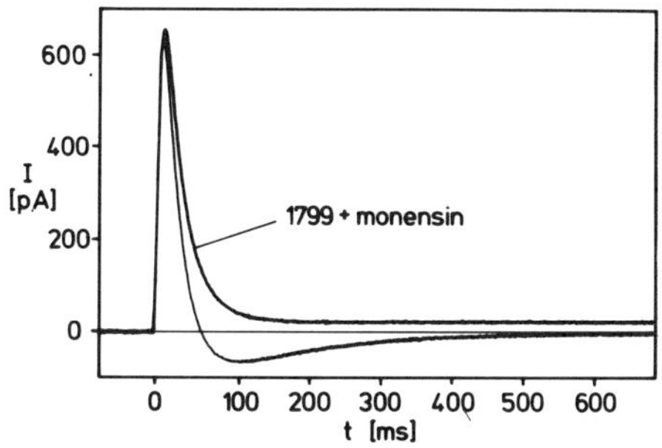

Figure 4: Short circuit currents in the absence and presence of ionophores.

With the data shown in figure 3 excellent fits with 3 exponential components were obtained. The slowest time constant τ_3 determined this way has to be assigned to the characteristic time constant τ_o of the equivalent circuit (Fendler et al., 1987). This is further supported by the experiment shown in figure 4. If the membrane conductivity is increased by the addition of ionophores, the slow negative component of the Na^+ current disappears.

The fast two time constants τ_1 and τ_2 reflect charge translocations during the reaction cycle of the enzyme. Because τ_3 is much larger than τ_1 and τ_2 distortion of the pump current by the network is small and will be neglected.

The two intrinsic time constants of the Na^+K^+-ATPase found in the photoinduced current suggest a two step first order reaction sequence with an electrosilent transition followed by an electrogenic one. The current transient generated by this model is given by :

The delay time D was introduced to account for the kinetics of ATP release from caged ATP. This function was fitted to the data at different ATP concentrations.

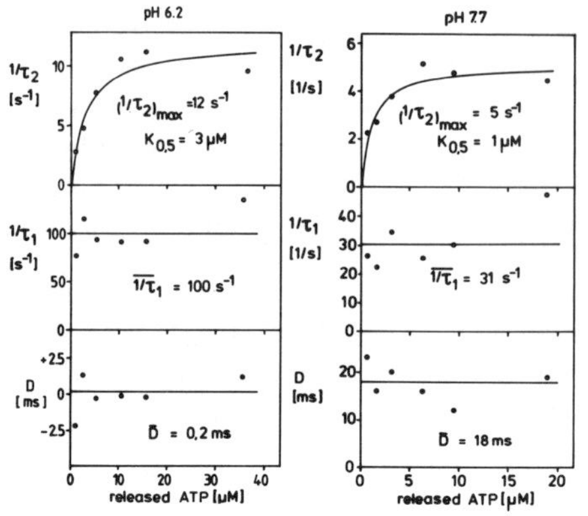

Figure 5: ATP dependence of the fit parameters of equation 1. The ATP dependence of the inverse time constants $1/\tau_1$ and $1/\tau_2$ and of the delay D (see equation 1) of the UV flash induced current was measured at pH 6.2 and pH 7.7. The electrolyte contained 130 mM NaCl, 20 mM KCl, 3 mM $MgCl_2$ different amounts of caged ATP and 25 mM imidazole at pH 6.2 and pH 7.7, respectively. For $1/\tau_2$ the solid line represents a fit of a Michaelis Menten type concentration dependence to the data. For $1/\tau_1$ and D the solid line is a horizontal line at the average value $1/\tau_1$ and D respectively.

The saturating behaviour of $1/\tau_2$ shows that substrate binding is involved in this step and that it consequently has to precede the reaction characterized by τ_1. The situation is, however, complicated by the fact that caged ATP is a competitive inhibitor of the Na^+K^+-ATPase (Forbush, 1985 and Nagel et al., 1987).

The arguments given above support a kinetic model consisting of a fast electrogenic step with rate constant 100 s^{-1} (pH 6.2) which is preceded by a non electrogenic slow process which takes place in approximately 80 ms and involves binding of the substrate. The similarity of the results in the absence and presence of K^+ suggests that the current is generated during the Na^+ dependent steps of the first half cycle:

$$E_1\text{+ATP} \rightarrow E_1\cdot\text{ATP} \rightarrow E_1P \rightarrow E_2P$$

As already discussed τ_2 is attributed to substrate binding. τ_1 must then be due to reactions of the phosphoenzymes. Assignment of the observed transition by comparison with the rate constants reported in the literature is not possible because of differences in enzyme provenience and electrolyte composition. The decision which of the phosphoenzyme transitions is described by τ_1 has therefore to remain open.

CONCLUSIONS

Detailed information about charge transport in ion translocating membrane proteins may be obtained with the method presented in this article. The peak currents observed in the Na^+K^+-ATPase disk-BLM system reflect early events in the reaction cycle, probably the E_1 -> E_2P half-cycle. This is illustrated for the Na^+K^+-ATPase by the arrow in figure 6.

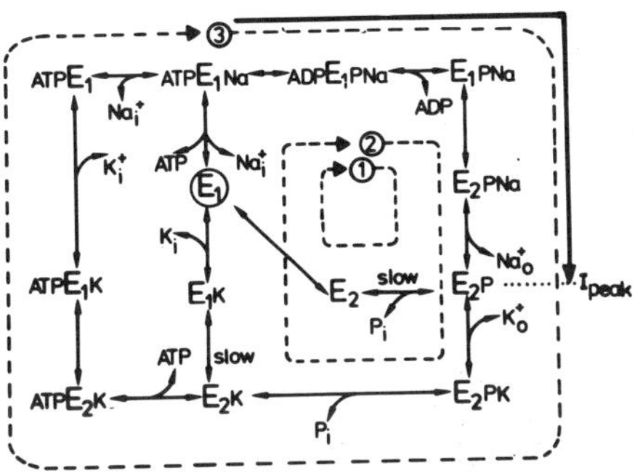

Figure 6: Reaction cycle of the Na^+K^+-ATPase (modified scheme from Karlish et al., 1978). Loop 1 shows the Na^+-ATPase cycle; Loop 2 the Na^+K^+-ATPase cycle at low ATP concentrations; loop 3 the physiological case, where the low affinity ATP binding site is saturated.

In addition, stationary currents give complementary information about slow steps of the cycle and the overall electrogenic activity of the enzyme. These experiments have been used to show that the Na^+K^+-ATPase functions as an electrogenic pump in the presence and absence of K^+ and that the H^+K^+-ATPase is not electrogenic.

From fast current measurements it is concluded, that the electrogenic step observed for the Na^+K^+-ATPase (rate constant 100 s^{-1}) is one of the phosphoenzyme transitions: E_1 -> E_1P -> E_2P. The preceding nonelectrogenic slow step (80 ms)[1] probably reflects ATP binding and exchange with caged ATP. No evidence has been found for a K^+ dependent electrogenic transport of ions back to the cytoplasmic side.

REFERENCES

Borlinghaus R, Apell HJ and Läuger P (1987) Fast charge translocations associated with partial reactions of the Na,K-Pump: I. Current and voltage transients after photochemical release of ATP. J Membrane Biol 97: 161-178

De Weer P (1986) The electrogenic sodium pump: Thermodynamics and kinetics. Progress in Zoology 33 (Suppl)

Dixon JF and Hokin LE (1980) The reconstituted (Na,K)-ATPase is electrogenic. J of Biol Chem 255: 10681-10686

Fahr A, Läuger P and Bamberg E (1981) Photocurrent kinetics of purple-membrane sheets bound to planar bilayer membranes. J Membrane Biol 60: 51-62

Fendler K, Grell E, Haubs M and Bamberg E (1985) Pump currents generated by the purified Na^+K^+-ATPase from kidney on black lipid membranes. EMBO J 4: 3079-3085

Fendler K, Grell E and Bamberg E (1987) Kinetics of pump currents generated by the Na^+K^+-ATPase. FEBS Letters, in the press

Forbush B (1985) Rapid ion movements in a single turnover of the Na^+ pump. In: "The Sodium Pump" (Glynn IM and Ellory C eds) Comp of Biologists Ltd, GB: 599-611

Goldshlegger R, Karlish SJD, Rephaeli A and Stein WD (1987) The effect of membrane potential on the mammalian sodium-potassium pump reconstituted into phospholipid vesicles. J Physiol 387: 331-355

Jørgensen PL (1974) Purification and characterization of (Na^++K^+)-ATPase. III. Purification from the outer medulla of mammalian kidney after selective removal of membrane components by sodium dodecylsulphate. Biochim Biophys Acta 356: 36-52

Kaplan JH, Forbush III B and Hoffmann JF (1978) Rapid photolytic release of adenosine 5'-triphosphate from a protected analogue: Utilization by the Na:K pump of human red blood cell ghosts. Biochemistry 17: 1929-1935

Karlish SJD, Yates DW and Glynn IM (1978) Conformational transitions between Na^+-bound and K^+-bound forms of (Na^++K^+)-ATPase, studied with formycin nucleotides. Biochim Biophys Acta 525: 252-264

Nagel G, Fendler K, Grell E and Bamberg E (1987) Na^+ currents generated by the purified Na^+K^+-ATPase on planar lipid membranes. Biochim Biophys Acta 901: 239-249

The Ion Pumps: Structure, Function, and Regulation, pages 233–238

VOLTAGE CONTROL OF THE SODIUM-POTASSIUM PUMP IN THE LOBSTER STRETCH RECEPTOR NEURONE

Wolfgang Grampp and Staffan Skogvall

Department of Physiology and Biophysics, University of Lund, Sölvegatan 19, S-223 62 Lund, Sweden

INTRODUCTION

In previous investigations it could be shown that in slowly and rapidly adapting lobster stretch receptor neurones an electrogenic Na-K pump is taking part in the regulation of impulse firing both by controlling intracellular ion concentrations and by producing a hyperpolarizing membrane current component (Gestrelius and Grampp, 1983; Edman et al., 1986; Edman et al., 1987). It could also be shown that the activation of the pump by intracellular Na^+ is describable by Garay-Garrahan kinetics (Garay-Garrahan, 1973) according to which the pump action is initiated after formation of three non-interacting Na^+-pump site complexes. Finally, it could be shown that the pump´s maximum Na^+ extrusion capacity is about two times higher in slowly than in rapidly adapting receptors. This last fact could be explained as an expression of different pump site densities in the two receptor types.

In all of the above investigations it was assumed that the pump function was independent of the transmembrane electric field. This assumption was founded on the notion that the pump current production was unaffected by voltage variations in a normal working range between -65 mV (resting voltage) and -45 mV (average voltage during repetitive impulse firing). However, from more recent investigations it has become evident that, when studied in a wider voltage range, the pump function may be distinctly voltage dependent both in living cells (Gadsby et al., 1985; Lafair and Schwarz, 1986; Eisner et al., 1987) and in lipid vesicles reconstituted with purified Na,K-ATPase (Goldschlegger et al., 1987). In view of these observations it was decided to re-examine the question of whether the Na-K pump, or its current production, is voltage dependent in the lobster stretch receptor neurone. Below it will be shown that such a voltage dependence seems to exist and that it may have implications for the regulation of the cell´s impulse activity.

METHODS

All experiments were performed on isolated slowly adapting lobster (Homarus gammarus) stretch receptor neurones. During the experiments the preparations were kept in an artificial sea water solution at 18°C. For electrophysiological measurements, the cells were impaled with double-barrelled microelectrodes connected to conventional electronic recording devices. The electrodes were filled with 3 M KCl or, alternatively, with 5 M NaCl in order to raise the intracellular Na^{+} concentration. For further technical details, see Edman et al. (1987).

RESULTS

For a determination of the relationship between membrane voltage and activity of the Na-K pump, estimations of the pump activity were obtained from measurements of a pump current production that could be either abolished by ouabain (5 mM) or enhanced by intracellular Na^{+} accumulation.

Tests Based on Pump Current Abolition

In tests using pump current abolition as a means of determining the voltage dependence of the pump function, cells were treated with tetrodotoxin (480 nM), tetraethylammonium (10 mM), 4-aminopyridine (0.5 mM), and cesium (2 mM) in order to eliminate gated membrane currents, and subjected to a series of 3 s long voltage-clamp pulses of various heights and either polarity with respect to resting voltage level. The initial and final values of the resulting membrane current responses were then substituted into so-called 'instantaneous' and 'delayed' current-voltage relationships. Relationships of this kind were determined both in control conditions and 5-7 min after ouabain application, and they were also obtained in two groups of cells of which one was impaled with KCl-filled, and the other with NaCl-filled microelectrodes. From the measurements it appeared (a) that the application of ouabain leads to a loss in outward membrane current (pump current); (b) that the outward current loss is greater in cells whose internal Na^+ concentration has been raised by impalement with NaCl-filled microelectrodes; and (c) that, independently of the microelectrode electrolyte, the loss in outward membrane current increases with membrane depolarization. This last fact is illustated in Fig. 1, in which there is also a theoretical relationship predicted by a previously designed receptor model (Edman et al., 1987) on the assumption that the Na-K pump has no voltage dependence and that, in agreement with previous findings (Edman et al., 1986), a 5-7 min pump blockage changes the internal Na^+ and K^+ concentrations by +10 mM, and -10 mM, respectively. It is evi-

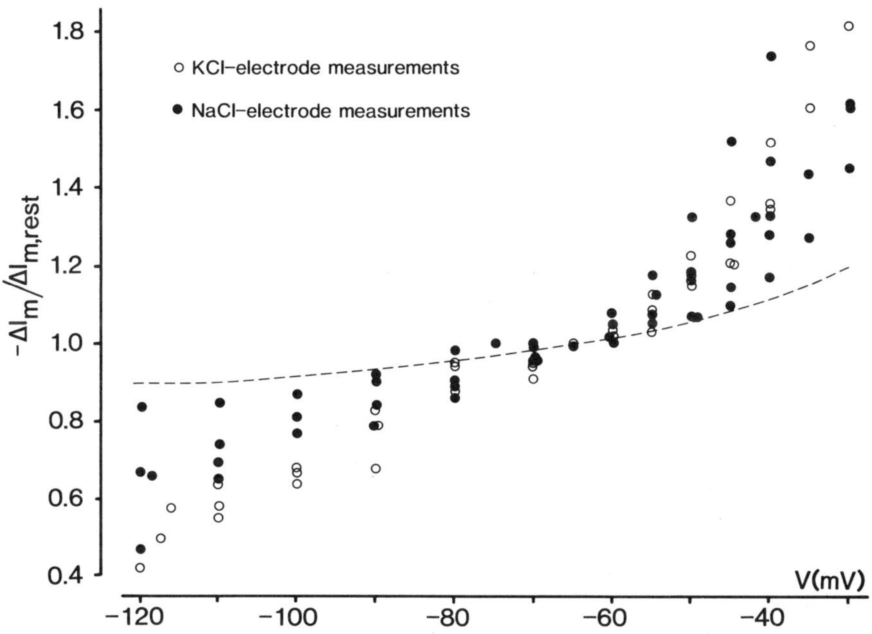

Figure 1. Relationship between membrane voltage, V, and ouabain induced loss of outward membrane current [relative to loss at resting membrane voltage (-65 mV)], $-\Delta I_m/\Delta I_{m,rest}$, in three cells impaled with KCl-filled microelectrodes and five cells impaled with NaCl-filled microelectrodes. The interrupted line represents a theoretical relationship based on assumptions as explained in the text.

dent that the computed relationship is significantly less steep than the measured relationships which, therefore, can be taken to indicate that the ouabain-sensitive pump current production is voltage dependent in the present preparation.

For an estimation of the function of time in the voltage dependence of the pump current production, ouabain-induced outward current losses inferred from 'instantaneous' and 'delayed' current-voltage relationships were compared with each other. It was found that there was no significant difference between the measured current losses. From this it may be concluded that the voltage dependence of the pump current production is either instantaneous or so fast that the bandwidth of the recording device (DC→500 Hz) is insufficient for its temporal resolution.

Tests Based on Pump Current Activation

In tests using pump current activation as a means of determining the voltage dependence of the pump function, cells were loaded with Na^+ by high-frequency (~50 Hz) impulse activation for 2 min. This resulted in an ouabain-sensitive tail of outward current (cf. Edman et al., 1986) whose height after 10 s, when all channel gating had come to an

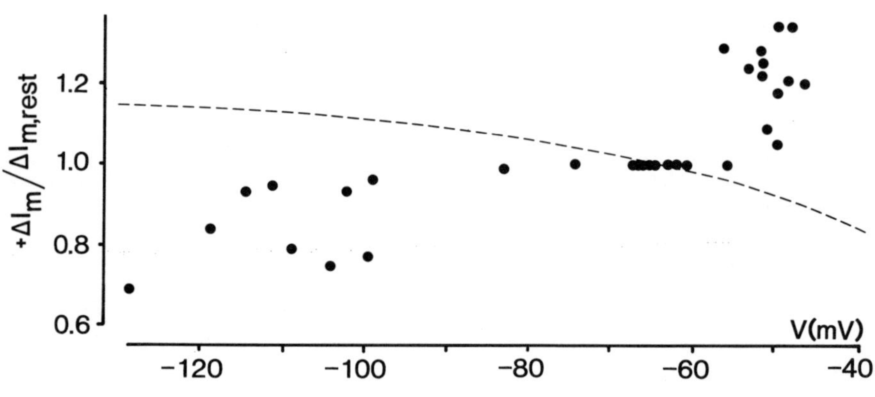

Figure 2. Relationship between membrane voltage, V, and impulse activation-induced increase in outward current (relative to increase at resting membrane voltage), $+\Delta I_m/\Delta I_{m,rest}$, in ten different cells. The interrupted line represents a theoretical relationship based on assumptions as explained in the text.

end, was measured at various voltage levels. The results of such measurements are shown in Fig. 2 together with a theoretical relationship predicted by the same receptor model as mentioned above on the assumption that, in agreement with the absolute values of the observed outward current increases, the internal Na^+ and K^+ concentrations are changed by +17.5 mM, and -17.5 mM, respectively. It is seen that the theoretical relationship and its empirical counterparts run at significantly different slopes, and that the experimental findings are thus in support of the notion that the pump current production is voltage dependent with a distinct tendency to increase with increasing membrane depolarization in the present preparation.

DISCUSSION

In the present preparation the voltage dependence of the Na-K pump implies that the pump current production is increased by a factor of about 2.5 in a voltage range from -130 to -30 mV. This factor is similar to that pertaining to pump current measurements in heart cells (Gadsby et al., 1987), but it differs distinctly from those pertaining to Na^+ extrusion measurements in oocytes (Eisner et al., 1987) or reconstituted Na,K-ATPase preparations (Goldschlegger et al., 1987). The latter might be because of fortuitous circumstances. However, it is also consistent with the possibility of a genuine difference in voltage dependence for Na^+ extrusion and pump current production due, for example, to the forward and backward reactions of the pump cycle being affected differently by the membrane voltage (Eisner et al., 1987).

The mechanisms underlying the voltage dependence of the Na-K pump in the lobster stretch receptor neurones are not clearly identified in the present investigation. However, as in reconstituted Na-K pumps (Goldschlegger et al., 1987), they might involve a voltage sensitivity of the rate constant of the $E_1P(3Na) \rightarrow E_2P3Na$ reaction. The fact that the voltage dependence of the pump current production is very fast should suggest that it is not associated with any voltage-induced conformational changes of the pump molecule. Nor does it seem to be related to the way Na^+ is bound to the pump molecule, since changing the intracellular Na^+ concentration by the alternate use of KCl and NaCl-filled microelectrodes turned out to have no effect on the realationship between membrane voltage and ouabain-induced pump current loss.

Functionally, the voltage dependence of the Na-K pump may be of importance for the receptor neurone, when it is subjected to long-term stimulation. This is evident from model studies which show that a 30% increase of the resting pump current production (as would result from an average membrane depolarization from -65 mV to -45 mV) shortens the time for the cell´s firing adaptation from about 10 min to 5 min. Hence, the voltage dependence of the Na-K pump may not only be an unavoidable consequence of the pump´s mode of operation but, also, a property which enables the cell to quickly adapt to new functional situations.

REFERENCES

Edman Å, Gestrelius S, Grampp W (1986). Transmembrane ion balance in slowly and rapidly adapting lobster stretch receptor neurones. J Physiol 377:171-191.

Edman Å, Gestrelius G, Grampp W (1987). Analysis of gated membrane currents and mechanisms of firing control in the rapidly adapting lobster stretch receptor neurone. J Physiol 384:649-669.

Eisner DA, Valdeolmillos M, Wray S (1987). The effect of membrane potential on active and passive sodium transport in _Xenopus_ oocytes. J Physiol 385:643-659.

Gadsby DC, Kimura J, Noma A (1985). Voltage dependence of Na/K pump current in isolated heart cells. Nature 315:63-65.

Garay RP, Garrahan PJ (1973). The interaction of sodium and potassium with the sodium pump in red cells. J Physiol 231:297-325.

Gestrelius S, Grampp W (1983). Impulse firing in the slowly adapting stretch receptor neurone of lobster and its numerical simulation. Acta physiol scand 118:253-261.

Goldschlegger R, Karlish SJD, Rephaeli A, Stein WD (1987). The effect of membrane potential on the mammalian sodium-potassium pump reconstituted into phospholipid vesicles. J Physiol 387:331-355.

Lafair AV, Schwarz W (1986). The voltage dependence of the rheogenic Na^+/K^+ ATPase in the membrane of oocytes of _Xenopus laevis_. J Membr Biol 91:43-51.

The Ion Pumps: Structure, Function, and Regulation, pages 239–246

HOW THE SODIUM PUMP USES ATP TO RENDER PUMPING OF BOTH Na AND K EFFECTIVE

Wilfred D.Stein

Dept of Biochemistry,
Hebrew University,
JERUSALEM, 91904, I S R A E L

IT IS NOT PUMP ASYMMETRY THAT DRIVES ACTIVE TRANSPORT

We all know (or think we know) that an ion pump works by loading its substrate at a high affinity site at the side from which pumping occurs, and discharging the substrate from a low affinity site at the other face of the membrane. The textbooks teach us that the energy from hydrolysis of ATP is used to bring about this change of the site affinities, and that this is what drives pumping. But Hill and Eisenberg (1981) showed us that this cannot be the whole story or even the major part of it. They showed that the release of the ATP's free energy can occur anywhere around the pumping cycle and yet pumping will still occur. Consider the conventional scheme for the Na/K ATPase (Fig 1):

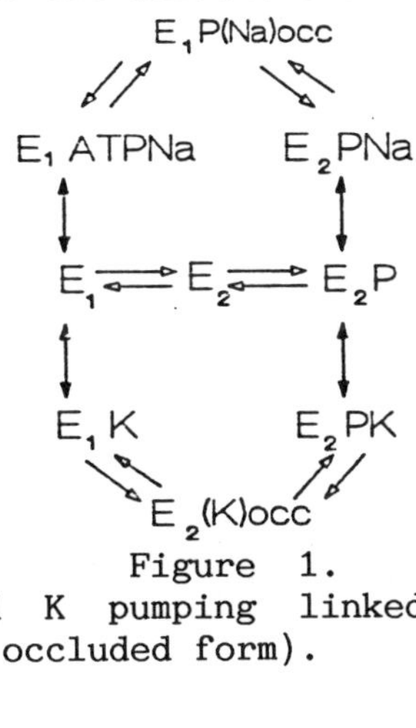

Figure 1. The conventional scheme for Na and K pumping linked to ATP splitting. ("occ" signifies an occluded form).

The upper half of the scheme show the pumping of sodium linked to the phosphorylation of the enzyme, the lower half, potassium pumping linked to dephosphorylation. Nowhere in the scheme as written is it necessary to state whether there is a change of pump affinity for either ion. Yet the scheme will pump sodium and potassium in opposite directions against their respective concentration gradients, if the splitting of ATP is tightly coupled to the ion movements. It is this tight coupling, Hill and Eisenberg emphasise, that drives the pump. We can now measure directly just how tight this coupling is in the case of the Na + K ATPase (Karlish and Stein, 1982,a). Jencks (1980) has shown us how the erection of appropriate energy barriers and their directed dismantling can ensure tight coupling.

A little reflection will convince one that pumping will occur, in a <u>strictly coupled</u> pump, even if the pump's affinities for ions are quite contrary to those ordinarily found. A pump can pump an ion which binds to it at a low affinity site at the side from which pumping is occurring (the "whence" side in what follows), and discharge it from a high affinity site at the side to which pumping is occurring (the "whither" side), provided only that the cycle of fig 1 continues to take place, that is, provided that the overall equilibrium between ATP, ADP and Pi, and the prevailing ion gradients are such as to allow ATP splitting to continue. It is thermodynamics that determines whether or not pumping will occur, not kinetics.

A SIMPLE CARRIER CAN BE ASYMMETRIC AND YET IT CANNOT PUMP

Indeed, one can easily demonstrate the falsity of the assumption that an asymmetry of affinities must bring about pumping. We now know that many simple carrier systems, which cannot bring about the active transport of their substrates, are asymmetric in their kinetics as well as in their structure. Two well-studied examples are the glucose transport system and the uridine transport system of the human red blood cell (Baker and Widdas, 1973; Cabantchik and Ginsburg, 1977). Interestingly, and revealingly, both these systems develop their asymmetry on cold-storage of the cells, and they do this without developing any capacity for pumping their substrates

(Weiser et al, 1983; Plagemann and Wohlhueter,1984). There is good evidence that it is the accumulation within the cells of some competitive inhibitor of transport that induces these asymmetries. Consider what will happen if a non-penetrating competitive inhibitor of transport is present at the cytoplasmic face of a carrier system. Within the cell, substrate and the inhibitor will compete. The effect on the efflux of the substrate is to raise the effective Km, leaving Vmax unchanged. For influx, on the other hand, substrate and inhibitor do not compete - they are on opposite sides of the membrane. The Km for influx is unchanged and the system becomes asymmetric in its kinetics - the original high affinity at the extracellular face, but a low affinity at the cytoplasmic surface. Vmax for influx is, however, reduced. The competitor binds some of the carrier at the cytoplasmic face and holds it there, trapping it, and taking it out of the pool of carriers that are available for substrate transport, thereby reducing the Vmax. Even at the highest concentrations of substrate, the effect of this inhibition at the cytoplasmic face cannot be overcome. In all circumstances, the ratio of Vmax to Km must be the same for efflux as it is for influx, and this occurs by the raising of the Km for efflux, while Vmax for influx is lowered.

The kinetic asymmetry of a carrier system depends on the difference in free energy between the carrier conformations that take part in transport. It is easy to show (Stein, 1986) that, for a carrier in which the loaded forms reorient across the membrane faster than do the unloaded forms (and this is true for the majority of the carrier systems), the asymmetry depends on the difference in free energy between the two conformations of the unloaded carrier. This is exactly true for the important exchange-only carriers. Since for all substrates that ride on the same carrier, the unloaded forms are the same, the kinetic asymmetry for all such substrates is identical, a strong result. This means that whatever are the intrinsic affinities for carrier and substrate at any face of the membrane, the ratio of affinities that a steady-state kinetic experiment will measure will be determined only by the free energy difference between the two forms of the free carrier. (In the case of a carrier plus competitive inhibitor system, the free energy difference takes into account the overall free energy change for the binding of competitive inhibitor to the carrier at the prevailing concentration of the inhibitor, as well as the free energy difference between the unloaded forms of the carrier.) Later we shall use an important corollary of the main result : If a carrier system behaves symmetrically in its kinetics, the two forms of its unloaded carrier do not differ in free energy i.e. they are iso-energetic.

WHAT PUMP ASYMMETRY DOES IS TO RENDER A PUMP EFFECTIVE

What does all this have to do with ion pumping? We have just shown that asymmetry does not in itself lead to pumping. Yet we know that ion pumps are asymmetric. Is this fact of significance for pumping? The answer is : Yes! Asymmetry increases the effectiveness of an ion pump. While it it true that a pump will pump with ion affinities low at the whence side and high at the whither, it will operate very slowly under these conditions. Much of the pump enzyme will be trapped unloaded at the whence side, unable to take part in the pumping process. Much of the time the pump will return loaded with its substrate from the whither side, having shuttled substrate fruitlessly across the membrane. (Both these situations waste pumping time not pumping energy (Honig and Stein, 1978). In a well-coupled pump, an ion that returns along the path from the whither to the whence side will resynthesise ATP by reversal of the splitting process. No energy is lost, only time.)

Time is of significance since our ion pumps are not isolated systems. They are embedded in natural membranes which are leaky to the pumped substrates. At the steady-state, the pumped substrate will leak back down the built-up concentration gradient just as fast as pumping occurs. This wastes the energy that was invested into ion pumping. A pump will be effective if it is designed to operate rapidly, and for this it must have that high affinity at the whence side and low affinity at the whither side that we have always known was important for pumping. We now see that this aspect of pump design is required to render ion pumping effective by speeding up the pump cycle, a kinetic problem not a thermodynamic one, and neglected in the Hill-Eisenberg analysis.

HOW IS ASYMMETRY MAINTAINED IN THE POTASSIUM HALF OF THE PUMP CYCLE?

Consider, first, the simpler, potassium half of the pump cycle. We have shown (Karlish and Stein, 1982 a,b) that, in the absence of sodium and hence of ATP splitting, the system behaves as a simple carrier for potassium ions. Net transport is slower than exchange transport so that the unloaded carrier reorients more slowly than does the potassium - loaded form. In the absence of ATP and Pi, the system behaves symmetrically, kinetically, with high affinity sites for potassium being demonstrable at both faces of the membrane. This must mean (Stein, 1986) that the two carrier conformations E1 and E2 must be more-or-less isoenergetic. But the intrinsic affinities of E1 and E2 for potassium are very different, being high at the extracellular face and low at the cytoplasmic face, by about two orders of magnitude. These affinities are just in the ratio that effective pump design would require (high at the whence side for potassium - the extracellular face; low at the whither side - the cytoplasmic face). How are these intrinsic affinities realised during potassium pumping, when both ATP and Pi are present?

First note that the forms E2 and E2P are not very different in free energy, the phosphate bond here being a low energy bond. The effective affinity of K for the system at side 2 is much the same whether or not Pi is present. Thus the asymmetry of potassium binding at its whence and whither sides does not depend on the difference of free energy between E1 and the phosphate form E2P, but rather on the differential binding of ATP to these two forms. It is clear that ATP binds to E2 forms with a low affinity, to E1 forms with a high affinity. Hence the addition of ATP stabilises the E1 form, allowing its low affinity for potassium to be realised. We have the operation of the thermodynamic "box" depicted in fig 2. Whether one travels from E2P to E1 by binding ATP poorly and K well, or from E1 to E2P by binding K poorly and ATP well, the initial and final states are the same, so the free energy difference is independent of route. Thus, if ATP binds poorly to E2P and well to E1, K must in like but reverse ratio, bind poorly to E1 and well to E2P. Essentially, ATP acts as a competitive inhibitor of potassium movement, its action being localised at the cytoplasmic face, just as if it was a non-penetrating competitive inhibitor as we discussed above. (Note that the ATP remains at the cytoplasmic face (side 1) although it is binding to forms E2P and E1 that bind K at, respectively, side 2 and side 1).

This role for ATP is merely a regulatory one, and it is not split in the process. (If ATP is not split, the energy of its hydrolysis cannot be used for pumping, thus the change of affinity - important for effective pumping - is not what brings about pumping.) The asymmetry in binding of potassium is given by the free energy difference for the binding of ATP to the two forms E1 and E2P of the potassium carrier (here the Na/K ATPase).

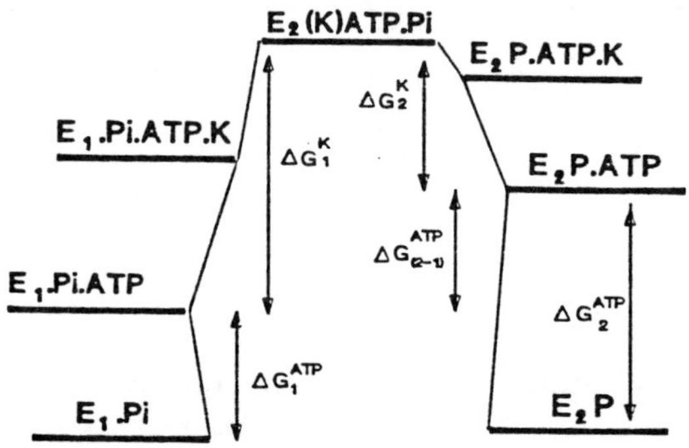

Figure 2. Energy levels for forms involved in the potassium half of the Na - K pump. Each horizontal line depicts the energy level (in the standard state) of the components written just above the line. The arrows with deltas are the free energy changes of the steps indicated and give the overall binding constants for the ion or for ATP at each face of the membrane.

HOW IS ASYMMETRY MAINTAINED IN THE SODIUM HALF OF THE PUMP CYCLE?

The cytoplasmic face of the pump is the whence side for sodium but the whither side for potassium. Thus we cannot use ATP binding to ensure that sodium binds more avidly at its whence face i.e., to the E2P form. We used up that trick already to stabilise the E1 form. The differential affinity of ATP for the E1 form means a lower affinity for ALL co-substrates to this form. Some other design principle is needed to generate a high affinity for sodium at the cytoplasmic face. The solution used is a kinetic one. The complex Na.ATP.E1 takes a different route to get to E2 than does the complex K.ATP.E1. The ion which is bound to E1.ATP determines the fate of the E1+ATP complex. Ion binding is the fate-determining step!

K.ATP.E1 merely undergoes a conformation change to yield the occluded form ATP.E2(K). In contrast, Na.ATP.E1 undergoes, of course, a chemical change with eventual phosphorylation of the enzyme and its reorientation to yield Na.E2.Pi + ADP. The reaction of phosphorylation followed by reorientation is heavily over to the side of Na.E2.Pi, so that the E2 form is stabilised over the E1 form, or E1 destabilised over E2 (Fig 3). The binding of Na to E1 is thus followed kinetically by a change which removes the Na.E1 complex as fast as it is formed. The binding of sodium to the carrier is heavily stabilised and the system behaves with a high affinity for sodium. The standard state free energy of E1 + ATP is, of course, far higher than that of E2P + ADP (Fig 3). Since, as we have seen, E1 and E2P are more-or-less isoenergetic, ALL of the chemical energy in splitting ATP to ADP + (Pi bound to E2) can be used to stabilise the binding of sodium to the pump from side 1. Sodium can thus bind with the required high affinity at its whence side.

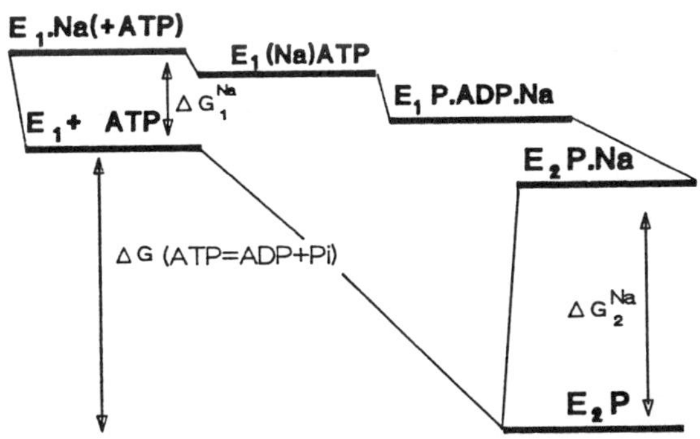

Fig 3. Energy levels in the sodium half of the Na-K pump cycle. Conventions as in Fig 2.

Splitting of ATP drives the whole pump cycle thermodynamically, but effective pump design ensures that the splitting of ATP takes place in a particular step which ensures the satisfaction of the kinetic requirement for a high affinity for sodium at the cytoplasmic face, the whence side for that ion. In contrast, binding of ATP without splitting reveals the high affinity for potassium at the extracellular face, the whence side for potassium.

REFERENCES

Baker, G F, Widdas, W F (1973). The asymmetry of the facilitated transfer system for hexoses in human red cells and the simple kinetics of a two component model. J Physiol (London) 231: 143-165.

Cabantchik, Z I, Ginsburg, H (1977). Transport of uridine in human red cells. Demonstration of a simple carrier. J Gen Physiol 69: 75-96.

Hill, T L, Eisenberg, E (1981). Can free energy transduction be localized at some crucial part of the enzymatic cycle? Quart Rev Biophys 14: 463-511.

Honig, B, Stein, W D (1978). Design principles for active transport systems. J Theor Biol 75: 299-305.

Jencks, W P (1980). The utilization of binding energy in coupled vectorial processes. Adv Enzymol 51: 75-106

Karlish, S J D, Stein, W D (1982a). Passive rubidium fluxes mediated by Na-K-ATPase reconstituted into phospholipid vesicles when ATP- and phosphate-free. J Physiol (London) 328: 295-316.

Karlish, S J D, Stein, W D (1982b). Effects of ATP or phosphate on passive rubidium fluxes mediated by Na-K-ATPase reconstituted into phospholipid vesicles when ATP- and phosphate-free. J Physiol (London) 328: 317-331.

Plagemann, P G W, Wohlhueter, R M (1984). Kinetics of nucleoside transport in human erythrocytes : alterations during blood preservation. Biochim Biophys Acta, 778: 176-184.

Stein, W D (1985) Transport and Diffusion across Cell Membranes, Academic Press, Orlando, Fla.

Weiser, M B, Razin, M, Stein, W D (1983). Kinetic tests of models for sugar transport in human erythrocytes and a comparison of fresh and cold-stored cells. Biochim Biophys Acta 727: 379-388.

The Ion Pumps: Structure, Function,
and Regulation, pages 247–254

THE GASTRIC H^+,K^+-ATPase

Pia Lorentzon, David Scott, Steve Hersey, Bjorn Wallmark, Edd Rabon and George Sachs

UCLA/Wadsworth VA, Emory U., Hassle A.B.

INTRODUCTION

The transport of hydrogen ions across biomembranes is catalyzed by specific peptides that recognize H^+ at their transport site. These peptides can be classified into passive and active porters, such as H^+/Na^+ antiport of eukaryotic plasma membranes which belong to the passive group, and the H^+-ATPases of several membrane systems that are active. In this latter category are multisubunit pumps such as the vacuolar ATPases of lysosomes, as well as the single subunit, phosphorylating ATPases of <u>Neurospora</u>, yeast and stomach. The countertransport properties of the H^+,K^+-ATPase of gastric mucosa has made this enzyme a model for correlation of kinetic and transport reactions and this short review will summarise some of what has been learnt about the mechanism of the H^+ and K^+ translocation reactions.

KINETICS

The activity of the ATPase is stimulated by the addition of monovalent cations in the sequence $Tl^+ > K^+ > Rb^+ > NH_4^+ \gg Cs^+ > Na^+, Li^+$, although there is also a basal Mg^+-ATPase activity that is about 5% of the maximal cation stimulated activity. The effect of these cations is biphasic with stimulation and then inhibition by increasing cation concentrations. Figure 1 shows the ATPase activity at 5 uM ATP as a function of changing K^+ or NH_4^+ concentrations. It can be seen that both cations give a biphasic effect, but that the NH_4^+ curve is shifted far to the right.

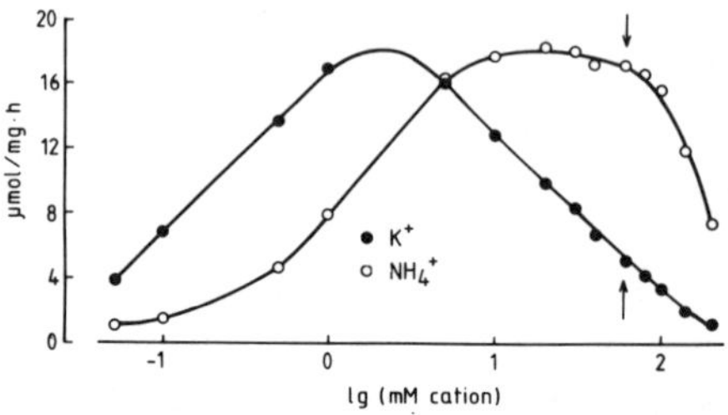

Figure 1. Activity of the H^+,K^+ ATPase as a function of varying K^+ or NH_4^+. The basal Mg^{2+} stimulated ATPase has been substracted as background activity.

There are several possible explanations of this phenomenon, but a likely explanation is that K^+ binds to the H^+,K^+-ATPase enzyme at an inhibitory site with higher affinity than the binding of NH_4^+, as first shown for the Na^+,K^+-ATPase (Post et al, 1972). Inhibition could be due to inhibition of phosphorylation or of dephosphorylation.

With the enzyme in ion-tight vesicles, the addition of K^+ hardly stimulates the enzyme. However, in the presence of ionophores such as gramicidin or nigericin, there is a large stimulation of as much as 50-fold in the most active preparations. Valinomycin gives significantly less stimulation at the optimal K^+ levels for activation by the other ionophores, due to the slow rate of penetration by the valinomycin-K^+ complex as compared to the K^+/H^+ exchange catalyzed by nigericin or gramicidin (Sachs et al, 1976). Thus, in vesicles prepared from hog gastric mucosa, access of K^+ to the luminal surface is rate limiting for ATP turnover. The luminal K^+ site is considered to be in the E_2 form, with cytosolic K^+ site, E_1.

In contrast to the ATPase reaction, pNPPase, a partial reaction of the ATPase does not show effects of either preincubation or ionophores. There appears to be no restriction for the site of K^+ activation of this reaction. The form of the enzyme that hydrolyzes pNPP can be deduced to be a form different from the $E_1 \cdot K$ form since K^+ is not inhibitory to the pNPPase reaction. It is therefore likely to be the $E_2 \cdot K$ conformer that can hydrolyze pNPP. The $E_2 \cdot K$ form does not occur in tight vesicles in the presence of ATP, otherwise the sidedness of K^+ action would not be observed. This leads to the further conclusion that the pathway of hydrolysis of ATP and pNPP might be different, for example, where pNPP does not phosphorylate the enzyme and phosphoenzyme does not allow formation of the $E_2 \cdot K$ form when K^+ is initially present on the cytosolic face.

The gastric ATPase is phosphorylated at an aspartyl phosphate. Labelling is faster at relatively acid pH values. This suggests that H^+ binding to a site on the enzyme is required for the phosphorylation reaction, acting like Na^+ or Ca^{2+} in the Na^+,K^+- or Ca^{2+}-ATPase. Furthermore, the reaction is also faster if ATP is premixed with the enzyme. This in turn indicates that the form of the enzyme that is phosphorylated is an ATP·enzyme conformer, presumably an E_1 configuration and, that prebinding of ATP induces this state of the enzyme (Wallmark et al, 1980). In ion-tight vesicles, the addition of K^+ does not stimulate dephosphorylation, nor does this ion enhance ATPase activity (Stewart et al, 1981), as shown in Figure 2. This result shows that K^+ inhibition is on a dephospho form of enzyme. These data also show that the rate limiting step in vesicular enzyme is K^+ access to the luminal face which is a requirement for rapid turnover of phosphoenzyme. Given that the enzyme exists <u>in situ</u> with about 120 mM K^+ at the cytosolic face, it might be that the enzyme would be inhibited under physiological conditions, but the effect of voltage on the K^+ limb of the pump, as well as the cytosolic ATP concentration, suggests that this is not the case.

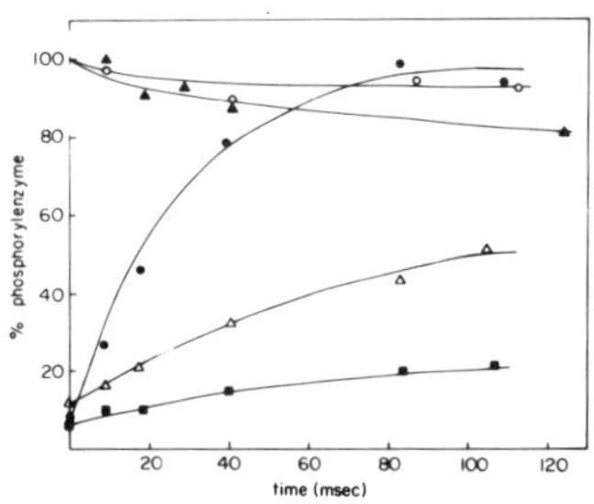

Figure 2. The effect of the addition of K^+ on phosphorylation of the H^+,K^+-ATPase in ion tight vesicles (where K^+ does not penetrate to the luminal face of the enzyme) by 5 uM ^{32}P-ATP, showing inhibition of the rate of phosphorylation by increasing K^+ concentrations.(△,■)

The rate of breakdown of phosphoenzyme in the absence of alkali cations is slow, but accounts for the basal Mg^+-ATPase activity of highly purified preparations. With the addition of K^+, for example, to the luminal face of the enzyme, there is a rapid breakdown of the intermediate, usually seen as a biphasic dephosphorylation with the initial fast rate constant several times greater than overall. The slower rate constant is about equal to the overall rate constant of about 200 per min, and may be assumed to be equivalent to the breakdown of the rate limiting intermediate, such as an H^+ occluded form of phosphoenzyme. A typical experiment is shown in Figure 3, where dephosphorylation is induced by K^+ only in the presence of gramicidin to allow K^+ luminal access.

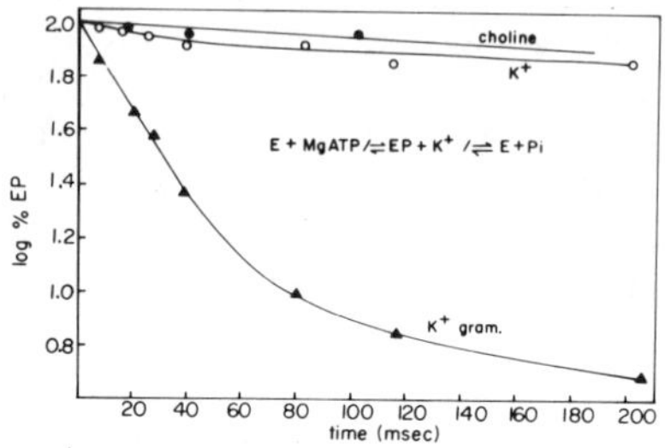

Figure 3. The effect of K^+ on preformed phosphoenzyme. In the upper curves, the spontaneous decay of E-P is shown also with the addition of K^+ in intact, ion-tight vesicles. The lower curve shows the effect of K^+ addition in the presence of gramicidin.

TRANSPORT

H^+ uptake is catalyzed by this enzyme when isolated in vesicular form, upon the addition of Mg^{2+}-ATP in the presence of KCl and valinomycin (Lee et al, 1973) or following preloading with a variety of K^+ salts (Sachs et al, 1976). It may be concluded that K^+ is required at the luminal face for transport of H^+, hence the pump catalyzes an H^+ for K^+ exchange reaction or simply requires K^+ at the luminal face without K^+ transport. The stoichiometry of H^+ transport can be measured by determining the disappearance of H^+ with the addition of Mg^{2+}-ATP at about pH 6.1. It has been shown that 2 H^+ are transported per ATP hydrolyzed, even through the physiologically attainable pH gradient is about 6.5 units (Rabon et al, 1982). Thermodynamically, therefore, the pump has to have variable stoichiometry, contrary to what is believed for other pumps of this class.

This enzyme transports K^+ from lumen to cytosol (Schackmann et al, 1977). By loading vesicles with KCl and measuring ATP turnover that is K^+ stimulated, it is possible to show that with time, ATPase activity falls to basal levels confirming K^+ transport by the ATPase (Rabon et al, 1982) and it is also possible to calculate that the stoichiometry of K^+ transported per ATP hydrolyzed is 2, even at pH 7.4. Thus, the ATPase as isolated in resting vesicles catalyzes a 2 H^+ for 2 K^+ exchange per ATP hydrolyzed. The enzyme also catalyzes a fairly rapid passive K^+/K^+ exchange sensitive to various inhibitors of the ATPase, such as vanadate and SCH 28080.

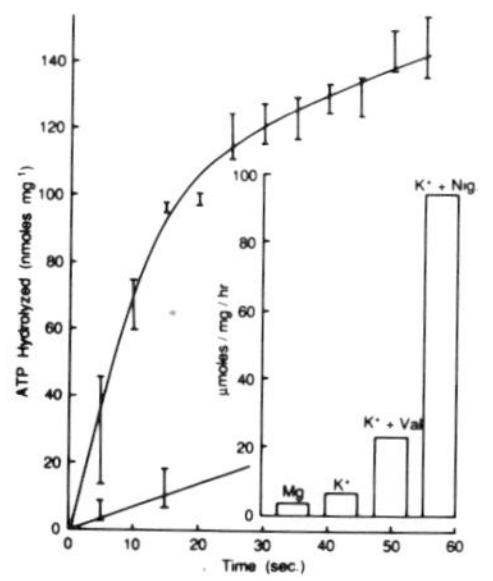

Figure 4. The hydrolysis of ATP by gastric vesicles. The inset of the figure shows that the hydrolysis is limited by the permeabililty of K^+, with nigericin giving a larger stimulation than valinomycin due to the slower rate of val-K^+ turnover. The hydrolysis of ATP when the vesicles are loaded with 150 mM KCl shows a rapid phase due to luminal K^+ stimulation and then returns to the basal, K^+-independent rate allowing an estimate of the K^+/ATP stoichiometry.

From the above, there are 2 possible types of regulation. One is regulation of KCl permeability of the ATPase membrane, and it has been shown that when the enzyme is isolated from stimulated stomach, there is no longer a requirement for either valinomycin or for preincubation in K^+ solutions for H^+ transport (Wolosin and Forte, 1985). It has been argued that this is due to an induction of parallel K^+ and Cl^- conductance, and it has been possible to show that there is the appearance of both K^+ and Cl^- conductances in the stimulated vesicles. However, the presence of a coupled K^+ and Cl^- pathway in parallel to a Cl^- conductance has not been excluded by studies to date. Another possibility is kinetic regulation of the enzyme by, for example, a protein activating factor isolated from gastric supernatants (Bandopadhyay et al, 1987). It is possible to determine the possible relative contribution of these two possibilities. One is to take advantage of the permeability of NH_3^+ across biomembranes and thus activation and transport uncoupling of the ATPase by the addition of NH_4^+. When glucose utilization is compared in unstimulated and stimulated rabbit gastric glands in the presence and absence of NH_4^+, it is found that NH_4^+ increases utilization in unstimulated glands almost to the stimulated level in regular medium. This shows that the major restriction on ATP utilization is indeed ion permeability.

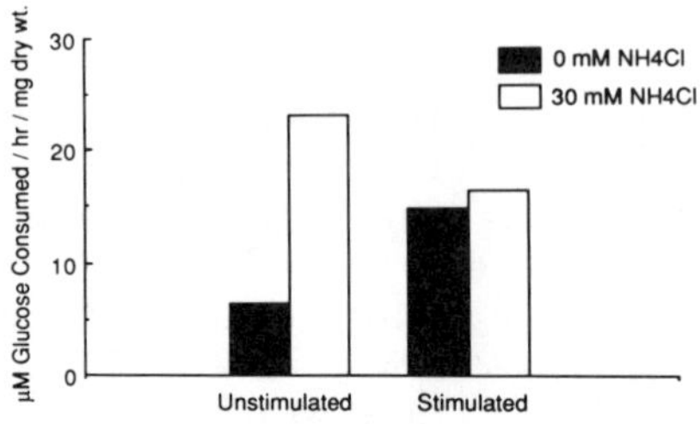

Figure 5. Glucose utilization by rabbit gastric glands in the absence and presence of stimulus, with and without the addition of 30 mM NH_4^+.

In a counter-transport ion pump, voltage dependence of net transport can arise in a variety of ways, dependent on the details of the ion transport pathway. If the stoichiometry of exchange is unequal, then charge will always be translocated by the pump. If the stoichiometry is equal, then net transport will always be electroneutral. However, the effect of voltage depends on whether the individual steps are charge carrying and as to which is rate limiting under the conditions of study. In addition, the pathway for ion transport can be either a single high resistance barrier, with free solution access, or a well preceding the high resistance barrier. In the former case, voltage may affect only the rate of movement of the ion, in the latter, the affinity of the ion for the site may also be altered by voltage (Mitchell and Moyle, 1974, Goldshlegger et al, 1986, Lauger, 1987). In the intact mucosa, it has been known for many years that voltage affects the rate of acid secretion (Rehm, 1965). In contrast, the equivalent stoichiometry of the H^+,K^+-ATPase in vesicles and the absence of voltage difference under transport conditions in vesicles has been difficult to reconcile with intact tissue data. If transport of either side of the enzyme is charge carrying, then applied voltage could affect transport. Kinetic evidence for this possibility can be derived from studies of the KCl dependence of ATPase activity in the presence of nigericin or valinomycin. As expected, from the earlier part of this paper, the ATPase activity is biphasic with respect to KCl concentration in the presence of nigericin, with peak activity seen at about 10 mM K^+. In contrast, in the presence of valinomycin, activity increases monophasically with K^+ concentration, without evidence of inhibition. However, when the protonophore, TCS, is added to abolish the interior positive potential due to valinomycin, inhibition of the ATPase activity is re-established. One interpretation of these data is that the interior positive potential increases dissociation of

K^+ from its inhibitory site on the enzyme, which may be analogous to the K^+ occluded form of the Na^+,K^+ ATPase (Lorentzon et al), Table 1.

TABLE 1	ATPase Activity um/mg/hr 10 mM K^+	175 mM K^+
Leaky vesicles + Val	55	18
Tight vesicles + Val	22	80
Tight vesicles + Val + TCS	60	33

If the sequence $E_2 \cdot K^+ \longrightarrow E_2(K^+) \longrightarrow E_1 \cdot ATP + K^+$ is voltage sensitive, it follows that this sequence is charge carrying and hence the other half of the cycle transporting H^+ is also charge carrying, and, therefore, also voltage sensitive. In the intact cell, the luminal face of the enzyme is positive with respect to the cytosolic face, and hence will be mimiced by valinomycin in the vesicle situation. Accordingly, the K^+ sequence of the pump will not be rate limiting. In intact tissue, therefore, if the H^+ transport steps are rate limiting, lumen negative voltage will accelerate and lumen positive voltage will slow acid secretion. A reasonable postulate for charge translocation in this enzyme would be a histidine residue protonated in the H^+ pumping direction and binding K^+ in the K^+ translocating direction. In partial support of this postulate, the histidine reagent, DEPC, has been shown to inhibit the enzyme by derivatizing a group with a pKa of about 7 (Saccomani et al, 1980).

A model of H^+ and K^+ translocation by the H^+,K^+-ATPase, showing net electroneutrality due to equal but opposite charge translocation by histidine residues is shown in the simplified kinetic scheme below, where the brackets denote rate limiting, occluded steps.

$$\begin{array}{ccccccc} & H^+.his.E_1.ATP & \longleftrightarrow & (H^+.his)E_1\text{-}P & \longrightarrow & E_2\text{-}P.his.H^+ & \\ K^+ \longleftarrow & \updownarrow & & & & \updownarrow & \longrightarrow H^+ \\ & K^+.his.E_1 & \longleftrightarrow & E_2(his.K^+) & \longleftrightarrow & E_2\text{-}P.his.K^+ & \end{array}$$

REFERENCES

Bandopadhyay S, Das PK, Wright MV, Bhattacharyay D, Ray TK (1987). Characteristics of a pure endogenous activator of the gastric H^+,K^+-ATPase system: evaluation of the role as a possible intracellular regulator. J Biol Chem

262:5664-5670.

Goldshlegger R, Karlish SJD, Rephaeli A, Stein WD (1986). Effect of membrane potential on the mammalian sodium/potassium pump reconstituted into phospholipid vesicles. J Physiol (Lond) 387:331-355.

Lauger P (1987). Voltage dependence of sodium calcium exchange. J Memb Biol 99:1-12.

Lee J, Simpson G, Scholes P (1974). An ATPase from dog gastric mucosa: changes of outer pH in suspensions of membrane vesicles accompanying ATP hydrolysis. Biochem Biophys Res Comm. 60:825-832.

Lorentzon P, Sachs G, Wallmark B (1987). Effects of cations on the gastric H^+,K^+-ATPase. J Biol Chem, (sub).

Mitchell P, Moyle J (1974). The mechanism of proton translocation in reversible proton translocating ATPases. In Bronk JR (ed): "Membrane adenosine triphosphatases and transport processes," New York: Biochem Society, pp 91-111.

Post RL, Hegavary C, Kume S (1972). Activation by adenosin triphosphate in the phosporylation kinetics of sodium and potassium ion transport adenosine triphosphatase. J Biol Chem 247:6530-6540.

Rabon E, McFall T, Sachs G (1982). The gastric H^+,K^+-ATPase stoichiometry. J Biol Chem 257:6296-6299.

Rehm WS (1965). Electrophysiology of gastric mucosa in Cl^- free solutions. Fed Proc 24:1387-1395.

Saccomani G, Barcellona ML, Sachs G (1980). Site directed modification of gastric H^+,K^+-ATPase. In Schulz I, Sachs G, Forte J, Ullrich KJ (eds): "Hydrogen ion transport in epthelia," New York: Elsevier, pp 175-184.

Sachs G, Chang HH, Rabon E, Schackmann R, Lewin M, Saccomani G (1976). A non-electrogenic H^+ pump in plasma membrane of hog stomach. J Biol Chem 251:7690-7698.

Schackmann R, Schwartz A, Saccomani G, Sachs G (1977). Cation transport by gastric ATPase. J Memb Biol 32:361-381.

Stewart HB, Wallmark B, Sachs G (1981). Interaction of H^+ and K^+ with the partial reactions of the H^+,K^+-ATPase. J Biol Chem 256:2682-2690.

Wallmark B, Stewart HB, Rabon E, Saccomani G, Sachs G (1980). The catalytic cycle of the gastric H^+,K^+-ATPase. J Biol Chem 255:5313-5319.

Wolosin JM, Forte JG (1985). K^+ and Cl^- conductances in the apical membranes from secreting oxyntic cells are concurrently inhibited by divalent cations. J Memb Biol 83:261-271.

The Ion Pumps: Structure, Function, and Regulation, pages 255–260

INTERACTIONS OF SCH 28080 WITH THE GASTRIC (H^{+}+K^{+})-ATPase.

D. J. Keeling, S. M. Laing and J. Senn-Bilfinger.

Smith Kline & French (Research) Ltd., The Frythe, Welwyn, Herts AL6 9AR, U.K. (D.J.K., S.M.L.). Byk Gulden Lomberg Chemische Fabrik GmbH, D-7750 Konstanz, West Germany. (J.S-B.).

INTRODUCTION

The gastric (H^{+}+K^{+})-ATPase (EC 3.6.1.36) is a membrane bound ion-translocating ATPase showing many similarities with the (Na^{+}+K^{+})-ATPase and the (Ca^{2+})-ATPase (see e.g. Petersen & Carafoli, 1987). Whilst the study of the (H^{+}+K^{+})-ATPase has to some extent followed that of these other two proteins, its central role in gastric acid secretion has provided the impetus towards identifying inhibitors of this ion pump as potential therapies for acid-related diseases in the gastro-intestinal tract.

In recent years two important classes of (H^{+}+K^{+})-ATPase inhibitor have been identified. The first class is that of the benzimidazole sulphoxides, such as omeprazole. These compounds are acid-activated sulphydryl agents, which inactivate the proton pump by reacting covalently with essential thiol groups accessible from extracellular side of the parietal cell membrane (Lorentzon et al., 1985, Keeling et al., 1985 and Keeling et al., 1987). The second class consists of a somewhat heterogeneous collection of hydrophobic amines which inhibit (H^{+}+K^{+})-ATPase activity reversibly and in a manner that is kinetically competitive with the activating cation, K^{+} (Nandi et al., 1983, Im et al., 1984). The activation of the (H^{+}+K^{+})-ATPase by K^{+} occurs at an extracellular (lumenal) site on the enzyme and results in the transport of K^{+} across the membrane. Whilst competitive inhibition of an enzyme is not proof that an inhibitor interacts at the same site as the substrate, we shall start from the simple assumption that this is, in fact, the case. These K^{+}-competitive inhibitors might therefore be considered lumenal "K^{+}-site" inhibitors.

SCH 28080 (8-benzyloxy-3-cyanomethyl-2-methyl-imidazo[1,2a]pyridine, Fig. 1) is a highly potent inhibitor of the gastric (H^++K^+)-ATPase, belonging to this second class (Beil et al., 1986). As such, it represents a useful probe for the study of the mechanism of the (H^++K^+)-ATPase. In this article, we discuss some of the properties of SCH 28080 in relation to those expected for a K^+-site inhibitor.

A

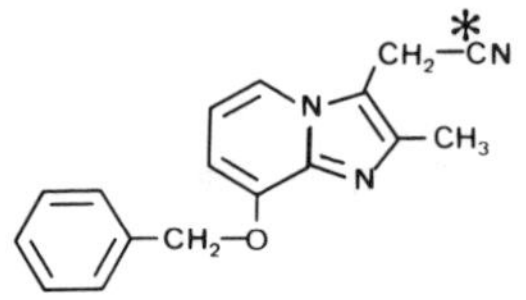

B

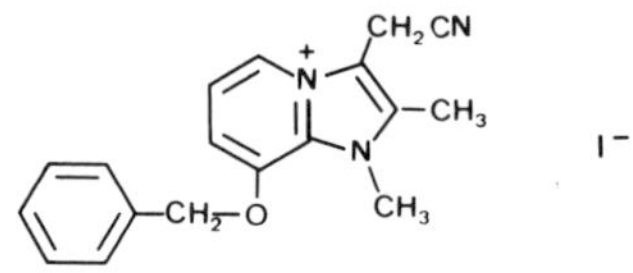

Figure 1. The structures of SCH 28080 (A) and its methylated analogue (B). The position of the radiolabelled atom in [^{14}C]SCH 28080 is shown (*).

METHODS

Intact gastric vesicles were prepared from pig stomach as described previously (Keeling et al., 1987). Where free access to the vesicle lumen was required intact vesicles were lyophilised (Keeling et al., 1985).

ATPase activity in lyophilised vesicles was determined by the liberation of inorganic phosphate (Yoda and Hokin, 1970) at 37°C, pH 6.1 to pH 8.0 in the presence of 2 mM Mg.ATP and a defined concentration of KCl (0.01 to 10 mM). Conventional kinetic methods were used to obtain the K_i value for the competitive inhibition of the K^+-stimulated ATPase activity by SCH 28080, or its methylated analogue.

Binding studies were performed on intact gastric vesicles at room temperature using [^{14}C-cyano]SCH 28080 (synthesised at SK&F). Osmolarity was maintained with 0.3 M sucrose. All incubations contained 2 mM $MgSO_4$. Bound material was isolated by rapid filtration and washing in ice cold buffer. Non-specific binding was defined as that binding that was observed in presence of 100 μM unlabelled SCH 28080 and was subtracted from the total binding.

Phosphoenzyme levels were determined at 0°C following incubation with 5 μM [γ-^{32}P]ATP and 2 mM $MgSO_4$. Covalently incorporated phosphate was isolated by rapid filtration of the TCA precipitable material.

RESULTS

It might be expected that an inhibitor binding to the K^+-site of the $(H^+ + K^+)$-ATPase would, like K^+, be positively charged. SCH 28080 is a protonatable weak base (pK_a = 5.6) and as such is approximately 4% protonated at pH 7. By following the potency (K_i) of SCH 28080 as a function of pH, we have attempted to determine whether the protonated form is indeed active. However interpretation of such data is often complicated by pH-induced kinetic changes to the enzyme. In order to correct for such changes, a close, methylated analogue of SCH 28080 (Fig. 1) was also studied. This compound could not alter its degree of positive charge with pH, due to its quaternary nitrogen. By expressing the potency of SCH 28080 relative to that of the methylated analogue the

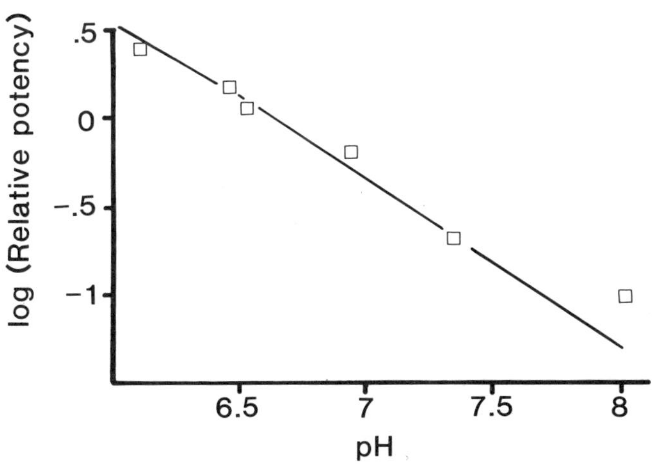

Figure 2. The effect of pH on the potency of SCH 28080 relative to that of its methylated analogue. Relative potency is expressed as ratio of the K_i values for the two inhibitors. The solid line shows the expected change in the protonated form of a compound with a pK_a of 5.6.

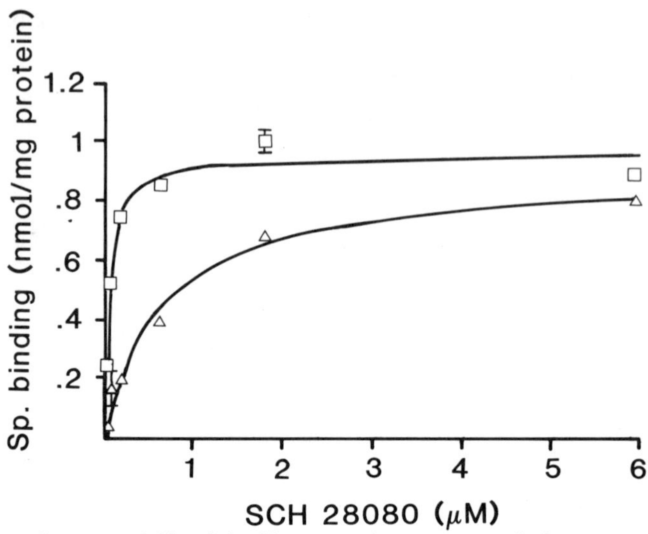

Figure 3. Specific binding to intact vesicles was determined in the presence (□) and absence (△) of 2 mM ATP.

effect of protonation on the inhibitory activity of SCH 28080 could be resolved from any pH-induced changes to the enzyme. Fig. 2 shows that the relative potency of SCH 28080 increased as the pH was lowered commensurate with the proportion in the protonated form. This finding is consistent with the effective inhibition of acid secretion by SCH 28080 *in vivo*, where the acidic conditions present on the lumenal side of the parietal cell membrane will lead to an accumulation of the protonated form of this compound.

With the aid of [^{14}C]SCH 28080, radiolabelled in the cyano- group, binding studies were undertaken. In the presence of 2mM ATP, SCH 28080 bound to intact gastric vesicles in a saturable manner (K_D= 45 $\pm$ 1 nM, SEM, n=4 experiments). The level of binding exceeded that of the catalytic phosphoenzyme by approximately 2-fold. Whether this indicated that there was more than one inhibitor binding site for each functional ATPase molecule, or that the level of phosphoenzyme did not properly reflect the amount of functional ATPase remains unclear. In intact gastric vesicles, K^+ was only able to displace SCH 28080 binding in the presence of the K^+ ionophore nigericin, consistent with binding at a lumenal K^+-site. In the absence of ATP, the binding affinity was reduced approximately 10-fold (Fig. 3). High affinity binding was restored by >1 μM ATP, 20 μM GTP but not by 200 μM [β,γ-CH_2]ATP. Since these low concent-rations of ATP were sufficient to cause phosphorylation of the enzyme, we suggest that, like K^+, SCH 28080 bound preferentially to the phosphoenzyme form of the (H^++K^+)-ATPase.

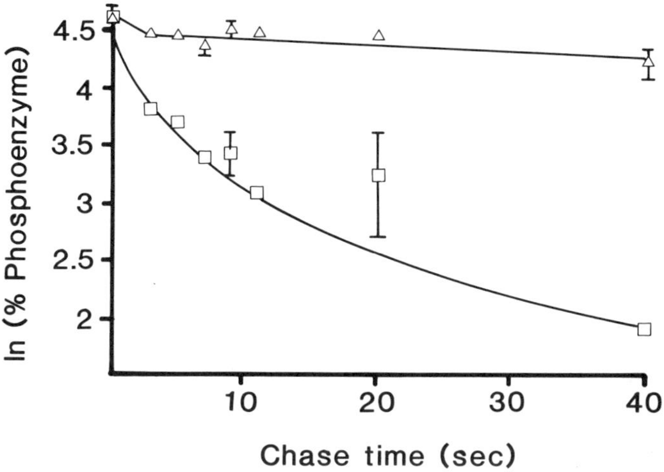

Figure 4. Phosphoenzyme was formed on ice by a 15 sec incubation with 5 μM [γ-^{32}P]ATP. Dephosphorylation was then induced by the addition of 0.5 mM KCl and 2 mM ATP in the presence (△) and absence (□) of 100 μM SCH 28080.

In common with the (Na^++K^+)-ATPase and (Ca^{2+})-ATPase, the catalytic cycle of the (H^++K^+)-ATPase alternates between a phosphorylated and dephosphorylated form. K^+ stimulates ATPase activity by increasing the rate of the dephosphoryl-ation reaction (Wallmark et al., 1980). It might be expected that a K^+-site inhibitor, such as SCH 28080, would block the K^+-stimulated dephosphorylation of the phosphoenzyme. However rapid reaction studies, using a concentration of SCH 28080 (10 μM) several times higher than the K_i value, have failed to show such an effect (Wallmark et al., 1987). By increasing the concentration of SCH 28080 to 100 μM a blockade of the K^+-stimulated dephosphorylation was observed (Fig. 4). A similar effect was found at lower (K_i) concentrations (50 nM) provided that SCH 28080 was allowed time to bind to the enzyme before the K^+ was added. Thus the previously reported absence of this effect may be explained by a relative slow association rate between SCH 28080 and the $(H^+ + K^+)$-ATPase.

CONCLUSION

SCH 28080 appears to act as a slow binding, lumenal K^+-site inhibitor, interacting preferentially in the protonated form with the phosphorylated form of the $(H^+ + K^+)$-ATPase.

REFERENCES

Beil W, Hackbarth I, Sewing K-Fr (1986). Mechanism of gastric antisecretory effct of SCH 28080. Br J Pharmacol 88:19-23.

Im WB, Blakeman DP, Mendelein J, Sachs G (1984). Inhibition of (H++K+)-ATPase and H+ accumulation in hog gastric membranes by trifluoperazine, verapamil and 8-(N,N-diethylamino)octyl-3,4,5-trimethoxybenzoate. Biochim biophys Acta 770:65-72.

Keeling DJ, Fallowfield C, Milliner KJ, Tingley SK, Ife RJ, Underwood AH (1985). Studies on the mechanism of action of omeprazole. Biochem Pharmacol 34:2967-2973.

Keeling DJ, Fallowfield C, Underwood AH (1987). The specificity of omeprazole as an $(H^+ + K^+)$-ATPase inhibitor depends upon the means of its activation. Biochem Pharmacol 36:339-344.

Lorentzon P, Eklund B, Brandstrom A, Wallmark B (1985). The mechanism for inhibition of gastric $(H^+ + K^+)$-ATPase by omeprazole. B.B.A. 817:25-32.

Nandi J, Wright MV, Ray T (1983). Mechanism of gastric antisecretory effects of nolinium bromide. Gastroenterology 85:938-945.

Petersen PL, Carafoli E (1987). Ion motive ATPases. TIBS 12: 146-150.

Wallmark B, Stewart HB, Rabon E, Saccomani G, Sachs G (1980). The catalytic cycle of gastric (H++K+)-ATPase. J Biol Chem 255:5313-5319.

Wallmark B, Briving C, Fryklund J, Munson K, Jackson R, Mendlein J, Rabon E, Sachs G (1987). Inhibition of gastric (H++K+)-ATPase and acid secretion by SCH 28080, a substituted pyridyl(1,2a)imidazole. J Biol Chem 262:2077-2084.

Yoda A, Hokin LE (1970). On the reversibility of binding of cardiotonic steriods to a partially purified (Na++K+)-activated adenosinetriphosphatase from beef brain. Biochem Biophys Res Commun 40:880-886.

The Ion Pumps: Structure, Function,
and Regulation, pages 261–268

MECHANISM OF ACTION OF THE MITOCHONDRIAL PROTON PUMPING ATPase IN ATP SYNTHESIS AND HYDROLYSIS

Harvey S. Penefsky

Department of Biochemistry, The
Public Health Research Institute,
New York, N.Y. 10016, U.S.A

INTRODUCTION

Recent advances in our understanding of the structure of proton pumping ATPases (F_1) are contributing substantially to an emerging description of the molecular mechanism of action of this family of enzymes. Of particular importance has been the elucidation of the amino acid sequence of the ATPase from *Escherichia coli* (Gay and Walker,1981) and a variety of other sources. The enzyme is composed of 5 different types of subunits in the stoichiometry $\alpha_3\beta_3\gamma\delta\varepsilon$. Although it appears that there are 6 adenine nucleotide binding sites on the enzymes from beef heart (Cross and Nalin, 1982), *E. coli* (Wise, et al, 1983) and perhaps chloroplast F_1 (Xue, et al, 1987), the early assignment of one binding site to each α subunit and one site to each β subunit (Cross and Nalin, 1982) is undergoing revision. Gromet-Elhanan and Khananshivili, 1984 were the first to show that subunit β (from R. rubrum) contains 2 adenine nucleotide binding sites. Two sites have since been observed in β subunits from beef heart (Cross, et al, 1987) and from chloroplast F_1 (Xue, et al, 1987). However, it is clear that isolated subunit α from E. coli also contains a nucleotide binding site (Dunn and Futai, 1980). These apparently conflicting observations could be reconciled if there were intersubunit "sharing" of a binding site. Whether or not such sharing exists has yet to be established.

Chemical modification and affinity label studies indicate that the catalytic site is on subunit β. Since this subunit is coded by a single gene (Gay and Walker,1981), it might be supposed that there are 3 catalytic sites per oligomer. Only 2 catalytic sites are directly demonstrable (Rosen, et al, 1979), (Grubmeyer, et al, 1981).The difficulty in showing the presence of 3 catalytic sites may be a consequence of asymmetries inherrent in the molecule. In fact, it seems clear that an understanding of the asymmetries in proton pumping ATPases will be a key element in an eventual understanding of mechanism. One of the interesting and unresolved asymmetries in F_1 is the manner in which the 3 minor subunits of the molecule (γ,δ and ε) interact with the 6 major subunits. There is no indication in the sequence of the molecule (Gay and Walker, 1981) that there can be equivalent interactions between the 6 major subunits and the 3 minor proteins.

RELEVANCE OF SINGLE SITE HYDROLYSIS TO THE MECHANISM OF ATP SYNTHESIS IN OXIDATIVE

Studies on the mechanism of ATP hydrolysis by both the soluble, homogeneous ATPase and the membrane-bound form in beef heart mitochondria have provided insights into the manner in which the enzyme could catalyze ATP synthesis in oxidative phosphorylation. The two key observations are first that the catalytic site exhibits a remarkably high affinity for substrate ATP ($K_A = 10^{12}\ M^{-1}$) and second, that the equilibrium constant for the catalytic step in hydrolysis is near unity. These observations developed from a study of the elementary steps in the mechansim of ATP hydrolysis by the soluble ATPase (Grubmeyer, et al, 1982 and Cross, et al, 1982) and were supported by further studies on the reaction mechanism for the membrane-bound enzyme,that is, submitochondrial particles (Penefsky, 1985a,b,c). The reaction mechanism for either form of the enzyme can be written as follows:

$$F_1 + ATP \underset{1}{\rightleftharpoons} F_1\ ATP \underset{2}{\rightleftharpoons} F_1\ ADP\ Pi \underset{3}{\overset{Pi}{\rightleftharpoons}} F_1\ ADP \underset{4}{\overset{ADP}{\rightleftharpoons}} F_1$$

The reaction shown is for hydrolysis at a single catalytic site on the enzyme. It is possible to study single site hydrolysis because of the very high affinity of the enzyme for substrate and because of negative cooperativity in binding of adenine nucleotides (Grubmeyer and Penefsky, 1981). Experimentally, a 3-fold molar excess of enzyme is incubated with $[\gamma\text{-}^{32}P]ATP$. Such conditions favor the binding of substrate in only one catalytic site. The enzyme-substrate complex is formed in Step 1 and the catalytic step is Step 2. The rate-limiting steps in single site catalysis are product release (Steps 3 and 4). As pointed out above, the two constants of most interest are the equilibrium constants for Steps 1 and 2. They support a model for ATP synthesis in oxidative phosphorylation which proposes that ATP is synthesized from bound ADP and Pi with little or no change in free energy (K_2 is near unity) and that the major requirement for energy in oxidative phsosphorylation is for the release of product ATP ($K_1 = 10^{12}\,M^{-1}$). In other words, the driving force for ATP synthesis is the free energy of binding of product ATP. The model is consistent with and supportive of the binding change mechanism of Boyer and coworkers (Gresser, et al, 1982) as well as suggestions put forth by Jencks (Jencks, 1975).

Several observations suggest that the high affinity catalytic site and other properties of single site catalysis may be widely distributed among the proton pumping ATPases. The values for enzymes from 3 different sources are summarized in Table I.

TABLE 1. Selected Constants for Single Site Catalysis by ATPases From Beef Heart, E. coli and Saccharomyces cerevisiae.

ENZYME*	K_1	K_2	k_3	k_4
Beef Heart				
Soluble	10^{12} M^{-1}	0.5	10^{-3} s^{-1}	10^{-4} s^{-1}
SMP	10^{12} M^{-1}	0.1-0.6	10^{-3} s^{-1}	10^{-4} s^{-1}
Yeast	10^{12} M^{-1}	0.3	10^{-3} s^{-1}	10^{-4} s^{-1}
E. coli	10^{11} M^{-1}	~0.5	10^{-3} s^{-1}	

* SMP: submitochondrial particles. The data are taken from: soluble F_1 (Grubmeyer, et al ,1982), SMP (Penefsky, 1985a,b,c),Yeast (Penefsky, 1987) and E. coli (Wise, et al ,1984 and Noumi, et al 1986). K_1 and K_2 are the equilibrium constants for steps 1 and 2. k_3 and k_4 are the forward rate constants for steps 3 and 4.

These enzymes also exhibit catalytic site cooperativity. That is, the rate of hydrolysis of [γ-^{32}P]ATP bound in the high affinity catalytic site is accelerated 10^5 to 10^6-fold (that is, to V_{max} rates) when substrate ATP in the form of a cold chase at mM concentrations is made available to additional catalytic sites on the molecule. The rate acceleration is about 10^6 for soluble (Cross, et al, 1982) and membrane-bound (Penefsky, 1985b) beef heart F_1 and for yeast F_1 (Penefsky, 1987) and 10^5 for E. coli F_1 (Wise, et al, 1984). Because substrate bound in the high affinity catalytic site can be hydrolyzed at normal rates, it was proposed that this is a normal catalytic site on the enzyme (Cross, et al, 1982). Recently, it was argued (Bullough, et al, 1987) that the high affinity catalytic site was not a normal site because the rate of hydrolysis of ATP bound in that site was not commensurate with the rate of hydrolysis of µM concentrations of ATP added as a cold chase. When cold chase experiments with µM concentrations of chase ATP are carried out in 1 ml reaction mixtures with rapid stirring or in a rapid mixing apparatus, the observed rate acceleration was the same as that reported earlier (Penefsky, 1987).

ENERGY-DEPENDENT DISSOCIATION OF ATP BOUND IN THE HIGH AFFINITY CATALYTIC SITE

If the reaction mechanism for single site hydrolysis of ATP described above is relevant to the mechanism of ATP synthesis by membrane-bound ATPase, it follows that ATP bound in the high affinity catalytic site of F_1 attached to the membrane of competent submitochondrial particles (that is, particles capable of catalyzing oxidative phosphorylation) should dissociate when the particles are energized by addition of an oxidizable substrate. An energy-dependent dissociation of bound ATP was observed. That is, addition of NADH or succinate to a complex of [γ-^{32}P]ATP and submitochondrial particles resulted in a dissociation of bound [γ-^{32}P]ATP that was inhibited by uncouplers of oxidative phosphorylation and prevented by inhibitors of respiration. The <u>rate</u> of dissociation was at least as fast as the rate of ATP synthesis measured under the same conditions. Thus, the high affinity catalytic site exhibited kinetic competence both with respect to maximum rates of ATP hydrolysis (as discussed above) and with respect to energy-dependent dissociation of bound ATP (Penefsky, 1985c).

LONG RANGE INTERACTIONS BETWEEN CATALYTIC SITES ON F_1 AND THE SUBUNITS OF F_o.

Incubation of submitochondrial particles with one equivalent of dicyclohexylcarbodiimide (DCCD), that is one mole of DCCD per mole of F_1, results in chemical modification of a single carboxyl group of subunit c in F_o, inhibition of ATPase activity, inhibition of proton translocation and of oxidative phosphorylation (Linnett and Beechey 1979). The distance between the DCCD binding site and the catalytic site (20 angstroms) was estimated by Montecucco and Azzi,1975, who inserted a spin-labeled carbodiimide in subunit c and Mn^{++} in the catalytic site of F_1 in beef heart submitochondrial particles. These observations are relevant to the finding that incubation of submitochondrial particles with one equivalent of DCCD or with oligomycin resulted in a profound decrease in the affinity of membrane-bound F_1 for ATP. Since the DCCD binding site and the catalytic site are clearly too far from each other for direct interactions to occur between them, it seemed reasonable to suggest that chemical modification of the glutamate carboxyl group in subunit c of F_o caused a conformatinal change in F_o, which, transmitted to F_1, caused a conformational change in F_1 that altered the nucleotide binding site and reduced the affinity of the site for ATP (Penefsky, 1985a). The possibility that long range conformational changes in F_o might be transmitted to F_1 has been raised by other workers.

It might be argued that the reduced binding affinity for ATP found after reaction of F_o with DCCD could be explained if binding of ATP were accompanied by the release of a proton and the requirement that the released proton enter the proton channel. Since the proton channel is blocked by DCCD (or oligomycin), interference with ATP binding could be expected. However, there is no clear evidence that insertion of a proton into the channel is a precondition for ATP binding and in fact there is a suggestion that a proton is released into the medium when ATP binds to chloroplast F_1 (Rumberg and Becher, 1984). Moreover, the E. coli mutant, in which the aspartate residue in position 61 of subunit c is replaced by a glycine residue, does not pump protons but does hydrolyze ATP. Thus, a blocked proton channel in the mutant does not seem to interfere with ATP binding and hydrolysis (Hoppe, et al,1982).

These observations support a more detailed model for the mechanism of ATP synthesis in oxidative phosphorylation which also invokes the possibility of long range interactions between F_o and F_1. It was proposed that protonation of the

DCCD-reactive carboxyl groups in F_o, or of other charged amino acid residues, presumably in response to proton flux through the system, caused a conformational change in F_o which was transmitted to F_1, reducing the affinity of catalytic sites on F_1 and permitting the dissociation of product ATP (Penefsky, 1985c).

REFERENCES

Bullough DA,Verburg JG, Yoshida M, Allison W (1987) Evidence for Functional Heterogeneity among the Catalytic Sites of the Bovine Heart Mitochondrial F_1-ATPase,J. Biol.Chem. 262: 11675-11683.

Cross,RL,Nalin,CM (1982),J.Biol.Chem.257:2874-2881.

Cross RL, Cunningham D, Miller CG, Xue Z, Zhou JM, Boyer PD (1987), Adenine nucleotide binding sites on beef heart F_1 ATPase. Proc Natl Acad Sci USA 84:5715-5719.

Cross RL, Grubmeyer C, Penefsky HS (1982) Mechanism of ATP Hydrolysis by Beef Heart Mitochondrial ATPase, J. Biol. Chem. 257:12101-12105.

Dunn SD, Futai M (1980), Reconstitution of functional coupling factor from the isolated subunits of E. coli F_1-ATPase J Biol Chem 255:113-18

Gay NJ, Walker JE (1981), The atp operon:nucleotide sequence of the region encoding the α subunit of E.coli ATP synthase, Nucleic Acids Res 9:2187-94

Gresser MJ, Myers JA, Boyer PD (1982) Catalytic SiteCooperativity of Beef Heart Mitochondrial F_1 ATPase,J. Biol. Chem. 257:12030-12038.

Gromet-Elhanan Z,Khananshvil(1984)Characterization of Two Nucleotide Binding Sites on the Isolated, Reconstitutively Active β Subunit of the F_o-F_1 ATP SynSynthase, Biochemistry 23:1022-1028.

Grubmeyer C, Penefsky HS (1981) The Presence of Two Hydrolytic Sites on Beef Heart Mitochondrial ATPase J Biol Chem 256:3718-3727.

Grubmeyer C, Cross RL, Penefsky HS (1982) Mechanism of ATP Hydrolysis by Beef Heart Mitochondrial ATPase, J. Biol. Chem. 257:12092-12100.

Hoppe J, Schairer HU, Friedel P, Sebald W, An Asp-Asn substitution in the proteolipid subunit of the ATPsynthase from E. coli leads to a non-functional proton channel, FEBS Lett. 145:21-24.

Jencks, WP (1975) Binding Energy, Specificity, and Enzyme Catalysis-The Circe Effect, Advances in Enzymol. 43:219-410.

Linnett PE, Beechey RB (1979) Inhibitors of the ATP Synthase System, Methods Enzymol. LV:472-518.

Montecucco C, Azzi A (1975) Molecular Interactions of ATPase with the Mitochondrial Membrane as Revealed by Spin Label Studies, J.Biol.Chem. 5050-5025.

Noumi T, Taniai M, Kanazawa H, Futai M (1986) Replacement of Arginine 246 by Histidine in the β Subunit of E. coli H^+-ATPase Resulted in Loss of Multi-site ATPase Activity, J. Biol. Chem. 261:9196-9201.

Penefsky HS (1985a) Mechanism of inhibition of mitochondrial adenosine triphosphatase by dicyclo hexylcarbodiimide and oligomycin:Relationship to ATP synthesis, Proc. Natl. Acad. Sci. USA 82:1589-1593.

Penefsky HS (1985b) Reaction Mechanism of the membrane-bound ATPase of Submitochondrial Particles from Beef Heart, J. Biol. Chem.260:13728-13734.

Penefsky HS (1985c) Energy-dependent Dissociation of ATP from High Affinity Catalytic Sites of Beef Heart Mitochondrial ATPase, J. Biol. Chem 260: 13735-13741.

Penefsky, HS (1987) Unpublished Experiments.

Rumberg B, Becher U (1984) Multiple pH Control of H^+-ATP Synthase Function in Chloroplasts, in Papa S, Altendorf K, Ernster L, Packer L (eds) H^+-ATPase (ATP Synthase): Structure,Function, Biogenesis, ICSU Press by Adriatica Editrice, Bari, Italy, 421-430.

Wise JG, Duncan TM, Latchney LR, Cox DN, Senior AE (1983),Properties of F_1-ATPase from the unc D412 mutant of E. coli, Biochem. J.215:343-350.

Wise, JG, Latchney LR, Ferguson AM, Senior AE (1984) Defective Proton ATPase of uncA Mutants of E. coli. 5'-Adenylyl Imidodiphosphate Binding and ATP Hydroysis, Biochemistry 23:1426-1432.

Xue Z, Zhou JM, Melese T, Cross RL, Boyer PD (1987) Chloroplast F_1 ATPase Has More Than Three Nucleotide Binding Sites and 2-azido ADP or 2-azido ATP at Both Catalytic and Non-catalytic Sites Labels the β Subunit, Biochemistry 26: 3749.

The Ion Pumps: Structure, Function, and Regulation, pages 269–277

BIOPHYSICAL EVALUATION OF LOCAL PROTON GRADIENT

Menachem Gutman, Esther Nachliel and Shimon Moshiach

Department of Biochemistry, Tel Aviv University, Ramat Aviv 69978, Israel.

The principal mechanism for ATP synthesis relies on enzymic transformation of available energy (reducing potential, or light) into directional proton (or charge) transfer across the membrane of specialized organelle. Other membranal enzymes conserve the stored energy by ATP synthesis.

Present studies are mainly focused on structure-function study of the enzymes involved in energy transforming reactions. The enzymes were isolated, purified and reconstituted in artifical membrane vesicles. Besides three dimensional structure analysis of these enzymes (see Witt et al. 1987 and references therein), the study is assisted by classical biochemical modifications (Zanotti et al., 1987) and genetic manipulations (Roa et al., 1987) of these proteins. What is common to these studies is their interest in the individual proteins without considering the novel properties emerging from the ensemble quality of the membranes. Insufficient interest is attributed to the packing form of the individual components in the membrane. In rat liver mitochondria (Schwermann et al., 1986), enzyme pairs which are involved in producing and consuming of $\Delta\mu H^{+}$ are 7-20 nM apart. Thus formulation of steady state dynamics of the oxidative (or photosynthetic) phosphorylation must consider the proximity of the reactants. One of the consequences of such dense packing is the appearance of local heterogenities of ion and charge concentrations on the membrane. Surface heterogenities with dimensions of 10-20 nM can already create microenvironments which may affect the enzymes within those domains. Indeed, it was demonstrated that lipid composition and surface charge can affect the activity of the chloroplasts ATP synthase (Pick et al., 1987) or the configuration of bacterial rhodopsin (Szundi & Stockenius, 1987). Such modulation can affect nearby enzymes leading to local control effects (for theoretical implications, see Vincent and Selengy, (1987)).

The complexity of reactions on the mebranes surface, especially under flux conditions, led to the appearance of many reaction models sharing the common concept of "local proton gradient". These models considered the aqueous phase, which interacts with all protogeneic sites on a membrane, not as a featureless continuum. The first generation models (Kell, 1979; Kell et al., 1981; Haraux & De Kouchkowsky, 1982) regarded the water as a biphasic matrix; the innermost layer, that is in contact with the membrane, had a special proton conductive capacity and was assumed to be partially insulated from the bulk water. Complex analog circuit diagrams were devised (Westerhoff et al., 1984) to simulate the experimental observations. The second generation models (Nagle & Dilly, 1986) already negate the simple two-phase concept. They prove its inadequacy and called for a network of conducting pathways stretched over the surface of the membrane.

The multiplicity of models and the fast emergence of a "second generation" formalism reflects the importance of this problem in modern bioenergetics. Yet, the fact that varieties of the same concept coexist indicates a lack of unambiguous methodology to monitor proton diffusion on a surface of a biomembrane.

A persistent state of acid-base disequilibrium between certain membranal domains and the bulk phase (Nagle and Dilly, 1986) necessitates a very special kinetic situation; the sites within the domain must exchange protons at a rate much faster than the emergence of protons out of the domain's boundary. In previous studies (Gutman , 1986) we demonstrated that surface groups react with protons or buffers in a diffusion controlled reaction. Thus, to comply with the above kinetic restrictions, the diffusion of protons on the membrane's surface must be faster than their diffusion to the bulk. There is a general belief (Kell et al., 1981) that proton diffusion on a surface will be faster than in bulk. This notion is based on the reduction of dimensionality principle of Adam and Delbruck (1968) (which is not applicable in this case) and the effect of water molecules' ordering. Tocanne and his co-workers (Prats et al., 1987) claimed that such ultrasfast proton diffusion on membrane water interface was indeed measured. Yet there are experimental peculiarities and theoretical objections (Kasionowicz et al., 1987) which call for reevaluation of this notion.

The best procedure to clarify this problem is to measure proton diffusion in a space made exclusively of the hydration layer of a phospholipid membrane. The results of these measurements are described in this study.

The preparation we selected for these studies, multilamellar liposomes, has been extensively investigated by Parsegian, Rand and their co-workers (Parsegian et al., 1979; Parsegian et al., 1986 and refs. therein). In their studies, multilamellar liposomes of known composition were subjected to external pressure and the width of the hydration layer (dw) separating the lipid double layers was measured by X-ray diffraction with accuracy of 0.05 nM.

The hydration layer is an extremely thin water layer - 1.5 nM deep - where the water molecules are ordered and oriented by long and short range interaction with the membrane (Israelachvilli & Mara, 1986). Due to this ordering, the hydration layer is an extension of the rigid structure of the body it wraps. The force needed to dislocate a water molecule from its energetically favored position varies exponentially with the distance to the surface. The innermost water molecules need ~ 1 Kcal/mol for displacement, at a distance of ~ 1 nM force is already smaller than the termal energy. The length constant of this exponential decay has the dimension of a water molecule (0.28 nM).

In our study, we swelled pure phosphatidyl choline in dilute aqueous solution of 8 hydroxypyrene 1,3,6 trisulfonate (pyranine), collected the vesicles by centrifugation and verified the closed multilamellar structures by electron microscopy.

The trapped compound, pyranine, is a mild base in its ground state (pK=7.7), but when excited to its first electronic singlet state it becomes a very strong acid (pK*=0.5). Following a synchronized excitation by a short (10 ns) laser pulse the pyranine population dissociates to excited anion RO^{*-} and free proton (see scheme I).

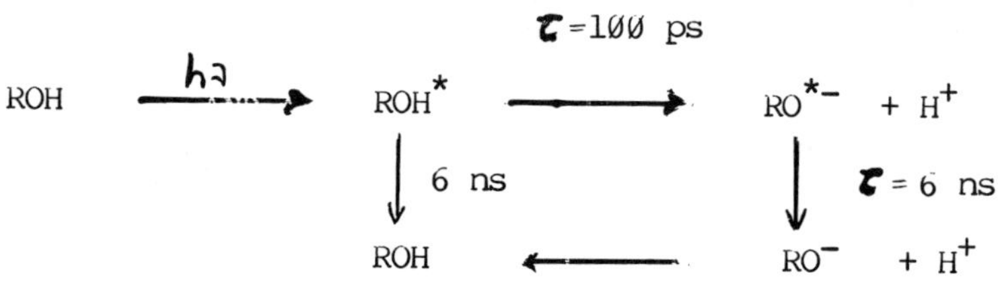

$$E_{450} = 25000\ M^{-1}cm^{-1}$$

A few ns later, the excited anion decays to the ground state, which has a strong absorbance at 450 nM. This anion recombines with H^+ and the absorbance of 450 nM is bleached. Thus we monitor directly the diffusion controlled reaction of a free H^+ and a target molecule dissolved in a hydration layer which is sandwiched between two membranes of known composition.

Figure 1 depicts kinetic measurement of proton-anion recombination taking place in the ultrathin water layers enclosed in a multilamellar liposome.

Immediately after the laser pulse the proton dissociates and RO^- population increases. The state of disequilibrium relaxes with apparent rate constant $\gamma = 3.2\ 10^6\ S^{-1}$ to a residual level. The latter decay at much slower rate, and represent a stable photoproduct of the dye which is not related to the acid-base reaction.

Addition of sucrose (1.1M) to the solution sucks the water out of the water shells and dw decreases to ~ 0.5 nM. In these compressed layers the recombination reaction is much slower (lines A and B, respectively).

Fig. 2 summarises the results of many experiments

carried out under the same conditions, except for one variable, the osmotic pressure applied on the vesicles. As the osmotic pressure increases, water molecules are sucked out from the interlamellar space and only those with potential higher than the threshold value will remain $\mu_{H_2O} = \Pi \cdot V_{H_2O}$ where Π is the osmotic pressure and V_{H_2O} is the molar volume of water molecule. This stripping removes the water layers which are in the "middle of the sandwich", i.e., those most remote from the two bordering on the lipid surfaces.

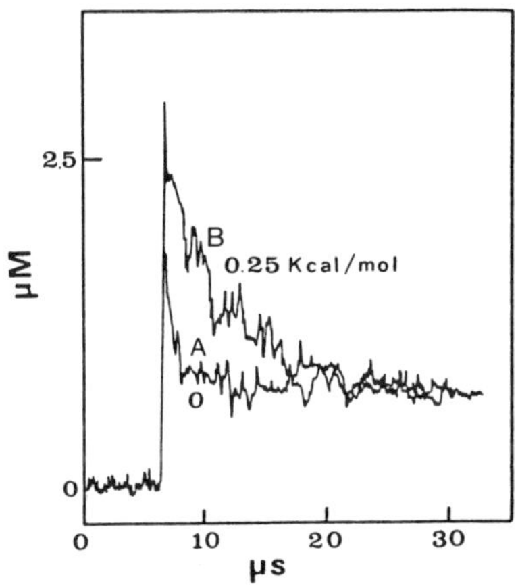

Fig.1

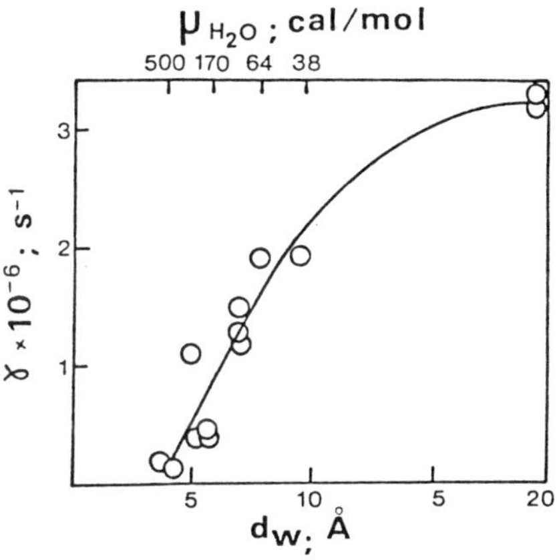

Fig.2

Figure 1. Time resolved dynamics of proton recombination with pyranine anion in the aqueous layer in multilamellar vesicles of phosphatidyl choline. Line A was measured in unstressed vesicles, line B under osmotic pressure which removes all water molecules with potential less than Ø.25 Kcal/mol from aqueous space.

Figure 2. The dependence of the rate constant of proton pyranine-anion recombination on the width of the hydration layer. The experiments described in Figure 1 were repeated under varying osmotic pressures and the width of the water layer was interpolated from the data of Parsegian et al. (1979).

As seen in Figure 2, narrowing of the water layer slows dramatically the rate of proton recombination. Thus we conclude that the water molecules which are adjacent to the phosphomoieties are lesser proton conductors than those which are in a more remote position. The 20-fold difference between rates measured in unstressed vesicles and fully compressed structures implies that under all conditions most of proton flux is carried out by those water molecules which are located in the middle of the water layer. This conclusion is crucial for the quantitative analysis of our measurement, as we can presume that all proton flux takes place in the two-dimensional surface of the highest conductive water layer.

This approximation legitimates the application of the Hardt (1979) equation for diffusion controlled reaction between non identical reactants trapped in a two-dimensional space.

According to Hardt equation the apparent rate constant of RO^- reprotonation (γ) will be a function of the proton's two dimensional diffusion coefficient D^H_2 and the surface concentration ($mole/cm^2$) of the pyranine anion (C^P_2).

$$\gamma = 2\pi N D^H_2 C^P_2 / \ln (9.6 \cdot 10^4 \sqrt{C^P_2})^{-1}$$

This relationship implies that, within a narrow range of C^P_2, where the logarithmic function will be practically constant, γ will vary linearily with C^P_2. This assumption was tested experimentally by modulating the intensity of the excitating laser pulse. The results of these measurements, presented as γ vs C^P_2 are shown in Figure 3. The mean value of D^H_2, calculated for unstressed phosphatidyl choline multilamellar structures is

$D^H_2 = (2.47 \pm 0.43) . 10^{-5}\ cm^2/sec.$

This value, corrected for the two dimensionality factor, corresponds with $D^H_3 = 3.7\ 10^{-5}\ cm^2/sec$. The diffusion coefficient of proton in bulk water, $9.3 . 10^{-5}\ cm^2/sec$, is definitely larger than the value estimated for the middle, conductive water layer, located 1.0 nm from a phosphatidylcholine surface. If we take this difference as significant, then we conclude that a water layer more than 1 nm above the phospholipid membrane has a proton conductivity which is lower than that of bulk water.

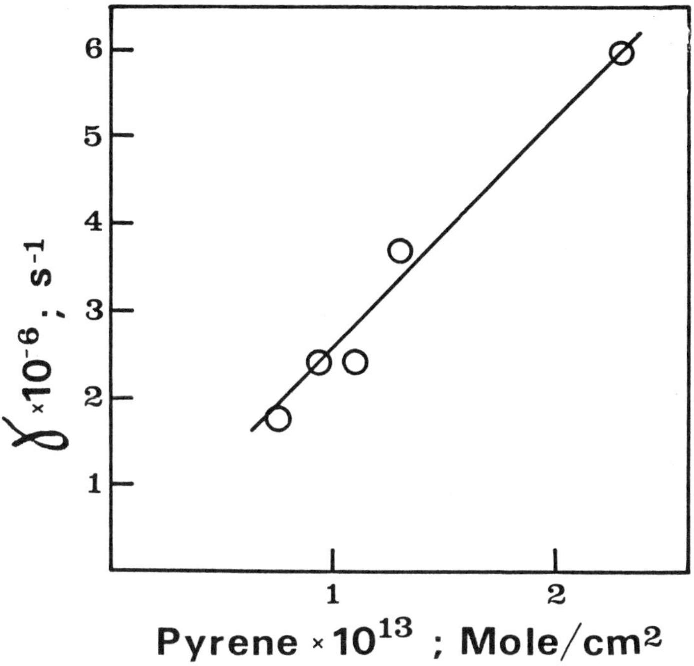

Figure 3. The dependence of the rate of proton recombination with the pyranine anion on the two dimensional concentration (mole/cm^2) of the reactants. Linear relationship complies with a bimolecular reaction and the slope is a function of the two dimensional diffusion coefficient.

Nagle and Dilly (1986) advanced theoretical proof that a local proton gradient mechanism must utilize special conducting pathways over the membrane's surface. We have already demonstrated that surface groups react either with protons, or with buffer in diffusion controlled reaction (Gutman, 1986). In the present study we demonstrated that the ordered water of the hydration layer is a poor proton carrier, and lateral proton diffusion is slower than its diffusion in bulk. Thus a proton released on the surface of a membrane will spontaneously diffuse into the bulk. Combination of these observations negates the physical existence of local proton gradient on a membrane's surface. If such models are to survive, a direct proton transfer between reactants must be adopted - in a space where the protons are excluded from contact with bulk phase water.

REFERENCES

Gutman M, (1986). Application of the laser induced proton pulse for measuring the protonation rate constants of specific sites on proteins and membranes. Methods in Enzymol 127: 522-538.

Haraux F, De Kouchkowsky Y (1982). Further investigation on the lateral and transversal proton currents at the thylakoid membrane level by hydrogen deuterium exchange. Biochim Biophys Acta 679: 235-247.

Hardt S, (1979). Rates of diffusion controlled reactions in one, two and three dimensions. Biophys Chem 10: 239-243.

Israelachvilli JN, Mara J (1968). Direct methods for measuring conformational water forces (hydration forces) between membranes and other surfaces. Methods in Enzymol 127: 353-360.

Kell DB, (1979). On the functional proton current pathway of electron transport phosphorylation - an electrodic view. Biochim Biophys Acta 549: 55-99.

Kell DB, Burns A, Clark DJ, Morris JG (1981). Proteinaceous proton pumps: A minimal model, some properties and their possible universality. Speculation in Science and Technol 4: 109-120.

Kasianowicz J, Benz R, Gutman M, McLaughlin S (1987). Lateral diffusion of protons along phospholipid membranes. J Memb Biol: in press.

Nagle JF, Dilly RA (1986). Models of localized energy coupling. J Bioenerg & Biomemb 18: 55-64.

Pick U, Weiss M, Gouraris K, Barber, J (1987). The role of different thylakoid glycolipids in the function of reconstituted chloroplasts ATP synthase. Biochim Biophys Acta 891: 28-39.

Parsegian VA, Fuller V, Rand RP (1979). Measured work of deformation and repulsion of lecithin bilayers. Proc Natl Acad Sci USA 76: 2750-2754.

Parsegian VA, Rand RP, Fuller NL, Rao DC (1986). Osmotic stress for the direct measurement of intermolecular forces. Methods in Enzymol 127: 400-416.

Roa R, Perlin DS, Senior AE (1987). The defective proton ATPase of unc-A mutants of E. coli ATP-binding and ATP-induced conformational changes in mutant alpha subunits. Arch Biochim Biophys 225: 309-315.

Shwermann K, Cruz-Orive LM, Eggman R, Sanger A, Weibel ER (1986). Molecular architecture of the inner membrane of mitochondria from rat liver, a combined biochemical and sterological study. J Cell Biol 102: 97-103.

Szundi I, Stoekenius W (1987). Effect of lipid surface charges on the purple to blue transition of Bacterorhodopsin. Proc Natl Acad Sci USA 84: 3681-3684.

Vincent JC, Selengy E (1987). Immobilized cyclic enzyme active transport system. pH feedback control. Biophys Chem 26: 45- 54.

Witt HT, Witt S, Gerken W, Saenger J, Dekkar P, Ranger M (1987). Crystalization of reaction center I of photosynthesis. FEBS Letters 221: 260-264.

Zanotti F, Guerieri F, Che YW, Scarfo R, Papa S (1987). Proton translocation by the proton ATPase of mitochrondria, effect of modification by monofunctional reagents of thiol residues of Fo polypeptides. Eur J Biochem 164: 517-524.

ACKNOWLEDGEMENT

This research was supported by the United States-Israel Binational Science Foundation (84-100).

The Ion Pumps: Structure, Function, and Regulation, pages 279–282

SURFACE CARBOXYLS AS A COLLECTING ANTENNA FOR PROTONS

Ruth Yam

Department of Biochemistry, Tel Aviv University, Ramat Aviv 69978, Israel.

The fluorescein-isothiocyanate adduct to bovine serum albumin was used as a model compound for studying the protonation kinetics of a chemically characterized site on a protein. The dynamics were measured by fast recording of light absorption of the protonable chromophore following a transient acidification by the laser induced proton pulse (Gutman, (1986)).

A typical transient is shown in Figure 1. Immediately after the pulse, the dye molecules are protonated, reaching a maximum at ~7 μs, then relaxes with time constant of ~100 μs.

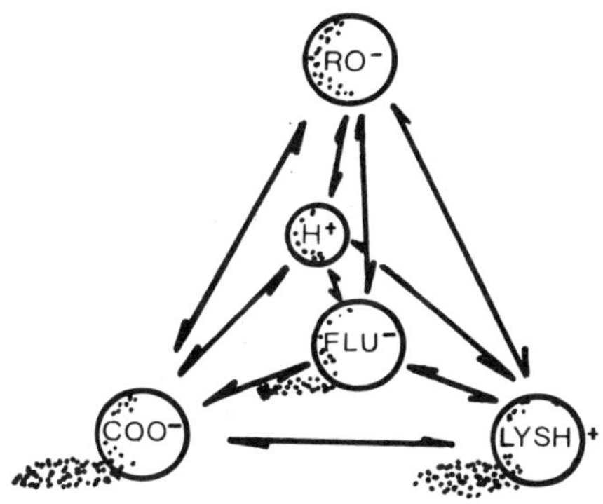

The analysis of this complex event, where 7 dye molecules (per protein) react with protons on a surface carrying 100 carboxyls (pK=5.0), 50 lysines (pK=9.9) and histidines (Tanford (1955)) calls for appropriate kinetic formalism, which is summarized in scheme I.

The circles represent the populations of the proton emitter (RO^-), Fluorescein (Flu^-), carboxyls ($1ØØ^-$) and lysine moieties ($LysH^+$), all reacting with protons (H^+). The line represents the partial reactions of proton transfer between the populations. The velocities of each partial reaction Vij=kij.Ci.Cj are combined by set of non linear, coupled differential equations as described elsewhere (Gutman, (1986)). These parametric equations are solved numerically, looking for a single set of rate constants which will reconstruct the experimental observation (the continuous line in Figure 1).

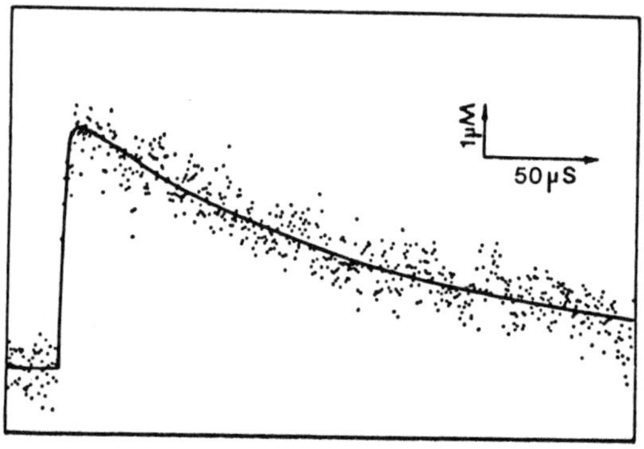

Figure 1

Transient protonation of fluorescein isothiocyanate - bovine serum albumin adduct. The protein solution, 75 uM with respect to the dye, supplemented by 1 mM of the proton emitter 2 naphthol 3,6 disulfonate, pH=8.Ø, was irradiated by 2Ø48 laser pulses. The continuous line is the theoretically predicted curve generated by the integration of the differential equation (Gutman, (1986)).

The solved system can reproduce the dynamics of all components participating in the combined event (Figure 2).

The dynamics of the proton emitter, RO^-m is depicted by line A. Immediately after the pulse is incremental concentration is maximal. Some of it (about 15%) is reprotonated within a very short time - this fast reaction is the phase where free protons, discharged in the bulk, react with RO^- and the protein.

The high concentration of the carboxyl groups renders them effective competitors for H^+. After 2ØØ ns, when the free proton concentration has decreased to the prepulse level, about 8Ø% of the discharged protons are

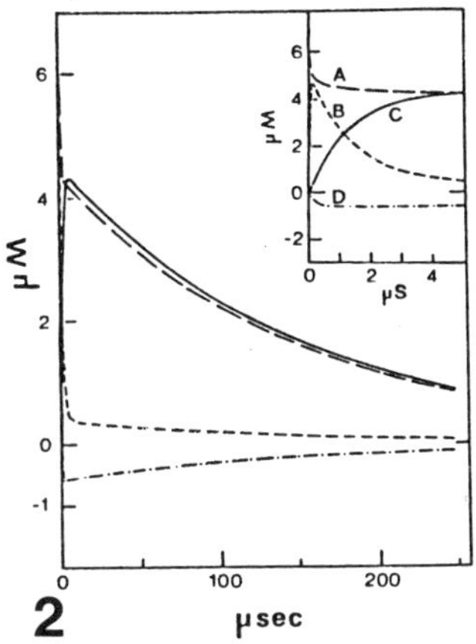

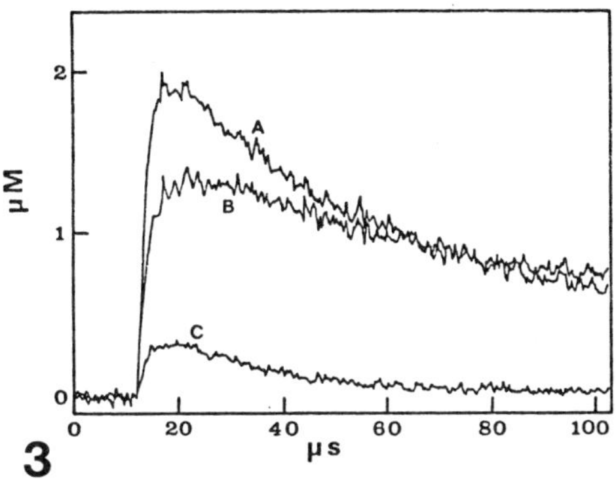

Figure 2

Transient protonation of the reactants participating in bulk-protein proton transfer. The lines were calculated for the conditions defined by the experiment shown in Figure 1. Line A (——) : the transient deprotonation of the proton emitter (RO^-). Line B (----): the protonation of the carboxyl groups of the proteion; Line C (——):transient protonation of the dye. this curve is identical with that shown in Figure 2. Line D (-.-): the deprotonation of the $-NH_3^+$ groups of the protein's lysine, the insert expands the first 5 μs of the main frame.

Figure 3

The effect of random blocking of surface carboxyls on the kinetics of fluorescein protonation. The experiment was carried out as in Figure 1 but with protein preparation where the carboxyl was blocked by-$CONH_2$ groups. A. Unblocked protein; B, 12% of carboxyls blocked; C, 45% of carboxyls blocked.

---

bound to the carboxyls of the protein (line B). From that time on, the bulk (RO^-) and protein surface (COO^-, dye and lysine) are in equilibrium with each other; the reprotonation of RO^- is now at the expense of protons dislodged from the protein.

The build up of COOH reaches its maximum at 200 ns and then, within a few microseconds, we observe a rapid shift of the protons from the carboxyl to the dye (line C). The efficiency of this transfer is high. As seen in the insert, within 5 us the decay of COOH and accumulation of protonated dye are coupled to each other, with only minor loss to RO^- (line A) or to $R\text{-}NH_2$ (line D).

The high efficiency of the carboxyl pathway is attributed to their high surface density. The average distance between carboxyls on the protein's surface is comparable to the Coulomb cage radius of a single charge ion in aqueous solution (7A). Thus in their non random assembly on the native protein, patches are formed where the proton can be transferred, with very high probability, between nearby moieties. Such a carboxyl rich domain in a vicinity of a fluorescein molecule will serve as a proton collecting antenna for the dye.

Any breakdown of the continuity in the "antenna" will reduce its connectivity with subsequent reduction of its efficiency. As seen in Figure 3, a random blocking of few carboxyls on the protein surface suffice to drastically reduce the pulse protonation of the dye.

ACKNOWLEDGEMENT

This research is supported by the United States-Israel Binational Science Foundation (84-100).

Ruth Yam is a recipient of a Buchman Fellowship.

REFERENCES

Gutman M (1986). The application of the laser induced proton pulse for measuring the protonation rate constants of specific sites on proteins and membranes. Methods in Enzymol 127: 522-538.

Tanford C, Swanson SA and Shore WS (1955). Hydrogen ion equilibria of bovine serum albumin. J Am Chem Soc 77: 6414-6421.

The Ion Pumps: Structure, Function, and Regulation, pages 283–287

PROTON BINDING DYNAMICS OF PHOSPHOLIPID MEMBRANE

Esther Nachliel
Dept. of Biochemistry, Tel Aviv University, Ramat Aviv 69978, Israel.

In a state of equilibrium, buffer is an inert component which does not affect the system. In dynamics, buffer is a dominating component which modulates the velocity of proton binding (Gutman & Nachliel (1986)). Thus the protonation dynamics of a given surface group will depend on the composition of the phosphohead groups of the membrane's lipids. This dependence was measured in the time resolved domain by the laser induced proton pulse (Gutman (1986)).

The experimental system consisted of three reactants: a proton emitter (ROH), a pH indicator (HIn) and phospholipids (PL). The latter two are adsorbed on supporting micelles, made of Brij-58, which is inert and does not participate in the acid-base reactions.

A short laser pulse excited the proton emitter to its first electronic singlet state which, due to its low pK ($pK^*=0.3$), discharges its proton. When the emitter relaxed to its ground state (less than 30 ns after the laser pulse), the $ROH \rightleftharpoons RO^- + H^+$ systems are poised in a state of disequilibrium. Both diffusion and collisional proton transfer propagate this perturbation to all other acid base pairs present in the system. The experimental observation is the sum of all these reactions.

Typical transient protonation of such system is demonstrated in Figure 1. The tracing represents the protonation of an indicator adsorbed to Brij-58 micelles, (line A) and how phospholipids affect the dynamics (lines B - D). Phosphatidyl choline increases the protonation while phosphatidyl serine and phosphatidic acid decrease it.

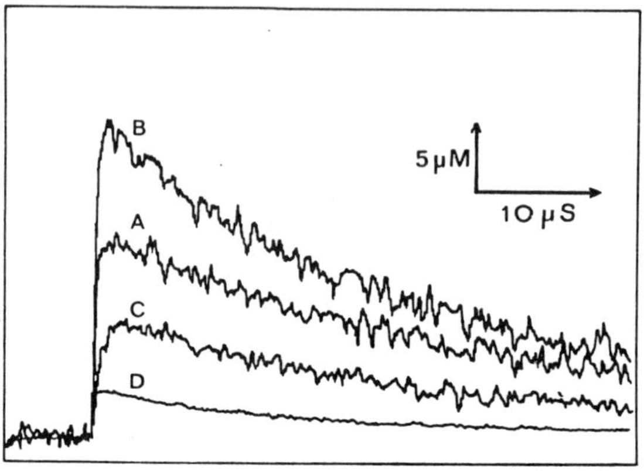

Figure 1:

Transient protonation of Bromocresol green adsorbed on mixed micelles. The reaction was carried out at pH = 7.3 + 0.1 in a solution containing 500 uM bromocresol green, 1 mM of the proton emitter 2 naphthol 3,6 disulfonate 500 μM micellar concentration of Brij-58 (40 mg/ml) and 3 mM of phospholipids. The transient protonation was measured at 633 nM through an optical path of 2.0 mm. A, control; no phospholipid added (averaging 256 events). B, Phosphatidyl choline (256 events). C, Phosphatidyl-serine (256 events). D, Phosphatidic acid (2098 events).

The protonation of a surface is a competition between two processes, recombination of protons with bulk anions vs. their reactions with membrane bound groups. The outcome depends on the pK of the surface groups. A surface covered with low pK groups will retain proton for a very short period. The dwell time of proton on a phosphatidyl-choline head group (pK=2.25) is about τ_{dw} = 30 ns. Within this time frame the PCH will dissociate and the proton, duffusing to the bulk, will react with RO^-. The proton dwell time on the carboxyl group of phosphatidyl-serine (pK= 4.6) is much longer, τ_{dw} = 4 μs.

A membrane's surface can be deprotonated also by collisional proton transfer to buffer. In presence of 1 mM buffer like imidazole, the time constants of collisional deprotonation of phosphatidyl choline and phosphatidyl serine are about τ_{pt}=2 μs. It is quite obvious that deprotonation of phosphatidyl-choline membrane will be mostly controlled by the dissociation of PCH, $\tau_{dw} \ll \tau_{pt}$. The deprotonation of phosphatidyl-serine membrane will be dominated by collisional proton transfer, $\tau_{dw} > \tau_{pt}$.

These figures imply that at a short time, the state of protonation of membrane regions enriched by phosphatidyl-serine or phosphatidyl-choline will differ and vary with the buffer concentration or composition.

A condensed representation of this property is shown in the simulated dynamics given in Figure 2, where we plotted the projected time dependence of RO^- (A), HIn (B) and PSH (C) for a system made of phosphatidyl-choline and phosphatidyl-serine.

A surface made of phosphatidyl-choline (plus 0.05 mole fraction of bromocresol green), is practically devoid of buffer capacity and ~ 60% of the discharged protons will rapidly recombine with RO^- (line 1 in frame A). The remaining protons will react with the indicator (frame B), except a minute fraction temporarily bound to PC (line 1 at bottom of frame A). Increasing the phosphatidyl-serine content of the membrane changes the scenario. The capacity of the membrane to compete for proton increases and reprotonation of RO^- is delayed (lines 1 - 6 in frame A). At X_{ps} = 0.5 nearly 90% of the discharged protons are bound to the membrane. Even low content of PS (X_{ps} = 0.05) already affects the protonation of the indicator (lines 1 and 2, frame B). The PS molecules increase the number of protons bound to the surface (line 2 in frame C) and the fast proton flux through the PC increases the protonation of the indicator. At high PS content (X = 0.5), nearly 66% of discharged protons react, within ~ 200 ns, with the PS. At such high content of phosphatidyl-serine the long dwell time of protons retards their mobility and the protonation of the indicator decreases (lines 5 and 6 in frame B). A phosphatidyl-serine rich membrane has a large proton reservoir; deprotonation of the indicator is compensated for by protons from PSH. As a result the life time of HIn is prolonged from 16 us in absence of PS to 22 μs.

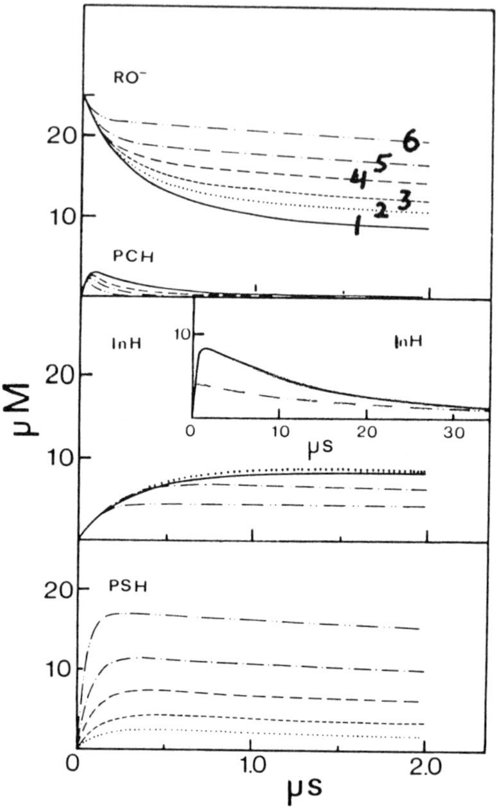

Figure 2:

Simulated dynamics describing bulk-surface proton transfer. The curves represent the state of protonation in a system containing a pH indicator adsorbed on a phospholipid membrane made of phosphatidyl-choline and phosphatidyl-serine at various proportions. The mole fraction of PS is marked in frame A. Line 1 (-) X_{ps}=0; 2.(.....)X_{ps}=Ø.Ø5; 3,(---) X_{ps}=Ø.Ø9;; 4,(--) X_{ps}=Ø.17; 5,(-.-) X_{ps}=Ø.28; 6,(-..-) X_{ps}=Ø.5.

Our experiments demonstrated that both time constants and protonation probability of a given group on a surface are grossly modulated by the nearby surface components. The mathematical analysis provided the physical rationale for these observations. The application of this reasoning in the study of membrane bound proton driven enzymes will help to elucidate the intricate mechanism of their catalytic activity.

ACKNOWLEDGEMENT

This research is supported by the United States-Israel Binational Science Foundation (84-100).

REFERENCES

Gutman M (1986). The application of the laser induced proton pulse for measuring the protonation rate constants of specific sites on proteins and membranes. Methods in Enzymol 127: 522-538.

Gutman M, Nachliel E (1985). Kinetic analysis of protonation of a specific site on a buffered surface of a macromolecular body. Biochemistry 24: 2941-2946.

Nachliel E, Gutman M (1984). Kinetic analysis of proton transfer between reactants adsorbed to the same micelle: The effect of proximity on the rate constants. Eur J Biochem 143: 83-88.

The Ion Pumps: Structure, Function,
and Regulation, pages 289–292

PROTON TRANSLOCATION AND ATP SYNTHESIS IN PHOTOSYNTHETIC ORGANISMS WITH μS ILLUMINATION

Girault G.*, Galmiche J.M.*, de Wolf F.**, Kraayenhof R.**. *DB/SBPH, CEN/Saclay 91191 GIF-SUR-YVETTE CEDEX, FRANCE. **Biological Laboratory, Vrije Universiteit, De Boelelaan 1087-1081 HV AMSTERDAM, THE NETHERLANDS.

INTRODUCTION

Changes of the proton flux and of the rate of ATP synthesis are related to the proton electrochemical potential difference across the membranes. Experiments were done with pea chloroplasts or photosystem I (PSI) enriched particles from spinach chloroplasts in conditions where the ATP synthetase is active.

RESULTS

In pea chloroplasts (Galmiche and Girault 1982), submitted to a train of flashes, the membrane potential remained constant but there was a proton accumulation inside the thylakoids (Fig. 1). Flash induced ATP synthesis paralleled the increase of the ΔpH observed (Fig. 2).

PSI particles (de Wolf et al., 1985), submitted to trains of flashes, synthetized ATP on the first flashes but the yield per flash increased with the number of flashes, as did the membrane potential (Fig. 3) until a limit value after around 10 flashes. No sizeable increase of the ΔpH could be measured.

With chloroplasts, ionic conduction depends on ΔpH. The proton efflux was measured by the slope of the dark decay of the absorbance change at 515nm. The proton efflux sharply increased at the optimum external pH for ATP synthesis activity in the presence of ADP plus Pi (Fig. 4).

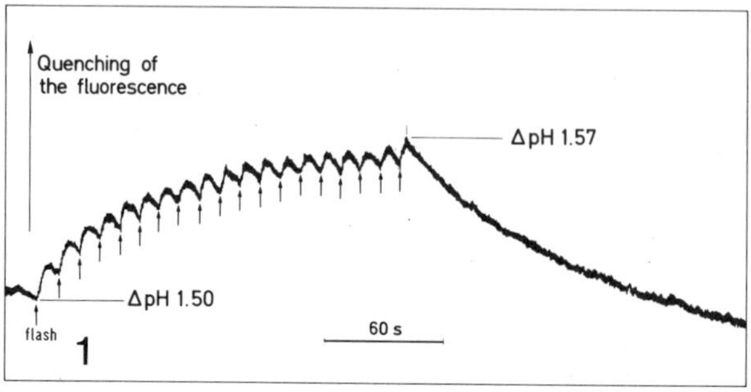

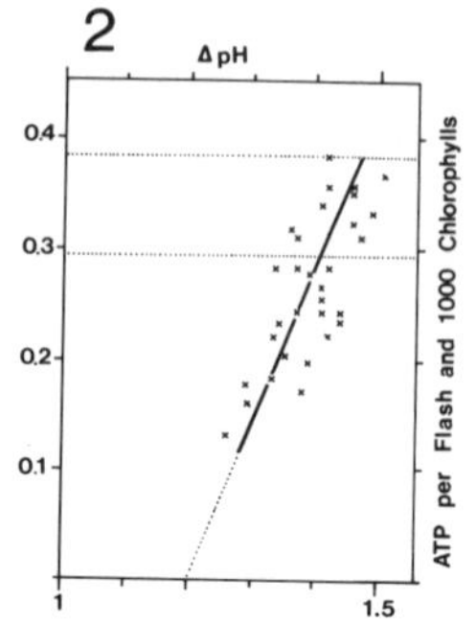

Figure 1. Flash induced increase of the ΔpH across the membrane.
Measurements were performed at 10°C in the following medium: 40mM Tricine pH8, 10mM $MgCO_3$, 2mM K_2HPO_4, 20mM $KHCO_3$, chloroplasts (20µg Chl/ml).

Figure 2. ATP synthetized/flash as a function of the ΔpH across the membrane.
Same conditions as on figure 1, supplemented for ATP measurements with 20µM ADP, 5µM diadenosylpentaphosphate, luciferin-luciferase and for pH measurements with 2.5µM atebrin.

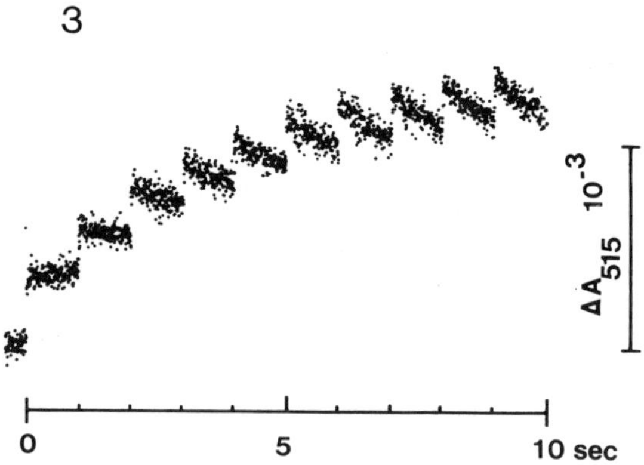

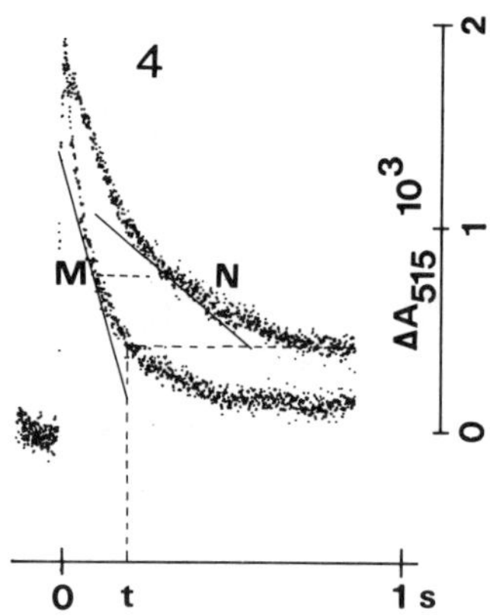

Figure 3. Flash-induced membrane potential in PSI particles monitored by absorbance change at 515nm.
PSI vesicles (50µg Chl/ml) were suspended in the same medium as on figure 1, supplemented with 5µM ferredoxin, 2mM NADPH.

Figure 4. Flash-induced membrane potential in pea chloroplasts monitored by absorbance change at 515nm. Conditions as on figure 1. Upper curve in the absence of ADP and Pi, lower curve in the presence of ADP and Pi.

With PSI particles ionic conduction depends on $\Delta\psi$. Above a limit $\Delta\psi$ threshold value, the ionic conductance increases sharply. The effect of $\Delta\psi$ is more pronounced when ATP is synthetized than in absence of ATP synthesis (Fig. 5). The difference between the two curves gives, for each flash, the extra proton flux caused by ATP synthesis (Fig. 6).

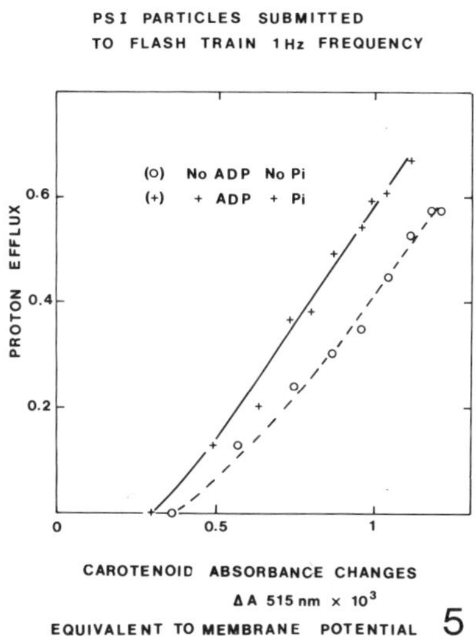

5

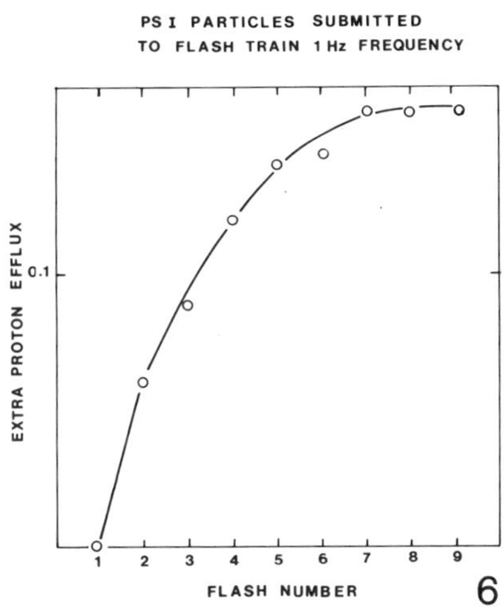

6

Figure 5. Proton efflux as a function of absorbance change at 515nm in PSI vesicles.
Conditions as on Fig. 3.

Figure 6. Extra proton efflux as a function of the flash group number in PSI vesicles.
Conditions and results are those of the experiment reported on Fig. 5. Extra proton efflux refers to the difference of the proton efflux in the absence or in the presence of ADP-Pi.

The ATP synthetase activity responds to $\Delta\tilde{\mu}H$ changes. With pea chloroplasts the flash induced membrane potential increases when the number of flashes per group increases. In the same conditions, the ATP synthetized per flash group increases and the ratio H^+/ATP decreases to ~ 2 (table 1).

ATP SYNTHESIS PER GROUP OF n FLASHES GIVEN AT 0.1 Hz FREQUENCY		
n	ATP PER 1000 CHLOROPHYLLS	RATIO H^+/ATP
1	0.47 0.34	3.23 4.96
2 10 ms apart	0.6 0.65	3.2 4
6 3 ms apart	1.36 1.72 1.51 2.75	3.11 2.26 3.8 1.65

Table 1. Flash induced ATP synthesis in pea chloroplasts. Conditions as on Fig. 2.

We propose that the limiting step of the reaction is, as in Boyer's hypothesis, the release of the newly formed ATP molecules from the ATP synthetase. Bound ATP, measured by luciferin-luciferase reaction, can be hydrolyzed before it is detected. The limiting step is more easily overcome at higher $\Delta\tilde{\mu}H$ when the turn-over of the ATP synthetase is faster.

REFERENCES

Galmiche J.M., Girault G. (1982). Synthesis of ATP induced in pea chloroplasts by single turnover flashes. FEBS Lett. 146, 123-128.

De Wolf F.A., Galmiche J.M., Kraayenhof R. and Girault G. (1985). Single turnover flash induced ATP synthesis in photosystem I-enriched subchloroplast vesicles. FEBS Lett. 192, 271-274.

The Ion Pumps: Structure, Function, and Regulation, pages 293–298

ELECTROGENESIS IN CHLOROPLAST MEMBRANES: Evidence for two slow components in the P515 signal after flash excitation

W.J. Vredenberg, J.J.J. Ooms and W.F. Buurmeijer

Laboratory of Plant Physiological Research, Agricultural University, Wageningen, the Netherlands

INTRODUCTION

Light-driven energization of the thylakoid membrane starts with a charge separation in the photosynthetic reaction centers of PS1 and PS2, giving rise to a transmembrane electric field. Secondary electron and proton transfer in the membrane-spanning photosynthetic electron transport chain is accompanied by electrochemical events at the thylakoid membrane. Amongst these is the generation of an electrochemical proton gradient which serves as a driving force for the ATP synthase. Ample evidence has accumulated that the electrochromic absorbance band shift of P515, often referred as the P515 response, is a powerful measure of these electrochemical events (Vredenberg, 1981).

It has been shown that the P515 response in well preserved chloroplasts upon flash illumination has a multiphasic kinetic profile (Schapendonk, 1980). So far at least two different electrochromic components have been identified and, at least partly, isolated. For the sake of briefness we will only consider the distinction between a fast component and a slow component. A third non-electrochromic component (Schapendonk et al., 1979) will not be dealt with here. There is general consensus that the fast component is a reflection of the transmembrane electric field generated by the primary charge separation in the reaction centers (Junge, 1977). However the so-called slow component so far has caused much debate with respect to its identification. In brief two opposing interpretations exist at the moment. The secondary slow rise in the overall response, observed in dark-adapted intact algae (Joliot and Delosme, 1974) and in chloroplasts under reducing conditions (c.f. Selak and Whitmarsh, 1982, Hope and Matthews, 1987) has been interpreted to be due to secondary electrogenic charge transfer in the so-called Q-cycle in which the Rieske-Fe -cyt b/f oxido-reductase complex is operative. In this interpretation one would expect equal rate constants of the dark decay of the slow and fast component (Vredenberg, 1981). Opposedly the slow component, called Reaction 2 (Schapendonk et al., 1979), when isolated from the fast component (Reaction 1) has been attributed to inner membrane electric fields which form upon charge stabilisation in localized membrane domains (Van Kooten, 1988). The distinction has been made mainly because the decay of the slow component is substantially slower than the decay of the fast component.

Here we will show that the flash-induced P515 response in chloroplasts consists of at least two slow components, which are clearly distinguishable on basis of their kinetics. One component with a rise time of about 40 ms and a dark relaxation time of 60-100 ms is a reflection of electrogenic secondary charge transfer and is called Reaction 1/Q. The other one, with a rise time of more than 100 ms and a dark relaxation time of about 500 ms, is called Reaction 2 as before. It is proposed to call the fast component reaction 1/RC, as it reflects primary charge separation in the reaction centers.

MATERIAL AND METHODS

Chloroplasts were isolated from freshly grown spinach. Growth conditions and isolation procedures were as described elsewhere in detail (Peters, 1986). Absorbance changes at 518 nm, induced by saturating single turnover light flashes were measured in a single beam absorbance difference apparatus as described before (Peters, 1986). Chloroplasts were taken from a concentrated stock suspension at $0^{o}C$ measured at a concentration of about 25 ug/ml chlorophyll in a medium containing 0.33 M sorbitol, 10 mM NaCl, 5 mM $MgCl_2$, 1 mM $MnCl_2$, 2mM EDTA, 5 mM ascorbic acid and 75 mM HEPES/KOH pH 7.5. They were routinely dark adapted for more than 2 hours before flash illuminations were given. Measurements were done at $5^{o}C$. Further details are given in the legends.

RESULTS AND INTERPRETATION

Figure 1A-D shows the P515 response of dark-adapted broken chloroplasts upon single (B,D) and double (A,C) flash illumination in the absence (A,B) and presence (C,D) of 1 um DBMIB. The net effect of the second flash under both conditions, obtained by subtracting the curves A and C, and B and D, respectively are shown in fig. 1E and F. Finally the effect of DBMIB on the response in the second flash, obtained by subtracting curves E and F, is shown in fig. 1G. This DBMIB-sensitive signal shows a slow rise completed in less than 40 ms and a decay which is completed in about 150 ms. The DBMIB effect on the first flash, conclusive with results of others (Selak and Whitmarsh, 1982, Hope and Matthews, 1987) is more complex and different from its effect in the second flash. One can easily see by

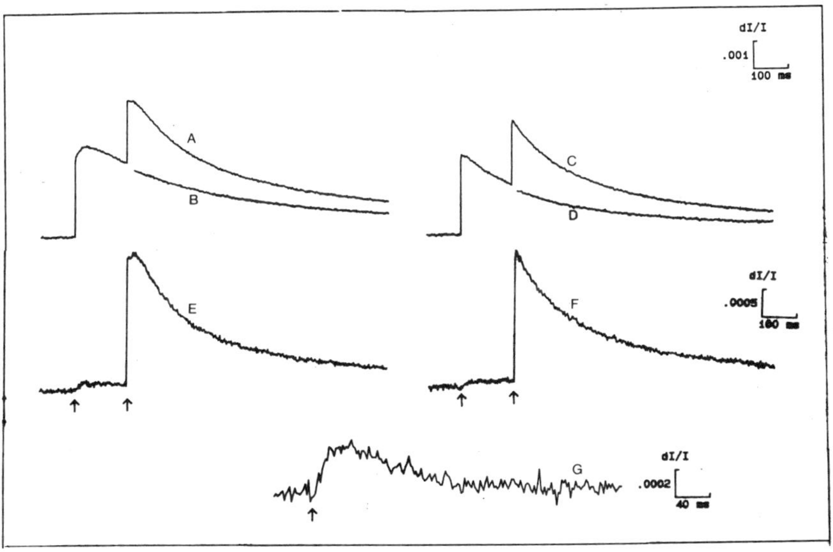

Figure 1. P515 response of intact chloroplasts, dark adapted for more than 2 hr, upon one (B,D) and two (A,C) single turnover saturating flashes in the absence (A,B) and presence (C,D), respectively of 1 μM DBMIB. Trace E(= A-B) and F (=C-D) represent the net effect caused by the second flash in the absence (E) and presence (F) of DBMIB. Trace G (= E-F) represents the effect of DBMIB on the response in the second flash. Note the different time scale in trace G.

subtracting figs. 1B and D that the difference (i.e. DBMIB-sensitive) signal in the first flash consists of a slow rise and a decay which is much slower than in the first flash and has not reached the zero level after 1 sec. We interprete these effects as follows. The DBMIB sensitive signal, isolated from the second flash, reflects the DBMIB inhibition of an electrogenic reaction, probably caused by secondary electron transfer through an endogeneous Q-cycle. This component is also present in the first flash. However, the non-electrogenic Reaction 2 component which has been shown to be abundant in the first flash, and much less or even absent in the second flash (Schapendonk, 1980), is also inhibited by DBMIB and consequently contributes to the DBMIB-sensitive difference signal. The low contribution of Reaction 2 in the second flash causes that the effect of DBMIB on the response in the second flash is mainly, if not exclusively due to a secondary reaction different from Reaction 2. The kinetics of this reaction would not violate a conclusion that it is associated with an electrogenic reaction in which an endogeneous Q cycle is involved. Therefore we call this component Reaction 1/Q.

The data and interpretations of fig. 1 are supported by the results of the following experiments. As shown before (Peters et al., 1984) CCCP at low concentration causes the inhibition of reaction 2. The DBMIB-sensitive signal in the presence of CCCP measured in a first flash is similar, if not identical to that of Reaction 1/Q, e.g. fig. 1G. An example is shown in fig. 2. This figure clearly shows the difference and distinction between the two slow components of the P515 response: Reaction 2, isolated here as the CCCP-sensitive part, and Reaction 1/Q, isolated as the DBMIB-sensitive part in the absence of Reaction 2.

As shown by others (Selak and Whitmarsh, 1982) a DBMIB-sensitive slow component can be induced by adding DQH_2 to a chloroplast suspension. The complementary signal which arises after the addition of DQH_2, as shown in fig. 3 has similar if not identical reaction kinetics as the DBMIB-sensitive signal in the absence of Reaction 2. It should be stressed that the decay below the zero level of the DQH_2 stimulated signal is variable for different samples, but significant. We presume that DQH_2 besides, stimulating a secondary electrogenic Q cycle, inhibits Reaction 2 to some extent. This inhibition gives rise to the observed undershoot. Indeed it was found that addition of DQH_2 in the presence of CCCP, causes a complementary signal similar, if not identical to that of reaction 1/Q (fig. 1G) with hardly, if at all an undershoot.

In conclusion our results suggest evidence for at least 3 distinguishable electrochromic components in the P515 signal in chloroplasts. These are listed in Table 1. The two slow components, of which only Reaction 1/Q is likely to be electrogenic because of its dark relaxation being equal to that of Reaction 1/RC, have different origin.

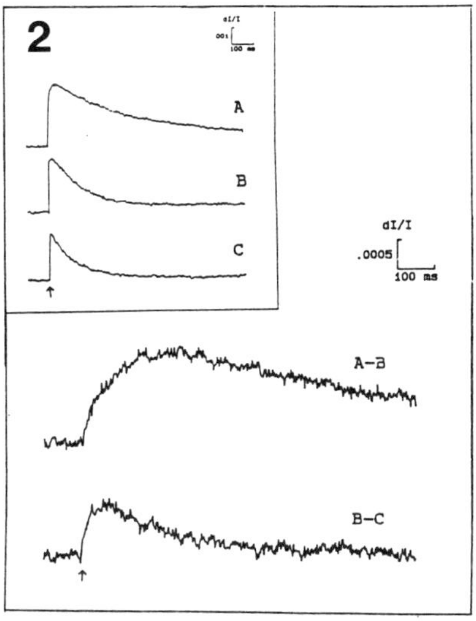

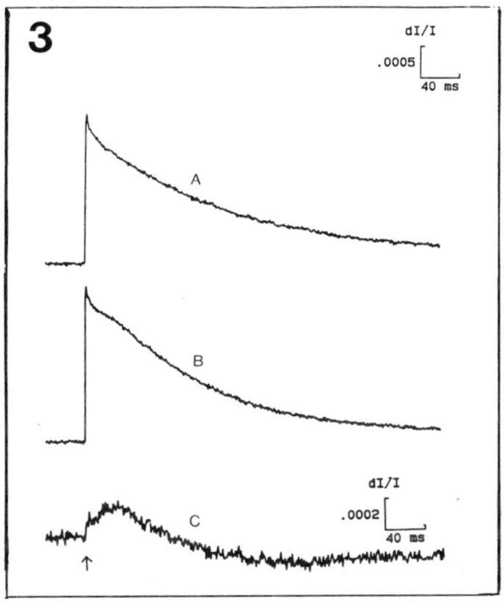

Figure 2. Insert: P515 response of dark-adapted intact chloroplasts in the absence (A), and presence of 0.4 uM CCCP (B,C) without (B) and with (C) 1 μM DBMIB. Curves A-B and B-C represent the CCCP-sensitive (reaction 2) and DBMIB-sensitive (reaction 1/Q) components of the responses (see further the text).

Figure 3. P515 response in broken chloroplasts which have been pre-illuminated (45s) in the presence of 300 μM ATP, 1 mM DTE and 60 μM MV, measured 3 min. after pre-illumination in the absence (A) and presence (B), respectively of 300 μM DQH_2 at 15°C. Trace C (=B-A) represents the DQH_2 stimulated signal.

Acknowledgement. This research was supported in part by the Stichting Scheikundig Onderzoek Nederland (SON), financed by the Netherlands Organization for the Advancement of Pure Research (ZWO)

TABLE 1.

P515-COMPONENTS

TYPE	(APPROX. REACTION TIME (ms) ON	OFF	INDICATOR
REACTION 1/RC (FAST)	<0.1	100	PRIMARY CHARGE SEPARATION RC
REACTION 1/Q (SLOW)	40	100	SECONDARY CHARGE SEPARATION Q-CYCLE
REACTION 2 (SLOW)	150	500	LOCAL INNERMEMBRANE ELECTRIC FIELDS

REFERENCES

Hope, A.B., Matthews, D.B. (1987). The slow phase of the electrochromic shift in relation to the Q-cycle in thylakoids. Aust. J. Plant Physiol. 14: 29-46.

Joliot, P., Delosme, R. (1974). Flash-induced 519 nm absorbance changes in green algae. Biochim. Biophys. Acta 357: 267-284.

Junge, W. (1977). Membrane potentials in photosynthesis. Annu. Rev. Plant Physiol. 28: 267-284.

Kooten, O. van (1988). Photosynthetic free energy transduction. Modelling electrochemical events. Thesis Agricultural University Wageningen (in press).

Peters, R.L.A. (1986). Electrochromic effects in relation to energy transduction and energy coupling in chloroplast membranes. Thesis Agricultural University, Wageningen.

Peters, R.L.A., Van Kooten, O., Vredenberg, W.J. (1984). The effect of uncouplers (F)CCCP and NH4Cl on the kinetics of the flash-induced P515 electrochromic bandshift in spinach chloroplasts. FEBS Letters 177, 11-17.

Schapendonk, A.H.C.M. (1980). Electrical events associated with primary photosynthetic reactions in chloroplast membranes. Thesis Agricultural University, Wageningen.

Schapendonk, A.H.C.M., Vredenberg, W.J., Tonk, W.J.M. (1979). Studies on the kinetics of the 515 nm absorbance changes in chloroplasts. FEBS Letters 100: 325-330.

Selak, M.A., Whitmarsh, J. (1982). Kinetics of the electrogenic step and cytochrome b6 and f redox changes in chloroplasts. Evidence for a Q cycle. FEBS Letters 150: 286-292.

Vredenberg, W.J. (1981). P515: A monitor of photosynthetic energization in chloroplast membranes. Physiol. Plant.: 53, 598-602.

The Ion Pumps: Structure, Function, and Regulation, pages 299–306

DETERMINATION OF THE FUNCTIONAL HOMOLOGY OF β SUBUNITS ISOLATED FROM VARIOUS F_1-ATPase COMPLEXES: Their Role in Catalysis and Coupled Proton-Translocation

Zippora Gromet-Elhanan[1]

Department of Biochemistry, The Weizmann Institute of Science, Rehovot 76100, Israel

INTRODUCTION

Electron-transport coupled ATP synthesis in respiratory and photosynthetic organisms is catalyzed by a reversible proton-translocating ATP synthase-ATPase complex which is composed of two distinct structures (Nelson, 1981; Futai and Kanazawa, 1983; Vignais and Satre, 1984). A readily solubilized catalytic ATPase sector (F_1), and an intrinsic membrane sector which is involved in proton fluxes (F_0). The F_1-ATPase from many different sources has a $\alpha_3\beta_3\gamma\delta\epsilon$ subunit structure and contains the catalytic site(s) for ATP synthesis, although in its soluble state it catalyzes only ATP hydrolysis. It is also involved in proton-translocation during catalysis, namely proton-coupled ATP synthesis and ATP-driven proton uptake, which are catalyzed only by the membrane-bound whole F_0F_1 complex. The mechanism of ATP synthesis and its coupling to proton-translocation are still unsolved. Their elucidation is dependent on a precise determination of the function of each of the individual subunits of the whole complex and on a detailed characterization of its catalytic site.

Studies aimed at the characterization of substrate binding sites on the F_1-ATPase and attempts to identify among them the catalytic binding site(s) have been carried out on a number of F_1-complexes using various affinity and photoaffinity analogs (Vignais and Satre, 1984). They have revealed up to six nucleotide binding sites on the F_1-complex that reside in its α and β subunits. Because of the subunit stoichiometry of $\alpha_3\beta_3$ it has been suggested that each molecule of α or β has one nucleotide binding site. They include both catalytic and noncatalytic sites, but their exact subunit location could not be clearly ascertained. A promising approach to

[1]Supported by a grant from the United States-Israel Binational Science Foundation (BSF) Jerusalem, Israel

the elucidation of this problem is the study of isolated, functionally active, α and β subunits.

Reconstitutively active β-subunits have been obtained from the photosynthetic bacterium *Rhodospirillum rubrum* (Philosoph et al., 1977), from various respiratory bacteria (Futai, 1977; Yoshida et al., 1977) and recently also from spinach chloroplasts (Richter et al., 1986). In the respiratory and chloroplast systems the β subunits have been isolated and purified from the respective F_1 complexes after their dissociation into a mixture of subunits. In the *R. rubrum* system the β subunit has been selectively extracted from the chromatophore membrane-bound F_0F_1 complex, leaving a β-less chromatophore preparation that lost all its ATP synthesis and hydrolysis activities. The *R. rubrum* β subunit (Rrβ) has been purified to homogeneity and optimal conditions were established for its reconstitution into the β-less crhomatophores, leading to full restoration of their lost activities.

The *R. rubrum* system offers a unique possibility to investigate both the isolated, reconstitutively active Rrβ and the resulting membrane-bound β-less RrF_0F_1 complex. On purified Rrβ substrate binding sites have been identified and their relationship to the catalytic site of the F_0F_1 enzyme complex has been clarified (Gromet-Elhanan and Khananshvili, 1984; Khananshvili and Gromet-Elhanan, 1984, 1985a, b). The β-less chromatophores have recently been used as a basis for formation of active hybrid F_0F_1 complexes through their reconstitution with purified β subunits that have been isolated from different sources (Gromet-Elhanan et al., 1985; Richter et al., 1986). Some new results of studies with the hybrid complexes will be summarized here, together with a short description of the experimental *R. rubrum* system.

RESULTS AND DISCUSSION

A. Characterization of the Experimental *R. rubrum* system

Fig. 1 illustrates the *R, rubrum* system used in these studies. Coupled chromatophores carry out light-induced electron transport, H^+-translocation, ATP synthesis and hydrolysis (Gromet-Elhanan, 1974; Binder and Gromet-Elhanan, 1974; Philosoph et al., 1977). Their extraction with LiCl resulted in complete loss of all ATP-linked reactions and this loss was correlated with the removal of the β subunit (Philosoph et al., 1977; Philosoph and Gromet-Elhanan, 1981). The β-less chromatophores retained, however, their capacity to take up protons during light-induced electron transport and form an electrochemical proton gradient (Binder and Gromet-Elhanan, 1974; Khananshvili and Gromet-Elhanan,

1982). Only a four subunit β-less F_1 complex composed of α,γ,δ, and ϵ, could be solubilized from these β-less chromatophores (Philosoph et al., 1977). Furthermore, an anti-β antibody raised against a purified preparation of Rrβ, was found to agglutinate coupled and reconstituted chromatophores but had no effect on the β-less ones (Philosoph and Gromet-Elhanan, 1981). The isolated purified Rrβ had no catalytic activity by itself, but its reconstitution into β-less chromatophores resulted in restoration of >90% of their ATP synthesis and hydrolysis activity (Philosoph et al., 1977; Khananshvili and Gromet-Elhanan, 1982). These results indicate that: (a) The β-less chromatophores have lost all their β subunit, but the other F_1 subunits remained attached to the membrane (Fig. 1), and (b) The extracted Rrβ does not act as a passive gate to proton-translocation through the F_0 channel, but is absolutely required for the catalytic activity of the RrF_0F_1.

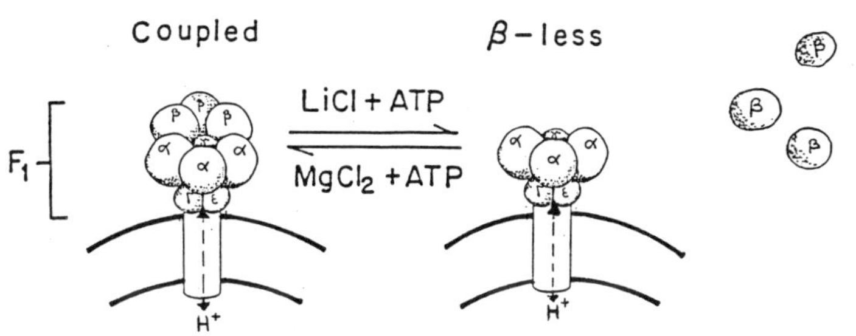

	Coupled	β-less	β
1. e⁻ Transport		+	–
2. H^+ Translocation		+	–
3. ATP Synthesis		–	–
4. ATPase — Ca^{2+}		–	–
4. ATPase			–
4. ATPase — Mg^{2+}		–	–

FIGURE 1. Summary of the reactions catalyzed by the native, coupled and β-less *R. rubrum* chromatophores and by the isolated β subunit. The procedures employed for extraction, reconstitution, purification of the extracted Rrβ, and for assays of activity were described in detail by Gromet-Elhanan and Khananshvili (1986).

Since Rrβ has no catalytic activity by itself we could measure directly its capacity to bind labeled ATP, ADP, and Pi. Such direct binding studies have revealed on Rrβ two binding sites for ATP and ADP and one for Pi (Gromet-Elhanan and Khananshvili, 1984; Khananshvili and

Gromet-Elhanan, 1984, 1985a). One of the nucleotide binding sites is a Mg-independent high-affinity site and one is a Mg-dependent low-affinity site. Pi was found to bind to Rrβ only in the presence of $MgCl_2$ and its binding was competitively inhibited by the presence of either ATP or ADP. ATP was, however, a much more potent inhibitor than ADP, thus suggesting that Pi binds at the site occupied by the γ-phosphate group of ATP, which remains empty in the presence of ADP.

Further evidence for this suggestion, as well as for the participation of the γ-phosphate subsite in the RrF_0F_1 catalytic activity, was obtained from studies with various modified Rrβ preparations (Khananshvili and Gromet-Elhanan, 1985b). Rrβ modified by the histidine reagent diethyl pyrocarbonate, or by the carboxyl group reagent Woodward's reagent K, was found to lose completely its Pi binding site as well as its capacity to bind ATP to the Mg-dependent site. On the other hand, binding of ATP to its Mg-independent site and binding of ADP to both sites were not affected by these modifications. Furthermore, Rrβ modified by either one of these reagents retained its capacity to rebind to β-less RrF_0F_1, but lost completely its ability to restore their catalytic activity. These results have demonstrated that Rrβ contains two substrate binding sites and have identified one of them, the Mg-dependent site, with the catalytic binding site of the RrF_0F_1 enzyme complex.

B. Evidence for the Functional Homology of β Subunits Isolated from Different Sources

The above described studies have confirmed earlier suggestions that the β subunit is the catalytic subunit of all F_0F_1 enzyme complexes, and demonstrated that the isolated, purified Rrβ is reconstitutively active, since its rebinding into β-less RrF_0-F_1 led to restoration of their ATP synthesis and hydrolysis activities. Assuming a similar mechanism of action for all F_0F_1 complexes one would expect their catalytic β subunits to exhibit a close structural and functional homology. A high degree of structural homology has been suggested on the basis of immunological studies (Philosoph and Gromet-Elhanan, 1981; Rott and Nelson, 1981) as well as from a comparison of amino acid sequences of $F_1\beta$ subunits from various photosynthetic or respiratory, eukaryotic and prokaryotic sources (Falk et al., 1985). The functional homology of various isolated β subunits has recently been investigated by employing the β-less *R. rubrum* chromatophores as a sensitive test system. Reconstitution of these chromatophores with either Ecβ (Gromet-Elhanan et al., 1985) or $CF_1\beta$ (Richter et al., 1986) has been found to result in restoration of their capacity to catalyze ATP-linked reactions, indicating the formation of active hybrid F_0F_1 complexes between the β-less *R. rubrum* chromatophores and the isolated $CF_1\beta$ or Ecβ. As in the native homologous system, the photophosphorylation activity of the hybrid systems was coupled to light-induced electron transport, being dependent on illumination and inhibited by the uncoupler gramicidin (Table 1).

TABLE 1: Restoration of light-induced ATP synthesis in β-less *R. rubrum* chromatophores by their reconstitution with Rrβ, Ecβ, or $CF_1\beta$

Reconstitution* System		ATP synthesis assayed under		
		illumination		darkness
Chroma-tophores	β-subunit	Gramicidin (-)	Gramicidin (+)	Gramicidin (-)
		(μ mol/hr/mg bacteriochlorophyll)		
Control	None	566	14	24
β-less	None	12	6	8
β-less	Rrβ	463	10	18
β-less	Ecβ	94	15	24
β-less	$CF_1\beta$	46	12	14

*Reconstitution and assay of photophosphorylation were conducted as described by Richter et al.(1986).

The native and hybrid systems showed, however, marked differences in a number of tests. Thus, they show large differences in sensitivity to various effectors (Table 2), which are clearly reflecting the different properties of the F_1 complexes from which the β subunits have been isolated. Sulfite, which has been reported to stimulate by several fold the Mg-ATPase activity of RrF_1 and CF_1 but not of EcF_1 (Webster and Edwards, 1977; Anthon and Jagendorf, 1983; Takeda et al., 1982), showed the same effect here. Also, tentoxin, which is a highly specific inhibitor of ATP-linked reactions in thylakoids (Selman and Durbin, 1978) blocked completely the $CF_1\beta$ hybrid, but did not affect any of the other systems. On the other hand the uncoupler gramicidin, stimulates only the native RrF_0F_1-ATPase. These results indicate that the activity of the hybrid complexes are indeed dependent on the presence of their specific β-subunits.

The absence of stimulation of both hybrid systems by gramicidin is probably related to the large differences observed in their ATP synthesis and hydrolysis activities. Whereas Rrβ did always restore both activities to the same extent (Philosoph et al., 1977; Khananshvili and Gromet-Elhanan, 1982), Ecβ and $CF_1\beta$ restored ATP hydrolysis much more efficiently than ATP synthesis (compare Tables 1 and 2). The Ecβ containing hybrid did even reach higher rates of ATP hydrolysis than those observed with control native *R. rubrum* chromatophores, when the reconstitution was carried out at pH 6.2 rather than the usual pH 8.0 (Gromet-Elhanan et al., 1985). The high rates of ATP hydrolysis (and low rates of ATP synthesis) were not due to any general uncoupling effect of Ecβ or $CF_1\beta$ on the chromatophore membrane, because: (a) ATP synthesis in the hybrids was as sensitive to gramicidin as in the native RrF_0F_1 (Table 1); (b) An identical light-induced quenching of quinacrine fluorescence was obtained with the hybrid and homologous chromatophores, and in both systems this light-induced quenching was completely eliminated by uncouplers (Gromet-Elhanan et al., 1985).

TABLE 2: Effect of sulfite, tentoxin and gramicidin on the Mg-ATPase activity of control and β-less *R. rubrum* chromatophores reconstituted with Rrβ, Ecβ, and $CF_1\beta$

Reconstitution system: Chromatophores	Reconstitution system: β subunit	Mg-ATPase activity assayed in presence of: no addition	sulfite	sulfite & tentoxin	sulfite & gramicidin
		(μmol/h/mg bacteriochlorophyll)			
Control	none	280	523	509	901
β-less	Rrβ	242	450	434	819
β-less	Ecβ	181	214	245	220
β-less	$CF_1\beta$	125	400	23	460

Surprisingly, the very efficient ATP hydrolysis activity of the hybrid F_0F_1 complexes is only very loosely coupled to proton translocation. Thus, in contrast to the native system, they have a very low capacity for ATP-induced quinacrine fluorescence (Gromet-Elhanan et al., 1985; Richter et al., 1986). The low rate of ATP synthesis and high rate of hydrolysis, which is only loosely coupled to proton-traslocation and is not stimulated by uncouplers, could all be due to a looser, less efficient interaction between Ecβ and $CF_1\beta$, as compared to Rrβ, and the other RrF_0F_1 subunits. Such a weaker interaction could enable the protons released during ATP hydrolysis to leak to the outside medium rather than move inside through the F_o sector (Gromet-Elhanan, 1987). The results presented here thus demonstrate (a) the functional homology of various F_1-β subunits in catalyzing ATP synthesis and hydrolysis, and (b) that these β subunits play an important role not only in catalysis but also in proton-translocation coupled to the catalytic activity.

REFERENCES

Anthon GE, Jagendorf AT (1983). Effect of methanol on spinach thylakoid ATPase. Biochim Biophys Acta 723: 358-365.

Binder A, Gromet-Elhanan Z (1974). Depletion and reconstitution of photophosphorylation in chromatophore membranes of *Rhodospirillum rubrum*. In Avron M (ed): "Proceedings of the Third Internaional Congress on Photosynthesis", Amsterdam: Elsevier, Vol. 2, pp. 1163-1170.

Falk G, Hamp A, Walker JE (1985). Nucleotide sequence of the *Rhodospirillum rubrum* atp operon. Biochem J 228: 391-407.

Futai M (1977). Reconstitution of ATPase activity from the isolated α, β, and γ subunits of the coupling factor F_1 of *Escherichia coli*. Biochim Biophys Res Commun 79: 1231-1237.

Futai M, Kanazawa H (1983). Structure and function of proton-translocating ATPase (F_0F_1): Biochemical and biological approaches. Microbiol Rev 47: 285-312.

Gromet-Elhanan Z (1974). Role of photophosphorylation coupling factor in energy conversion by depleted chromatophores of *Rhodospirillum rubrum*. J Biol Chem 249: 2522-2527.

Gromet-Elhanan Z, Khananshvili D (1984). Charaterization of two nucleotide binding sites on the isolated, reconstitutively active β subunit of the F_0F_1 ATP synthase. Biochemistry 23: 1022-1028.

Gromet-Elhanan Z, Khananshvili D, Weiss S, Kanazawa H, Futai M (1985). ATP synthesis and hydrolysis by a hybrid system reconstituted from the β subunit of *E. coli* F_1-ATPase and β-less chromatophores of *R. rubrum*. J Biol Chem 260: 12635-12640.

Gromet-Elhanan Z, Khananshvili D (1986). Selective extraction and reconstitution of F_1 subunit from *R. rubrum* chromatophores. Methods Enzymol 126: 528-538.

Gromet-Elhanan Z (1987). Characterization of the catalitic site on the β subunit of the H^+-ATP synthase-ATPasecomplex and demonstration of functional homology of β subunits isolated from different sources. In Biggins J (ed): "Progress in Photosynthesis Research", Dordrect: Martinus Nijhoff Publishers, Vol. 3, pp 63-69.

Khanashvili D, Gromet-Elhanan Z (1982). Isolation and purification of an active γ subunit of the F_0F_1 ATP synthase from chromatophore membranes of *R. rubrum*. J Biol Chem 257: 11377-11383.

Khananshvili D, Gromet-Elhanan Z (1984). Demonstration of two binding sites for ADP on the isolated β subunit of the *R. rubrum* RrF_0F_1 ATP synthase. FEBS Lett 178: 10-14.

Khananshvili D, Gromet-Elhanan Z (1985a). Characterization of an inorganic phosphate binding site on the isolated, reconstitutively active β subunit of F_0F_1 ATP synthase. Biochemistry 24: 2482-2487.

Khananshvili D, Gromet-Elhanan Z (1985b). Evidence that the Mg-dependent low-affinity site for ATP and Pi demonstrated on the isolated β subunit of the F_0F_1 ATP synthase is a catalytic site. Proc. Natl. Acad Sci USA 82: 1886-1890.

Nelson N (1981). Proton-ATPase of chloroplasts. Curr Top Bioenerg 11: 1-35.

Philosoph S, Binder A, Gromet-Elhanan Z (1977). Coupling factor ATPase complex of *R. rubrum*. Purification and properties of a reconstitutively active single subunit. J Biol Chem 252: 8747-8752.

Philosoph S, Gromet-Elhanan Z (1981). Antibodies to the F_1-ATPase of *R. rubrum* and its purified native β subunit. Eur J Biochem 119: 107-113.

Richter ML, Gromet-Elhanan Z, McCarty RE (1986). Reconstitution of the H^+-ATPase complex of *R. rubrum* by the β subunit of the chloroplast coupling factor 1. J Biol Chem 261: 12109-12113.

Rott R, Nelson N (1981). Purification and immunological properties of proton-ATPase complexes from yeast and rat liver mitochondria. J Biol Chem 256: 9224-9228.

Selman BR, Durbin RD (1978). Evidence for a catalytic function of the coupling factor 1 protein reconstituted with chloroplast thylakoid membranes. Biochim Biophys Acta 502: 29-37.

Takeda K, Hirano M, Kanazawa H, Nukiwa N, Kagawa Y, Futai M (1982). Hybrid ATPase's formed from subunits of coupling factor F_1's of *E. coli* and thermophilic bacterium PS3. J Biochem 91: 695-701.

Vignais PV, Satre M (1984). Recent developments on structural and functional aspects of the F_1 sector of H^+-linked ATPases. Mol Cell Biochem 52: 801-824.

Webster GD, Edwards PA, Jackson JB (1977). Interconversion of two kinetically distinct states of the membrane-bound and solubilized H^+-translocating ATPase from *R. rubrum*. FEBS Lett 76: 29-35.

Yoshida M, Sone N, Hirata H, Kagawa Y (1977). Reconstitution of adenosine triphosphatase of thermophilic bacterium from purified individual subunits. J Biol Chem 252: 3480-3485.

The Ion Pumps: Structure, Function, and Regulation, pages 307–314

RAPID KINETICS MEASUREMENTS OF Mg^{2+} AND Ca^{2+} IONS RELEASE FROM CF_1-ATPase

R. Hiller and C. Carmeli*

Department of Biochemistry, Tel Aviv University
Tel Aviv, Israel

INTRODUCTION

Subunits interaction is a major regulatory process in the H^+-ATPase of mitochondria, bacteria and chloroplasts. Negative cooperativity in nucleotide binding results in faster catalysis (Cross, 1981), while positive cooperativity in the binding of divalent metal ions causes a slowdown in presteady state catalysis (McCarty and Carmeli, 1982). Stroop and Boyer (1987) suggested that the interaction among the three catalytic subunits is essential for a sequential order of catalysis. Quantitative evaluation of the cooperativity was studied by measurements of binding and release of Mn^{2+} from CF_1. Mn^{2+} is a component of the MnATP complex which is the substrate of ATPase activity and therefore could be used as a substrate analogue. In these studies we have found that CF_1 contains three positively cooperative strong Mn^{2+} binding sites (Hiller and Carmeli, 1985). These were suggested to be the active sites because covalent modification of an arginine residue by naphthylglyoxal which inhibits catalysis (Takabe et al, 1982) also inhibits the cooperativity among the binding sites. Direct measurements by x-ray absorption (EXAFS) of the metal reveals the existence of tertiary complexes of enzyme, Mn^{2+} and ATP at these sites (Carmeli et al, 1986).

Assuming that the cooperativity interactions among the subunits are expressed as a change in the binding constants of ligands, the kinetics of ligand binding could

* This work was supported by a grant from BSF/87.

monitor the conformational changes. Measurements of the kinetics of Mn^{2+} binding to CF_1 supported this assumption (Carmeli et al, 1987). Kinetic analysis of the data could be best fitted to a model which assumes sequential changes in conformation on binding and release of the metal ions (Hiller et al, 1988). The rate constants were increased by an order of magnitude on activation of the latent enzyme and were of the same order as those of the presteady state kinetics of ATPase activity (Carmeli et al, 1981). In the present work we extended the analysis to two important metal ions. One of which is Mg^{2+} which functions under physiological conditions and the second is Ca^{2+} which catalyzes only hydrolysis of ATP. The kinetic behavior of Mg^{2+} was found to be essentially similar to that of Mn^{2+} and was also accelerated following the modification of the enzyme by naphthylglyoxal. The binding and release of Ca^{2+} were much faster than those of the two former ions thus pointing to a possible reason for its incapacity to catalyze ATP synthesis.

PROCEDURE

CF_1 was isolated from lettuce chloroplasts, stored as ammonium sulfate precipitate, desalted on G-50 Sephadex columns, heat activated and assayed for ATPase activity as previously described (Hiller and Carmeli, 1986). Naphthylglyoxal modified CF_1 was prepared according to the method of Takabe et at (1982). Measurements of the kinetics of cation release were done by rapid mixing of 80 uM CF_1 preincubated with up to 500 uM of the indicated cation with a chelator dye. The absorption changes were monitored by an Aminco DW-2 spectrophotometer equipped with a stopped flow apparatus (dead time 5 ms). Quin-2 and Arsenazo III were used to monitor the release of Ca^{2+} while Eriochrome black T was used for Mg^{2+} and Mn^{2+} ions. The binding experiments were conducted at pH 8.8 where the dissociation constants of the dyes is lower than 1 uM and the enzyme retains more than 75% of the activity. The kinetic model previously developed by us for the analysis of Mn^{2+} binding to CF_1 (Hiller et al, 1988) was used for this analysis. The model is constructed from a set of nonlinear rate equations solved with the aid of DVERK subroutine from IMSL library based on Runge-Kutta algorithm (Hull et al,1976).

RESULTS AND DISCUSSION

Cation chelator dyes are used for determination of concentration and as indicators of kinetic changes in cation concentrations in biological systems (Tsien, 1983). At pH 8.8 Eriochrom black T (EBT) has a dissociation constant lower than 1 µM and a rather large extinctiong coefficient at the range of 0-0.3 mM Mg^{2+} (Fig.1). The binding of Mg^{2+} and Mn^{2+} as determined by stopped flow measurements was completed within 200 ms at this concentration range. EBT could therefore be used for measurements of the release of these ions from CF_1 having a K_d of 5 µM and a rate of release slower than 200 ms.

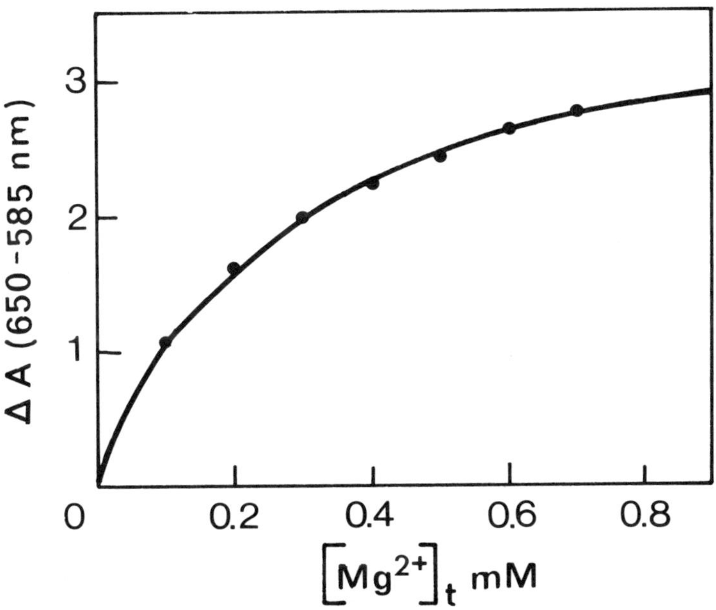

Fig.1. The effect of Mg^{2+} on the absorbance of Eriochrome black T (EBT).The absorbance difference at 650-585 nm of 200 uM EBT in 5mM HEPES-NaOH, pH 8.8 was measured in response to the change in Mg^{2+} concentration.

The release of Ca^{2+} was measured either with Arsenazo III or with Quin-2 both of which have a K_d of 0.1 µM and a binding rate constant faster than the mixing time of the stopped flow (5 ms). In all the experiments 5mM HEPES buffer was used as higher concentrations proved to interfere with the rate of cation binding. The rate of the releaseof bound Mg^{2+} was dependent on the concentration of the bound ion however it assumed a high kinetic order (Fig.2). In these experiments the concentration of the free

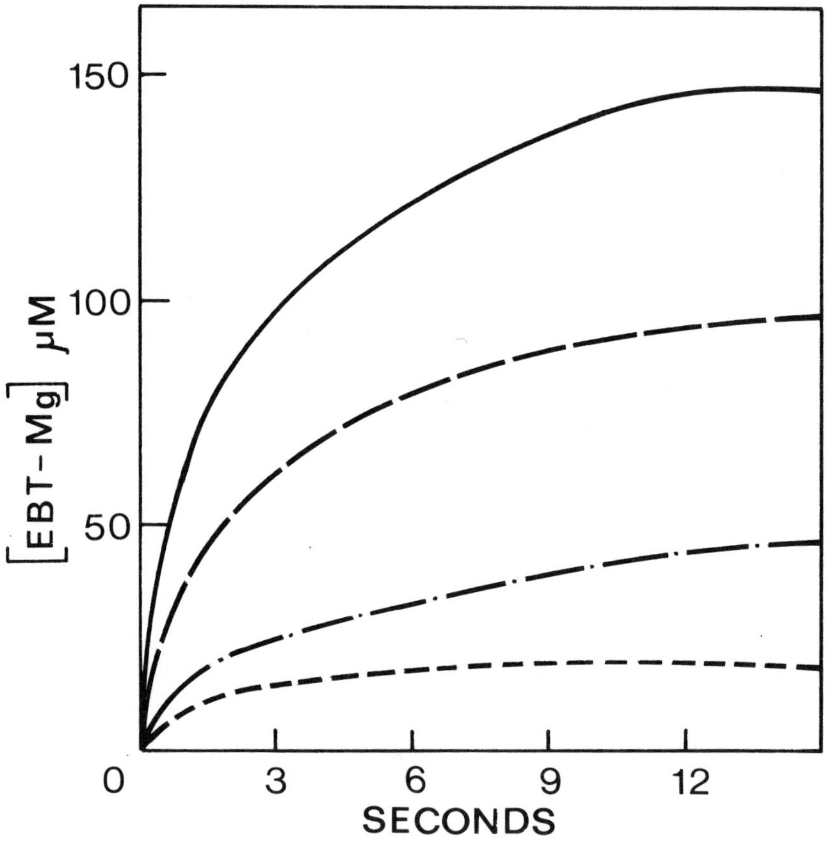

Fig.2. Time course of the release of Mg^{2+} from CF_1. Equal volumes of solutions containing 80μM latent CF_1, 5mM HEPES pH 8.8, and in Mg^{2+} concentrations of 40(- -),100 (-·-),200 (— —) or 300 μM (——) were mixed with 400μM EBT, 5mM HEPES pH 8.8 in a stopped flow apparatus. Increase in the absorption difference at 650-585 nm was converted to the concentration of Mg^{2+}.

ion was negligible and therefore is not apparent in the recording of the data. Modulation in the activation state of the enzyme resulted in changes in the rate of ion release. The rate increased when the latent enzyme was activated as was previously shown (Hiller et al, 1988) for Mn^{2+} ions. The faster rate was achieved with the naphthylglyoxal modified CF_1 (Fig.3). Similar kinetics was observed when Mn^{2+} release was measured with EBT (data not shown).

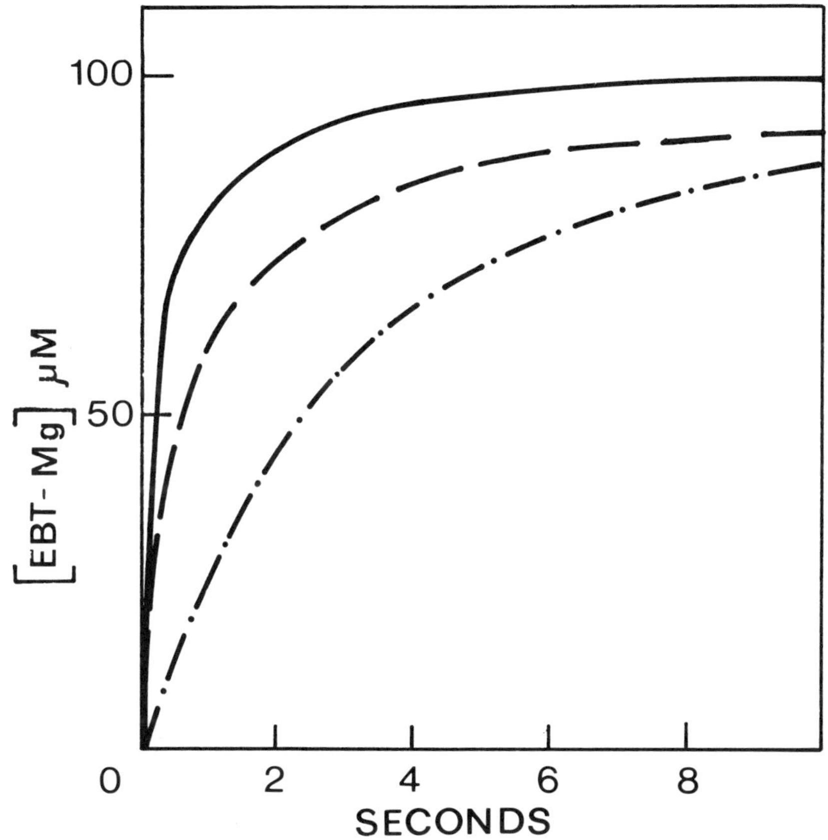

Fig.3. Release of Mg^{2+} from CF_1 at various states of activations. 80uM CF_1 preincubated with 200μM $MgCl_2$ were mixed with 400μM EBT as described under Fig.2. CF_1 was either latent(-•-), heat activated(- -) or chemically modified by naphthylglyoxal(——).

The data did not fit a first or second order kinetics. However, it could be fitted to a model developed by us for kinetic analysis of cooperative binding of Mn^{2+} to CF_1 (Hiller et al, 1988). The analysis is based on a model suggested by Adair (1925) assuming a sequential binding to three interacting sites as presented in scheme 1.

$$\text{Scheme 1: } E{+}M \underset{k_{-1}}{\overset{k_1}{\rightleftharpoons}} \overset{*}{E}M{+}M \underset{k_{-2}}{\overset{k_2}{\rightleftharpoons}} \overset{**}{E}M_2{+}M \underset{k_{-3}}{\overset{k_3}{\rightleftharpoons}} \overset{***}{E}M_3$$

E and E^* represent the enzyme in its native form and after it underwent a conformational change on binding of metal ion (M), respectively. The scheme is expressed as three differential mass law dependent rate equations. These were numerically solved and searched for solutions that best fitted the experimental data as described under "Procedure". The calculated kinetic constants for binding and release of Mg^{2+} for each of the three sites are given in Table 1.

Table 1. Kinetic rate constants of Mg^{2+} binding to CF_1

The kinetic constants were calculated from the experimental data as presented in Fig.2 using analysis as decribed under "Procedure". The random (r),the ordered (o) sequential and the noncooperative (n) models were used for calculation. The constants are given for the latent (L), heat activated (H) and the naphthylglyoxal modified (NG) CF_1.

Treatment/	Mg^{2+} binding (10^4.mole^{-1}.sec^{-1})			Mg^{2+} release (sec^{-1})		
	k_1	k_2	k_3	k_{-1}	k_{-2}	k_{-3}
LCF_1 (r)	8.00	4.00	0.08	0.20	0.11	0.05
LCF_1 (o)	8.00	8.00	0.23	0.60	0.21	0.05
HCF_1 (r)	11.00	7.00	3.00	0.70	1.95	0.19
HCF_1 (o)	11.00	14.00	9.00	1.85	3.85	0.19
$NGCF_1$(n)	2.90	0.12	0.10	1.28	1.02	0.04

The thermodynamic dissociation constants (not shown) were calculated for the latent enzyme to be 7.5, 2.6 and 19 µM for the three sites respectively. In the chemically modified enzyme the average Kd increased by an order of magnitude. The data could be fitted to both random and ordered sequential versions of the Adair model. Similar values were obtained in both cases, however a sequential model fits better the known asymmetrical properties of this enzyme. The asymmetry is examplified by the fact that under given conditions only one of the beta subunits binds ADP or becomes chemically modified (McCarty and Carmeli, 1982).In the case of naphthylglyoxal modified enzyme a non interacting model was used for analysis of the kinetics as no cooperativity was shown in the equilibrium metal binding studies for this form of the enzyme (Hiller and Carmeli, 1985). Activation of the enzyme cause a large increase in the kinetic rate constants with hardly any change in the Kd of the binding. A faster rate of conformational changes is expected to produce such an effect if it is the rate limiting factor in binding. The suggestion that the transmition of cooperative interaction might be the rate limiting step in catalysis is also in harmony with this observation. Yet, even the binding constants of the activated enzyme are more than an order of magnitude slower than the rate of catalysis. This is not surprising as the activated enzyme reaches full rate only after presteady state acceleration in the presence of ATP.

Almost 90% of the enzyme bound Ca^{2+} was released within the mixing time of the instrument as measured with either Arsenazo III or with Quin-2. Indicating that Ca^{2+} kinetics is an order of magnitude faster than the other two ions tested. The release of Ca^{2+} is much faster than that of Mg^{2+} and Mn^{2+}. About 90% of this ion were released from the latent CF_1 within the mixing time of the instrument giving a kinetic rate constant not smaller than 200 S^{-1}. In this case the conformational changes can not be a rate limiting step in the release of the ions. The lack of capacity of Ca^{2+} to bind in a manner which is affected by the conformational changes might be the reason why these ions can catalyze only hydrolysis of ATP but not its synthesis.Support for this suggestion is drawn from the observation that the rate of Mg^{2+} release from two of the binding sites in naphthylglyoxal modified enzyme, in which the cooperativity is inhibited, is faster than in any other form of CF_1 tested.

ACKNOWLEDGMENT: Our thanks to Dr. Gutman for helpful discussion in data analysis.

REFERENCES

Adair GS (1925). The haemoglobin system. VI. The oxygen dissociation curve of haemoglobin. J Biol Chem 63:529-545.

Carmeli C, Lifshitz Y, Gutman M (1981). Modulation by divalent metal ions of the autocatalytic reactivity of adenosinetriphosphatase from chloroplasts. Biochemistry 20:3940-3944.

Carmeli C, Huang JY, Mills DM, Jagendorf AT, Lewis A (1986). Extended x-ray absorption fine structure of Mn^{2+} and Mn^{2+}.ATP complex bound to coupling factor 1 of the H^+-ATPase from chloroplasts.

Carmeli C, Hiller R, Boyer PD (1987). Kinetics of Mn^{2+} to CF_1. In Biggins J (ed): "Progress in Photosynthesis Research", Dordrecht: Martinus Nijhoff, Vol III pp 123-126.

Cross RL (1981). The mechanism and regulation of ATP synthesis by F_1-ATPase. Ann Rev Biochem 50:681-714.

Hiller R, Carmeli C (1985). Cooperativity among manganese-binding sites in the H^+-ATPase of chloroplasts. J Biol Chem 260:1614-1617.

Hiller R, Carmeli C (1988). Kinetic analysis of cooperative intractions induced by Mn^{2+} binding to the chloroplast H^+-ATPase. (Submitted for publication).

Hull TE, Enright WH, Jackson KR (1976) "User's Guide for DVERK - A Subroutine for Solving Non-Stiff ODE's" University of Toronto.

McCarty RE, Carmeli C (1982). Proton translocating ATPase of photosynthetic membranes. In Govindjee (ed): "Photosynthesis, Energy Conversion by Plants and Bacteria," New York: Academic Press, pp 647-695.

Stroop SD, Boyer PD (1987). Catalytic and regulatory effects of light intensity on chloroplast ATP synthase. Biochemistry 26:1479-1484.

Takabe T, Debenedetti, Jagendorf AT (1982). Inhibition of chloroplast coupling factor by naphthylglyoxal. Biocim Biophys Acta 682: 11-20.

Tsien RY (1983). Intracellular measurements of ion activities. Ann Rev Biophy Bioeng 12: 91-116.

The Ion Pumps: Structure, Function,
and Regulation, pages 315–320

CUPRIZONE-INDUCED GIANT MITOCHONDRIA: A USEFUL SYSTEM FOR STUDIES ON ION PUMPS?

Valeria Petronilli and Mario Zoratti

C.N.R. Unit for the Physiology of Mitochondria and Institute of General Pathology, University of Padova, Italy

INTRODUCTION

"Giant" mitochondria have attracted considerable interest in the past (Tandler and Hoppel, 1986) because they have been associated with a variety of diseases, nutritional deficiencies, hormonal or drug treatments as well as with aging. Enlarged organelles are generally found to coexist with normal-size ones. Many of the studies concerned primarily ethiological and morphological aspects. When functional studies were carried out, they were performed almost invariably on subpopulations of mitochondria of near-normal size, which could be more easily purified. The assumption was made that the results would apply to the larger organelles as well.

Moving from a different point of view, giant mitochondria have been exploited as material for experiments involving micromanipulations of single organelles, impossible to carry out with normal mitochondria. Thus, giant mitochondria have been impaled with microelectrodes in an attempt to measure a transmembrane potential (Maloff et al., 1978), and used to measure diffusion coefficients of membrane proteins by the FRAP technique (Hochman et al., 1985). Most recently, they have become the material of choice for ongoing efforts to apply the patch-clamp technique to the study of the conductance properties of the outer (Tedeschi et al., 1987) and inner mitochondrial membranes. The latter approach holds obvious promise for studies on the ion pumps.

Among the various methods available to obtain giant

mitochondria, isolation from the liver of mice fed a diet containing cuprizone has so far always been chosen for work aiming to clarify the properties of mitochondria in general. It is obviously important to establish whether cuprizone liver mouse mitochondria (CMLM) are suitable material for this purpose, i.e., whether their membranes and membrane enzymes have the same properties as those of normal mitochondria. That this might not be so might be suspected from the fact that CMLM, unlike normal mitochondria, frequently come into contact within the cell, suggesting possible variations in surface charges. Also, they have been varously reported to possess a lowered (Hoppel and Tandler, 1973) or a normal (Wagner and Rafael, 1977; Hochman et al., 1982) respiratory activity. Our study aims to resolve this discrepancy and to establish whether the properties of the individual CMLM depend on their size.

RESULTS

One important feature of the production of CMLM, not emphasized in the literature, is the great variability of the effects of cuprizone. In an apparently casual manner, some mice developed the expected symptoms (lack of weight gain, weakness, low motility) and eventually died: the liver of these mice generally presented an abundance of large mitochondria, with the characteristics described below. Other mice, even within the same litter, showed no such symptoms, and survived indefinitely on the cuprizone-containing diet. Their livers contained mitochondria of near-normal size, which behaved like control mitochondria. Occasionally we encountered intermediate situations, with (statistically) somewhat enlarged mitochondria whose properties fell between the two extremes. This did not depend on the length of time the mice were fed the cuprizone-containing diet.

We routinely checked the homogenate of each liver separately, and pooled only the ones showing an abundance of large organelles. CMLM preparations were then obtained by various differential centrifugation protocols. These CMLM were extremely fragile. While the homogenate showed more or less normal respiratory control ratios (RCR), this parameter declined rapidly along the purification steps. The preparation procedures of Hochman et al. and of Wagner and Rafael in our hands produced uncoupled organelles. This characteristic fragility was not a consequence of the size of the mitochondria. We often obtained two or more fractions of different size distribution simply by centrifuging the same suspension at the same centrifugal force for different times: even the smallest mitochondria, with size distributions comparable to those of normal MLM, were rapidly uncoupled by procedures such as centrifugation at 1000 g and gentle resuspension by swirling.

We found respiration to be inhibited at least at the second site of the mitochondrial respiratory chain. The extent of inhibition varied between 20 and 50%, presumably reflecting the variability in the effects of cuprizone already mentioned. Table I shows data from a typical experiment, together with data on the ATPase activity. We found the latter to depend markedly on whether the mitochondria had been centrifuged on a discontinuous sucrose density gradient (Hochman et al., 1985) or not. Contact with 0.5 M sucrose caused large increases of ATPase activity in both CMLM and controls (see also Table II).

More meaningful information could be obtained from a comparison of pump activities per unit of membrane surface area. Work is in progress to obtain reliable data of this type. Furthermore, it would be highly desirable to ascertain whether the observed inhibition is due to a lower density of pumps in the membrane or to a modification of the enzyme properties. Information on the number of bc_1 complexes can be obtained by extrapolation of the titration curves giving respiratory activity vs. amount of antymicin added (Kroeger and Klingenberg, 1973; Wagner and Rafael,

Table I. Respiratory and ATPase activities of CMLM and control mouse liver mitochondria(*).

	Respiration (natoms O $xmg^{-1}min^{-1}$)		ATP Hydrolysis (nmols $xmg^{-1}min^{-1}$)
	NADH	Succinate	
CMLM	174	112	656
Controls	166	146	691

(*) The mitochondria were subjected to hypotonic shock to achieve maximal rates of respiration and ATP hydrolysis.

1977). Titrations of this type are shown in Fig.1.

To determine whether the size of the CMLM had a bearing on the properties of their membranes, in a few experiments we obtained two or more subpopulations of CMLM from the same pool of liver homogenates, determined their statistical size distributions using a Zeiss-Kontron IBAS 2000 image enhancement/processing apparatus, and measured the relevant biochemical parameters. The mitochondria were assimilated to spheres, without applying corrections for the presence of cristae. Table II presents the results of two such experiments. The higher rates of ATP hydrolysis in experiment n.1 are due to the use of a discontinuous sucrose gradient for that preparation. Within the limits of the assumptions mentioned, our data suggest that no significative differences exist in the activities, normalized to unit of membrane surface, of the respiratory chain or of the ATPase of CMLM obtained from the same liver, regardless of size.

DISCUSSION

The study of the effects of cuprizone administration and of the properties of CMLM is complicated by temporal

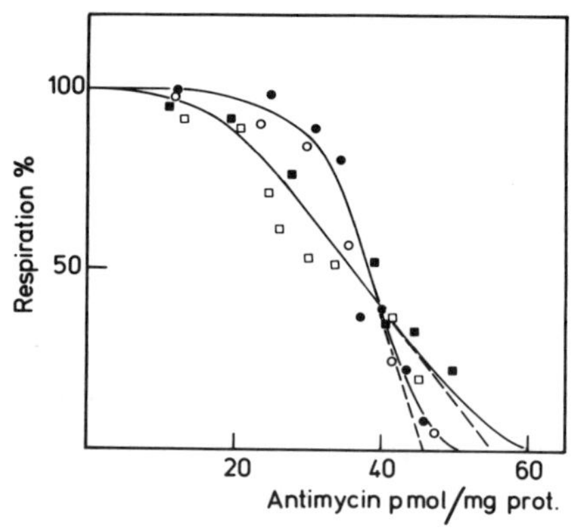

Fig.1. Antimycin titrations of succinate respiration by hypotonically shocked preparations of CMLM (dots) and of control MLM (squares). Extrapolations: 55 and 46 pmols Antimycin/mg.protein.

Table II. Surface-normalized rates of respiration and ATP hydrolysis in CMLM preparations of different size distribution(*).

	Mean surface (μ^2)	Respiration (g.atoms O xμ^{-2}min^{-1}) x10^{17}	ATP Hydrolysis (moles xμ^{-2}min^{-1}) x10^{17}
Expt.1	15.8	0.73	9.0
	9.4	0.81	10.3
Expt.2	16.2	0.82	6.1
	10.9	0.79	5.9

(*) The two preparations within each experiment were obtained from the same livers.

patterns (Tandler and Hoppel, 1986) and by the variability of effects (Suchy and Cooper, 1974). We could not pinpoint any particular cause for this variability: weaning age does not appear to be the only factor. Possibly the general state of health of the mouse, or whether it was well nourished by the mother, are important determinants. Some of the contrasting results in the literature might well be due to this heterogeneity of the effects: in most cases studies were carried out pooling together a number of livers, some of which might well have contained near-normal mitochondria. The tendency to use fractions containing small organelles only, together with this variability, may account for reports of coupled preparations (Hochman et al., 1985; Maloff et al., 1978). On the other hand, our observations, as well as others (Suchy and Cooper, 1974), indicate that at least a large fraction of the CMLM have normal coupling properties <u>in situ.</u> The degree of coupling of the organelles seems then to depend critically on the the amount of manipulation they are subjected to.

The properties of the mitochondrial membrane after isolation are of importance for electrophysiological studies. A high basal conductivity of the CMLM membranes is to be expected a priori, unless extreme care is exerted during isolation. The reasons for this fragility, once size itself is discounted, are speculative. The lipid composition of the membranes has been reported to be approximately normal (Wagner and Rafel, 1977; Hochman et al., 1985).

Our data show that indeed cuprizone treatment results in a decrease of the activity of the respiratory chain (Table I). This appears to be due at least in part to a reduction of the number of bc_1 complexes per mg. protein (Fig.1). Further work is however needed, and is being carried out, to safely conclude that the pumps of the CMLM, although reduced in number in at least one case, behave normally and are therefore suitable for study. After normalization for membrane surface, we could find no significant dependence on size of any of the properties of CMLM (Table II). The lowered respiratory activity seems therefore to be a consequence of the cuprizone treatment itself, and not a by-product of the process leading to the formation of the giant organelles.

REFERENCES

Hochman J, Ferguson-Miller S, Schindler M (1985) Mobility in the mitochondrial electron transport chain. Biochemistry 24:2509-2516.

Hoppel CL, Tandler B (1973) Biochemical effects of cuprizone on mouse liver and heart mitochondria. Biochem Pharmacol 22:2311-2318.

Kroeger A, Klingenberg M (1973). Further evidence for the pool function of ubiquinone as derived from the inhibition of electron transport by Antimycin. Eur J Biochem 39:313-323.

Maloff BL, Scordilis SP, Reynolds C, Tedeschi H (1978). Membrane potentials and resistances of giant mitochondria. Metabolic dependence and effects of Valinomycin. J Cell Biol 78:199-213.

Suchy J, Cooper C (1974). Isolation and respiratory meaurements on a single large mitochondrion. Exp Cell Res 88:198-202.

Tandler B, Hoppel CL (1986). Studies on giant mitochondria. Ann N Y Acad Sci 488:65-81.

Tedeschi H, Mannella CA, Bowman CL (1987). Patch clamping the outer mitochondrial membrane. J Memb Biol 97:21-29.

Wagner T, Rafael J (1977). Biochemical properties of liver mitochondria induced by chloroamphenicol or cuprizone. Exptl Cell Res 107:1-13.

The Ion Pumps: Structure, Function, and Regulation, pages 321–326

LIGHT INDUCED INTRAMOLECULAR PROCESSES OF BACTERIORHODOPSIN MONITORED BY TIME RESOLVED FTIR (FOURIER TRANSFORM INFRARED) SPECTROSCOPY

Klaus Gerwert, Benno Hess

Max-Planck-Institut für Ernährungsphysiologie,
Rheinlanddamm 201, 4600 Dortmund, FRG

The retinal protein bacteriorhodopsin transduces lightenergy into electro chemical energy by a vectorial proton transport across the membrane (1). In order to understand the mechanism of the protonpump on a sub-molecular level the reactions of intramolecular groups during the photocycle of lightadapted bacteriorhodopsin (BR) are investigated by FTIR-difference-spectroscopy. The initial photochemical step leads to J_{610} (J) which decays thermally through the intermediates K_{610} (K), L_{550} (L), M_{412} (M) and O_{640} (O). The pathway of the photocycle, especially the M decay is not unequivocally be determined (2).

The intermediates K, L and M can be stabilised at deep temperature. Therefore BR-K, BR-L and BR-M difference spectra can be obtained by static deep temperature FTIR spectroscopy. The results of these FTIR experiments gave evidence for protonation changes of internal aspartic acids and single bond isomerization of the chromophore retinal (3,4,5). Thereby, the transduction of lightenergy into phototransfer reaction can be explained by a mechanism in which the chromophore acts as a proton switch between two internal carbocyclic acids: light-induced isomerization around the $C_{13} = C_{14}$ double and C_{14}-C_{15} single bond breaks the salt bridge (6) between the asp 2 counterion and the positive chromophore charge and the proton is transported near a protonated

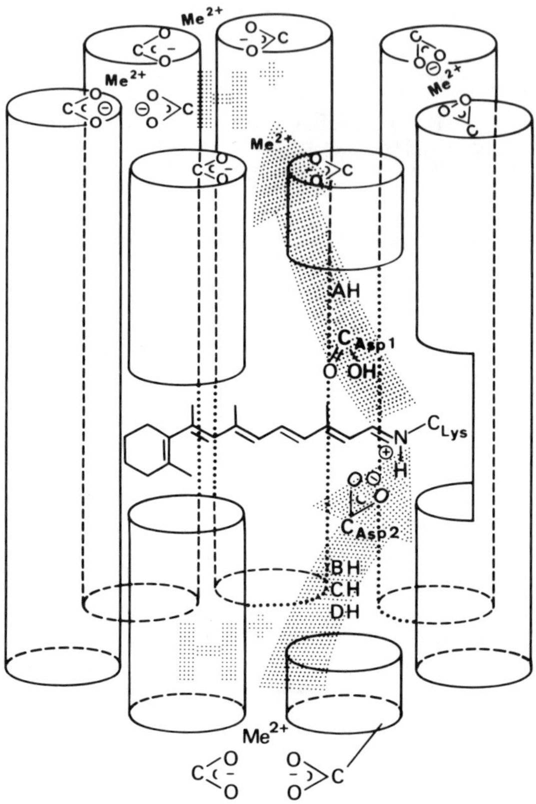

Fig. 1: The chromophore transported by a light-induced isomerization its positive charge near Asp 1 and initiated thereby a proton transfer to the outside of the protein

aspartic acid (Asp 1) (Fig. 1). In order to stabilise the positive chromophore charge again Asp 1 is deprotonated during the K to L transition. Thereby the proton transfer in the protein is initiated. For details see (5,7).

In the current investigation the proton transfer from the surface of the protein to the active

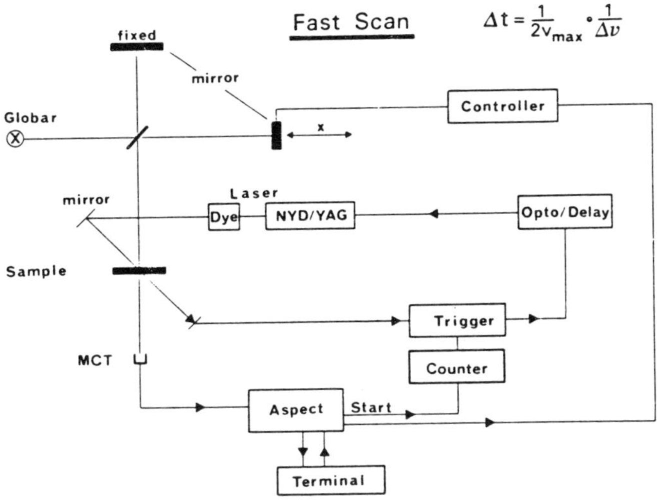

Fig. 2a: Apparatus for time resolved FTIR measurements

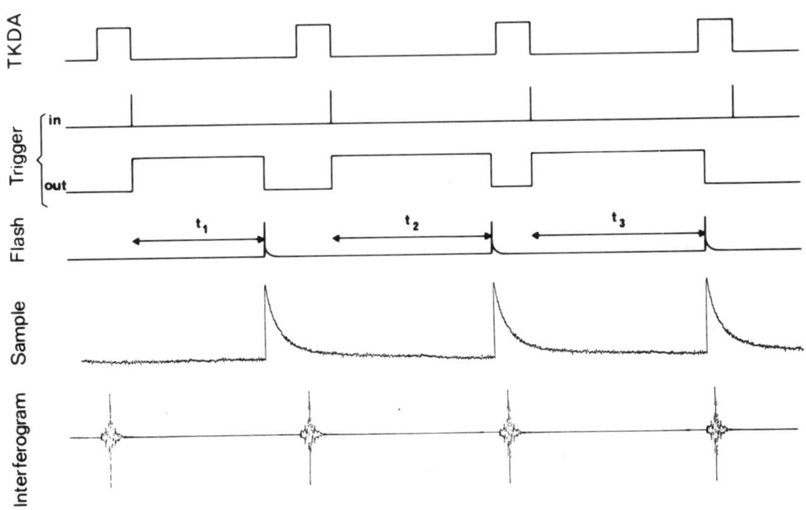

Fig. 2b: Time course of the apparatus.

center will be elucidated. Because the species rising in the M to BR reaction cannot be stabilised at deep temperature, an apparatus for time resolved FTIR measurements was developed (8). The apparatus is shown in Figure 2a and the instrumental time course in Figure 2b. When data of the FTIR apparatus are taken, a pulse of the aspect computer triggers a laser flash after a variable delay time ΔT_i. By this method spectra of different states of chromoproteins can be measured at room temperature. The time resolution Δt is determined by the velocity v_{max} of the scanner, the scanlength Δx and the spectral resolution Δv:

$$\Delta T = \frac{1}{2v_{max}} \cdot \Delta x = \frac{1}{2v_{max}} \cdot \frac{1}{\Delta\nu}$$

At v_{max} = 6.3 cm/sec a time resolution Δt = 10 ms at $\Delta\nu$ = 8 cm^{-1} spectral resolution is achieved.

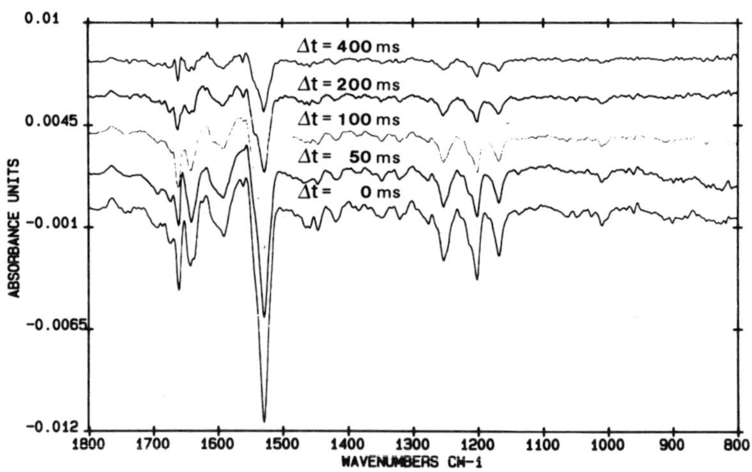

Fig. 3: Time resolved difference spectra at different delay times of Fe-bR

Fig. 3 show time resolved difference spectra with $\Delta v = 4\ cm^{-1}$ taken at different delay times ΔT after the laserflash of Fe-BR.
If cation free blue bacteriorhodopsin (9) was titrated by Fe-cations back to purple the Fe-bacteriorhodopsin shows a slowed down photcycle.
In Fig. 3 no positive bands are rising in the fingerprint region indicating that no O-Intermediate can be observed. The difference bands at about 1640 cm^{-1} and 1550 cm^{-1} can be interpreted as small conformational changes of the protein backbone involving a few groups only. Also a deprotonation of an internal carbocyclic acid, indicated by the negative band at 1740, is seen.

Fig. 4a shows the transmission change in the visible at 570 nm of a bacteriorhodopsin sample indicating the photocycle of BR_{570} In Fig. 4b the log (Δ Absorbance) of the C=C stretching vibration at 1525 cm^{-1} for different delay times is shown in comparison to the log (Δ Absorbance) at 570 nm of the same sample. The results show within the experimental error the same experimental results for the absorbance change at 570 nm and at 1525 cm^{-1}.

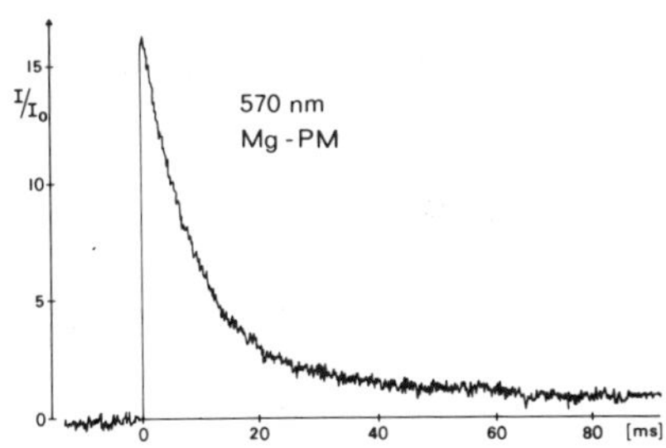

Fig. 4a: Flash photolysis signal at 570 nm

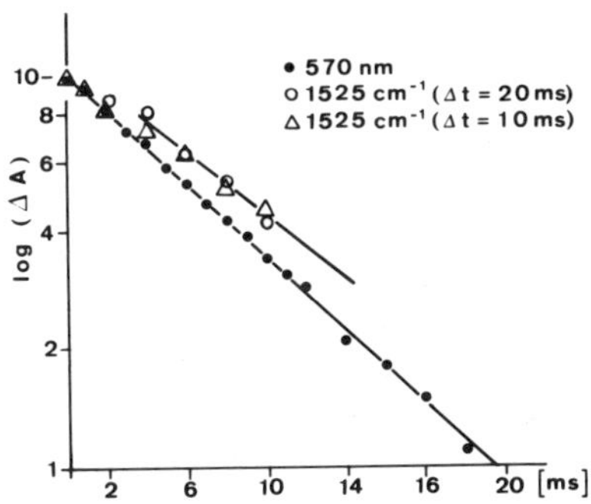

Fig. 4b: log (Δ Absorbance) of measurement in the visible and IR

With this method simultaneously the dynamics of the chromophore and protein groups can be monitored. Currently, the kinetics of the different groups are analysed by multipotential fit procedures. Thereby, the internal dynamic of the protein will be determined.

Acknowledgement

We would like to thank R. Eichas-Nell and M. Schubert for their help with the measurements.

References

(1) W. Stoeckenius, R.A. Bogomolni, R.A. (1982) Ann. Rev. Biochem. 51, 587-616
(2) L.A. Parodi, R.H. Lozier, S.M. Bhattacharjee and J.F. Nagle (1984) Photochem. Photobiol. 40, 501-512
(3) K. Gerwert, W. Kreutz, W. Mäntele, F. Siebert, M. Engelhard and B. Hess (1984) Photochem. Photobiol. 39, 275
(4) M. Engelhard, K. Gerwert, B. Hess, W. Kreutz, F. Siebert (1985) Biochem. 24, 100
(5) K. Gerwert and F. Siebert (1986) EMBO 5, 805
(6) B. Honig. T.G. Ebrey, R.H. Callender, U. Dinur and M. Ottolenghi (1979) Proc. Natl. Acad. Sci. U.S.A. 76, 2503-2507
(7) K. Gerwert and B. Hess (1987) "Intramolecular Protontransfer in Bacteriorhodopsin Monitored by Fourier Transform Infrared Difference Spectroscopy", in: Interactions of Water in Ionic and Nonionic Hydrates, ed. H. Kleeberg, Springer-Verlag, Berlin Heidelberg
(8) K. Gerwert and B. Hess (1987) Microchemica Acta, in press
(9) K. Gerwert and B. Hess, FEBS Lett. (1987) 212, 39-44

The Ion Pumps: Structure, Function, and Regulation, pages 327–330

THE ION BINDING BY DARK ADAPTED RHODOPSIN

Eugenia Chirieri-Kovács, Alexandru Dinu and Vasile Vasilescu

Department of Biophysics, Medical Faculty, Bucharest 76241, Romania

INTRODUCTION

One of the most intriguing features of the rod photoreceptor cell is that, in the dark-adapted state, there is a large difference existing between Ca^{2+} concentration in the intradiscal space (∼20 mM) and in the cytoplasmic fluid (∼1 μM) (Fatt, 1982). There must be a Ca^{2+} pump in the disk membrane which maintain this difference in calcium content. The functional significance of the difference in Ca^{2+} concentration across the photoreceptor membrane is not known (Stryer, 1986). Despite the reach information we have at present concerning the c-GMP cascade as a transmitter of information from light-excited rhodopsin to the cell membrane, we don't know which side of the structural change of rhodopsin is triggering the c-GMP cascade, i.e., the nature of the structural signal of rhodopsin is not known. In this paper our attention is focussed on the structural changes of rhodopsin induced by light.

METHODS

The results are obtained by electrical conductivity measurements on rhodopsin extracted from frog photoreceptor membranes. 48 hour dark-adapted frogs were decapitated; retinas were excised from isolated eyeballs and rhodopsin extracted by technique described by Fong et al. (1982). Two mild, non-ionic detergents were used for rhodopsin extraction: digitonin and Triton X-100. pH of the extracts was not adjusted since we avoided to introduce ions into the solution. All conductivity measurements were thus performed at pH 5 which proved to remain constant throughout the experiment.

Rhodopsin extracts were succesively diluted by triple distilled and deionized water (pH 5, γ = 6,7 µS) and electrical conductivities of the obtained solutions were measured.

RESULTS AND DISCUSSION

It was found that at concentrations of the extract at which rhodopsin is released from detergent micelles (below CMC of the detergent) the behaviour of illuminated and dark-adapted rhodopsin is quite different (Figs. 1A and 2); while conductivity of illuminated rhodopsin increases with progressing dilution, conductivity of dark-adapted extracts drops, reaching (in Triton extracts) values far below that of the distilled water ! (Fig. 2). This tendency is much attenuated if the solvent is D_2O (Vasilescu et al., 1986) (Fig. 1B).

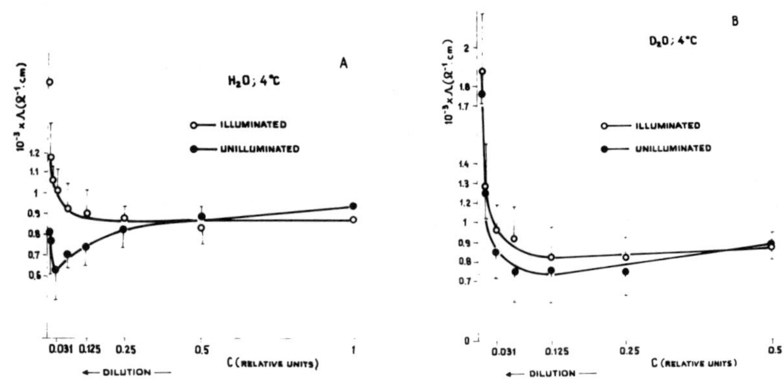

Figure 1. A plot of conductance vs. concentration of the rhodopsin extract diluted by H_2O (A) and by D_2O (B). •, dark-adapted rhodopsin; o, bleached rhodopsin. A concentration of 3.2×10^{-5} M rhodopsin is taken as unity on the abscissa.

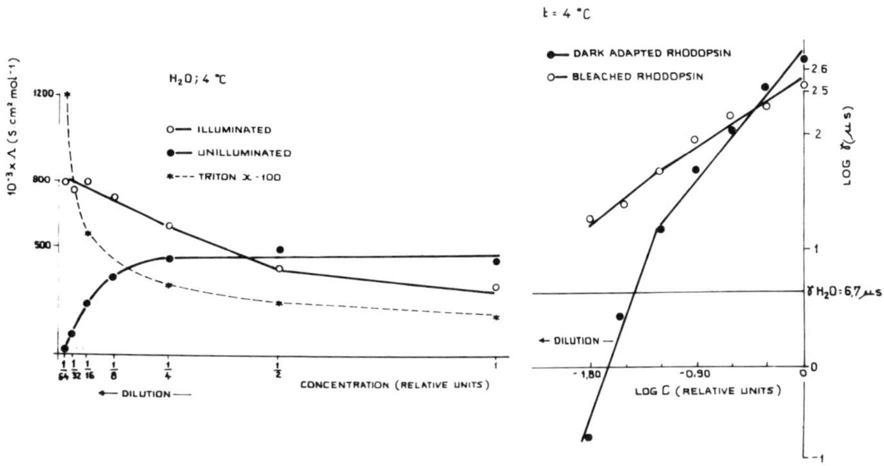

Figure 2. A plot of conductance (left) and a logarithmic plot of specific conductance (right) vs. concentration of the rhodopsin extract. •, dark-adapted rhodopsin; o, bleached rhodopsin; *, Triton X-100. Solvent: H_2O. A concentration of 3.2×10^{-5}M rhodopsin/1 % Triton X-100 is taken as unity on the abscissa.

The only available explanation of the observed phenomenon is that, when released from detergent micelles, dark-adapted rhodopsin molecules aggregate; the new formed aggregates bind or trap virtually all ions in solution, causing thus the sudden drop of the electric conductance.

Which is the biological significance of the above physico-chemical findings ? We have found that, in extracts, the dark-adapted rhodopsin has a greater capacity of binding or trapping ions than the bleahed one. If that would be true _in situ_ it would be enough for triggering the enzymatic events which lead to the excitation of the cell membrane.

Our previous electrophysiological studies are suggesting that the property of rhodopsin we observe _in vitro_ might have a physiological correspondence (Chirieri-Kovács and Vasilescu, 1982). It was found, for instance, that the latency of the ON response of retinal receptors is not affected by deuteration of retina, while the response itself is vanishing very quickly (Fig. 3). That could mean that rhodopsin molecules which have not been yet deuterated trigger in time the cell excitation but getting fewer with progressing deuteration, they generate a progressively smaller receptor potential.

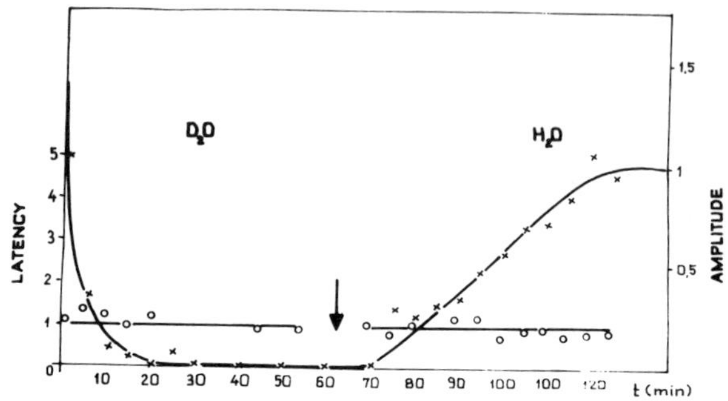

Figure 3. Deuteration effect on the amplitude (x) and latency (o) of the ON response of retina. Arrow marks the change of the immersion solution. The initial values are taken as units.

If releasing ions is a property of illuminated rhodopsin which trigger the generation of receptor potential, inhibition of that property by D_2O could explain the effect of deuteration on retina biopotentials.

REFERENCES

Chirieri-Kovács E, Vasilescu V (1982). Water involvement in the mechanisms of the retina electrical activity. Physiol. Chem Phys 14: 281-291.

Fatt P (1982). An extended Ca^{2+}-hypothesis of visual transduction with a role for cyclic GMP. FEBS Letters 149: 159-166.

Fong SL, Tsin A, Bridges CDB, Lion GI (1982). Detergents for extraction of visual pigments. In Packer L (ed): Methods in Enzymology 81: Acad. Press, pp 133-140.

Stryer L (1986). Cyclic GMP cascade of vision. Ann Rev Neurosci 9: 1-51.

Vasilescu V, Dinu A, Aricescu I, Chirieri-Kovács E (1986). On the electrical conductivity of rhodopsin solutions. Biochim Biophys Acta 849: 172-174.

The Ion Pumps: Structure, Function, and Regulation, pages 331–336

METAL ION NMR STUDY OF CATION TRANSPORT THROUGH THE GRAMICIDIN CHANNEL

J.F. Hinton, D.C. Buster, P. Easton and T. Privett

Department of Chemistry and Biochemistry, University of Arkansas, Fayetteville, Arkansas 72701

The transport of ions across biological membranes plays an important role in many different physiological processes. Most systems involved in cation transport, such as the Na^+ and K^+ channels of nerve axon membranes, are large protein complexes. Voltage gated, ion-permeable channels in the membranes of excitable cells have the general characteristics of being cation selective for the monovalent cations and being able to maintain a high transport rate through the channel.

It has been shown that relatively small antibiotics have the ability to induce selective transport of cations across membranes. These include the carrier-type antibiotics, such as valinomycin, and channel formers, such as gramicidin. The gramicidin family of antibiotics has been chosen for study since it represents a biologically viable channel system of related peptides of known structure in which specific changes in structure and the placement of polarizable substituents can be correlated with channel selectivity, stability, lipid dependence and transport. Gramicidin-A is a linear polypeptide whose dimeric form has been shown to induce monovalent cation permeability in membranes (Hladky and Haydon, 1970; Meyers and Haydon, 1972; Andersen, 1983; Levitt, 1978). It has a number of characteristics similar to those of physiological channels, such as monovalent cation selectivity and blocking, saturation effects, concentration dependence of permeability ratios, multivalent anion screening effects and divalent damping of single channel currents (Sandblom et al., 1977; Apell et al., 1977; Bamberg and Lauger, 1977).

The movement of an ion through a gramicidin channel involves 5 distinct steps: (1) diffusion through the aqueous phase up to the channel; (2) association with the channel entrance; (3) translocation through the channel; (4) dissociation from the channel and; (5) diffusion away from the channel. These steps are shown in Fig. 1 (Etchebest and Pullman, 1985). An understanding of each step is necessary for the characterization of the selectivity and transport of a channel system, such as gramicidin.

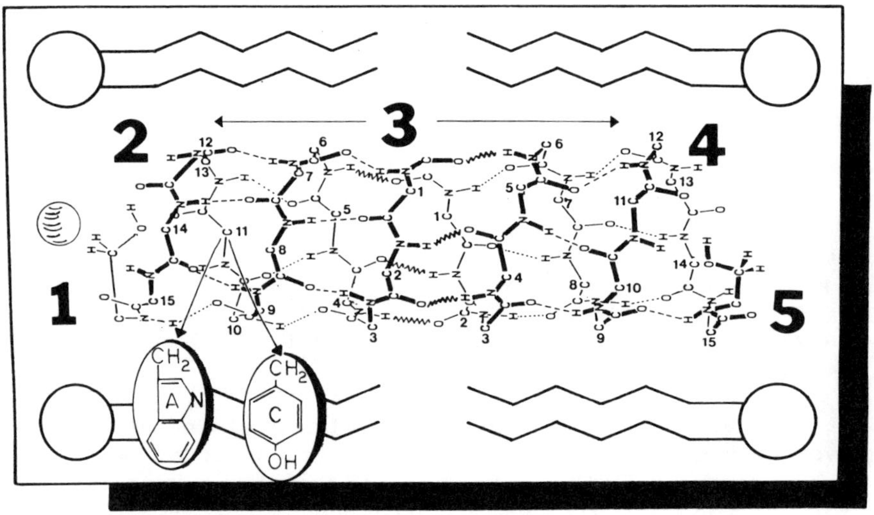

Fig. 1. Gramicidin dimer in lipid environment

Metal-ion NMR spectroscopy has been used to study the gramicidin facilitated transport of monovalent cations across the lipid bilayers of large unilamellar vesicles under equilibrium conditions. Through the use of a chemical shift reagent, such as dysprosium(III) tripolyphosphate (Gupta and Gupta, 1982), placed in the external aqueous environment, one can observe the NMR signals of the internal and external monovalent cation (e.g., Li^+, Na^+, K^+, etc.) pools. Once the internal and external NMR signals are separated, changes in them may be used to study the kinetics of transmembrane ion transport. We report here the preliminary results of the study of the gramicidin facilitated transport of Na^+ and Li^+ across the bilayers of large unilamellar vesicles under ionic equilibrium conditions. The technique presented is potentially very useful in probing a number of the properties of the gramicidin channel.

The equilibrium experiment allows one to study: (1) the equilibrium exchange rate of the monovalent cations through the gramicidin channel, (2) bi-ionic flux ratios of different cations, (3) the effect of single amino acid substitution in gramicidin on the transport rate.

The spectrum at the top of Figure 2 shows the Na^+-23 NMR signals of the internal (small) and external (large) sodium pools in the aqueous vesicle system under ionic equilibrium conditions.

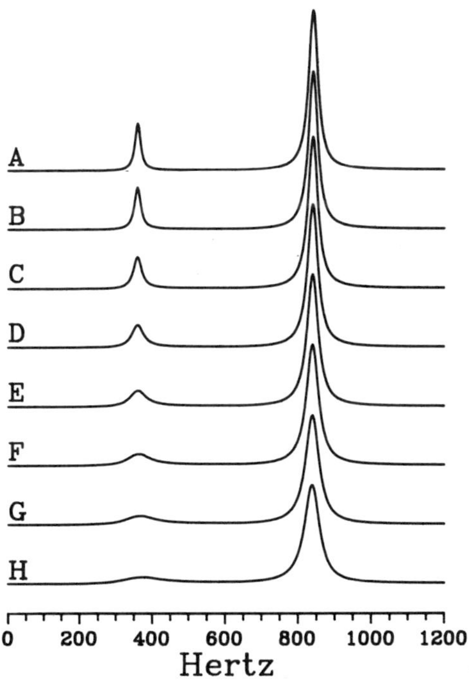

Fig. 2. Na-23 spectra with A(0), B(1.4), C(2.8), D(3.2) E(4.6), F(6.0), G(7.4) and H(9.0) μM gramicidin

The incorporation of increasing amounts of gramicidin-A into the vesicles produces very rapid exchange of the cations between the internal and external aqueous environments. The increasing exchange rate manifests itself in the linebroadening of the two NMR signals, particularly the internal Na^+-23 signal. The linewidth at half-height is proportional to the transport rate through the gramicidin channel.

Figure 3 shows the relationship between the gramicidin-A concentration and simultaneous transport of the Na^+-23 and Li^+-7 cations in the same vesicle system. The transport rate ($1/\tau$) is seen to be a linear function of the square of the gramicidin concentration for both cations.

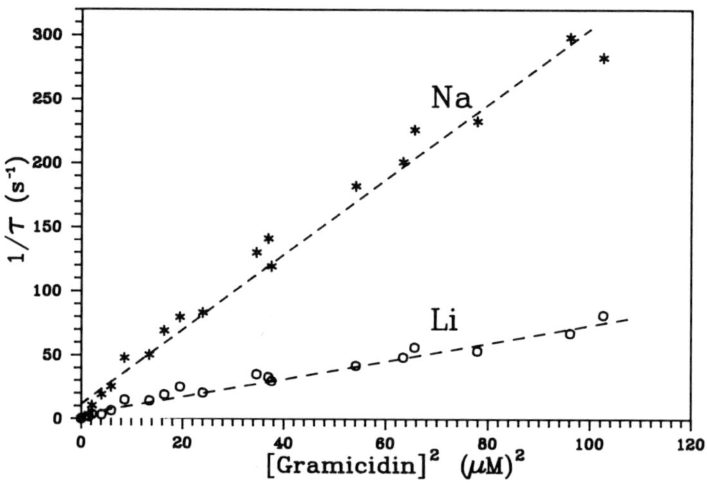

Fig. 3. Transport rate vs. $[\text{gramicidin}]^2$

A kinetic plot of the log of the transport rate as a function of the log of the gramicidin concentration for both cations gives a line whose slope is equal to the order of the reaction with respect to the gramicidin concentration. The slope of each line in Fig. 4 is 2. This second order dependence of transport rate on gramicidin concentration is consistent with the fact that the viable channel is a dimer. It is also very clear that the transport rate of the Li^+ ion is less than that of Na^+.

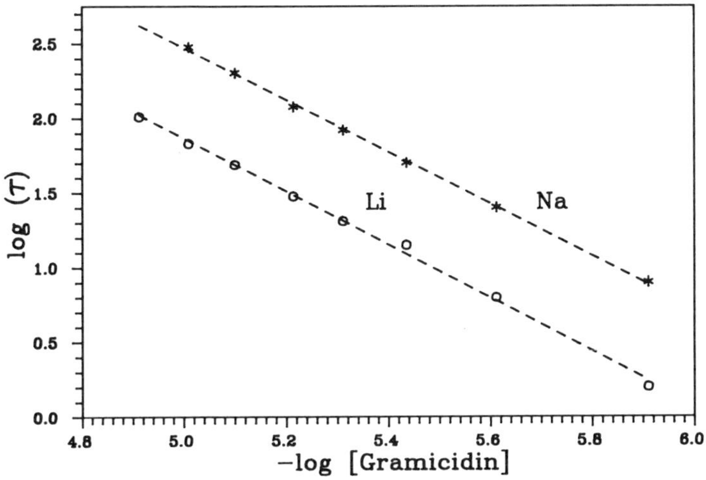

Fig. 4. Log (transport rate) vs. -log [gramicidin]

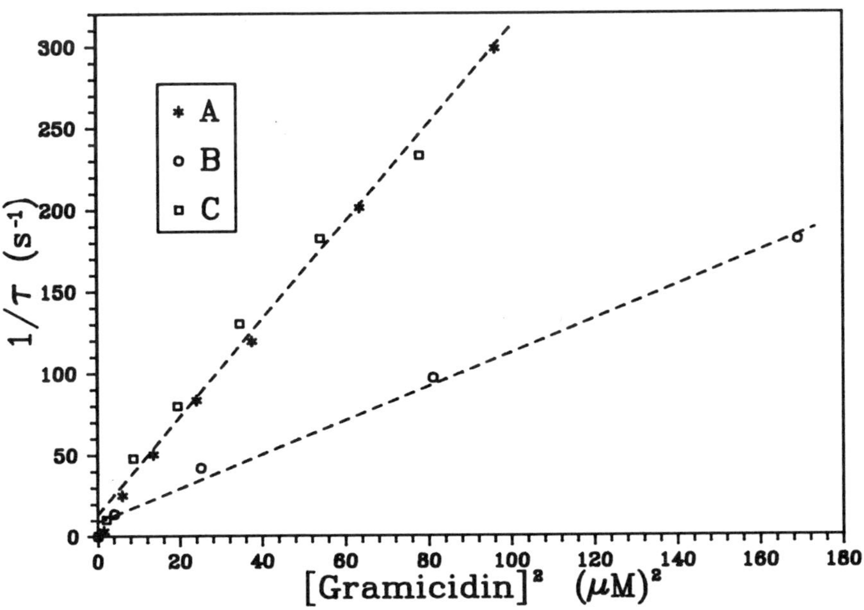

Fig. 5. Transport rate vs. $[\text{gramicidin}]^2$ for gramicidin A, B and C analogs

The transport rate of Na^+ through gramicidin-A and gramicidin-C is about the same but is much less with gramicidin-B, which has phenylalanine at position 11, as shown in Fig. 5. Thus, the equilibrium NMR technique appears to provide a relatively easy method for investigating the effect of single amino acid substititution on monovalent cation transport.

REFERENCES

Andersen O (1983). Ion movement through gramicidin channels. Biophys J 41:119.

Apell HJ, Bamberg E, Alpes H, Lauger P (1977). Formation of ion channels by a negatively charged analog of gramicidin. J Membr Biol 31:171.

Bamberg E, Lauger P (1977). Blocking of the gramicidin channel by divalent cations. J Membr Biol 35:351.

Etchebest C, Pullman A (1985). The effect of amino-acid side chains on the energy profile for ion transport in the gramicidin-A channel. J Biomole Struct Dynamics 2:859.

Gupta R, Gupta P (1982). Direct observation of resolved resonances from intra- and extracellular sodium-23 ions in NMR studies of intact cells and tissues using a paramagnetic shift reagent. J Magn Reson 47:344.

Hladky SB, Haydon DA (1970). Discreteness of conductance changes in biomolecular lipid membranes in the presence of certain antibiotics. Nature 225:451.

Levitt D (1978). Electrostatic calculations for an ion channel. Biophys J 22:209.

Meyers VB, Haydon DA (1972). Ion transfer across lipid membranes in the presence of gramicidin A. Biochim Biophys Acta 274:313.

Sandblom J, Eisenman G, Neher E (1977). Ionic selectivity, saturation and block in gramicidin A channels. J Membr Biol 31:383.

The Ion Pumps: Structure, Function, and Regulation, pages 337–342

CONTINUOUS MONITORING OF ATP IN THE MICROENVIRONMENT OF IMMOBILIZED ENZYMES BY FIREFLY LUCIFERASE.

Claude Aflalo and Marlene DeLuca[+]

Department of Chemistry, University of California at San Diego, La Jolla, California, 92093.

(+) Deceased on November 18, 1987.

INTRODUCTION

The study of enzymes sequestered in artificial or biological systems is generally conducted by indirect methodology through measurements of reactants in the bulk medium. While these macroscopic data reflect catalytic events, they include physical steps as diffusion, transfer, and partition of metabolites between the enzymic phase(s) and the bulk solution (Goldstein, 1976) which are not related to the intrinsic catalytic mechanism of the enzymes. Thus, molecular mechanisms proposed for occluded enzymes action and regulation are often derived from macroscopic measurements, in terms of steady state kinetics, implicitly assuming a large excess of reactants over the enzymes and a homogeneous distribution of freely diffusible intermediates and reactants. These assumptions are very unlikely to hold true in heterogeneous systems.

We introduce a new approach to study heterogeneous catalysis using firefly luciferase (FL) in the presence of luciferin (LH_2) to monitor [ATP] directly in the microenvironment of colocalized enzymes which produce or consume ATP. At low enough [ATP], the light output from the FL reaction is constant (very slow consumption of ATP by FL) and proportional to ATP, so that FL represents a non-interacting probe. The principles for [ATP] quantitation and the interaction of diffusion and catalysis in these systems are presented and discussed in detail elsewhere (Aflalo & DeLuca, 1987). This communication describes qualitatively the accumulation or depletion of ATP in the microenvironment of pyruvate kinase (PK) and hexokinase (HK), respectively.

EXPERIMENTAL

Enzymes were immobilized on CNBr-activated Sepharose-4B beads (Weinhausen et al., 1984). The enzymes were incubated at room temperature for 1 min in 1 ml of a reaction mix containing (in umol): LH_2, 0.1; K-tricine, pH 8, 50; K-EDTA, 2; $MgCl_2$, 10; and BSA, 0.5 mg. ATP, 0.1 pmole; glucose, 10 umol; PEP, 1 umol: ADP, 1 umol; or HK, 2 IU (containing 10 umol glucose) were rapidly injected in small volumes. The light output was monitored using an LKB luminometer equipped with a constant stirring device, and the signal recorded on a strip chart recorder. Velocity constants were estimated by regression analysis of the time course of the light output.

THEORY

Diffusion control with enzymes evenly distributed in spherical supports (radius=R)

Upon addition of a low concentration of substrate ($S_m << K_m$), a steady state is rapidly attained between diffusion and catalysis in the support. Applying Flick's second law of diffusion, at a radius r we get:

$$(1)\quad \left(\frac{dS}{dt}\right)_{diff} = -\left(\frac{dS}{dt}\right)_{cat} \quad \text{OR} \quad D\left(\frac{d^2S}{dr^2} + \frac{2}{r}\frac{dS}{dr}\right) + \frac{V_{max}}{K_m} S$$

Neglecting external diffusion, we can take $S=S_o$ at r=R, and the analytical solution is

$$(2)\quad S = S_m \frac{R}{\sinh(\alpha R)} \frac{\sinh(\alpha r)}{r} \qquad \text{where} \quad \alpha = \left(\frac{V_{max}}{K_m D}\right)^{1/2}$$

The disappearance of the substrate from the medium (at the surface) is given by (Flick's first law of diffusion)

$$(3)\quad -\frac{dS_m}{dt} = \frac{1}{V_m} D\, 4\pi R^2 \left(\frac{dS}{dr}\right)_R = \frac{4\pi RD}{V_m}\{\alpha R\coth(\alpha R) - 1\}\, S_m$$

Where V_m represents the volume of bulk medium. For the same amount of soluble enzyme the velocity would have been

$$(4)\quad v = \frac{V_{max}}{K_m} \frac{4/3\pi R^3}{V_m} S_m$$

We may define the catalytic efficiency of the immobilized enzyme as the ratio of (3) to (4)

$$(5)\quad \mu = \frac{3}{\alpha^2 R^2}\{\alpha R \coth(\alpha R) - 1\}$$

Thus, diffusion control will strongly depend on high values of αR, and so the heterogeneous distribution of the subsrate in the support.

Firefly luciferase as a probe for ATP in the microenvironment of immobilized enzymes

For soluble luciferase (FL), the light output (I) at steady state is proportional to ATP concentration ($[ATP] << K_m$)

$$(6)\quad I = \frac{dh\nu}{dt} = \frac{k_{cat}}{K_m}[FL]\, V_m\, [ATP] = C\,[ATP]$$

For FL immobilized in a sphere of radius R

$$(7)\quad dI = C\,[ATP]\, 4\pi r^2\, dr \qquad \text{and} \qquad I = 4\pi\, C \int_0^R [ATP]\, r^2\, dr$$

Due to the extremely slow ATP utilization (at low [ATP] and [FL]) we expect that it must be evenly distributed in the support ($\alpha \rightarrow 0$ and $\mu \rightarrow 1$, see Fig. 1) and that the enzyme will act essentially as in solution. Thus,

$$(8)\quad I = C\,[ATP]_m \frac{V_b}{V_m} \qquad \text{where } V_b \text{ represents the volume of the support.}$$

With a coimmobilized enzyme utilizing ATP, we may substitute [ATP] in (7) according to (2) and after rearrangement, we get

$$(9)\quad I_{ss} = C\frac{3}{\alpha^2 R^2}\{\alpha R\coth(\alpha R) - 1\}\,[ATP]_m \frac{V_b}{V_m} = C\,\mu\,[ATP]_m \frac{V_b}{V_m}$$

and the catalytic efficiency μ may be determined as the ratio of light intensities with (9) and without (8) active coupled enzyme.

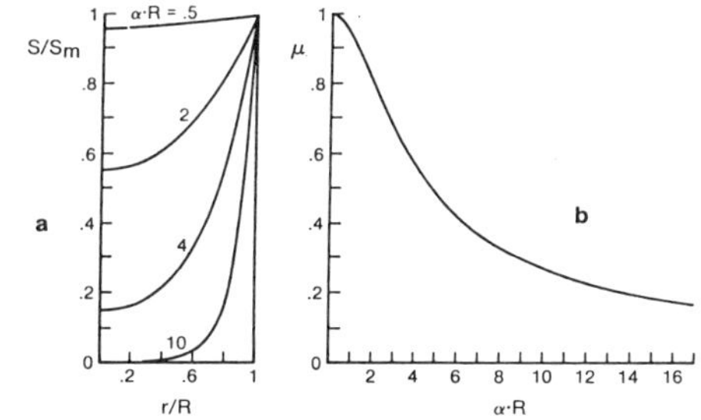

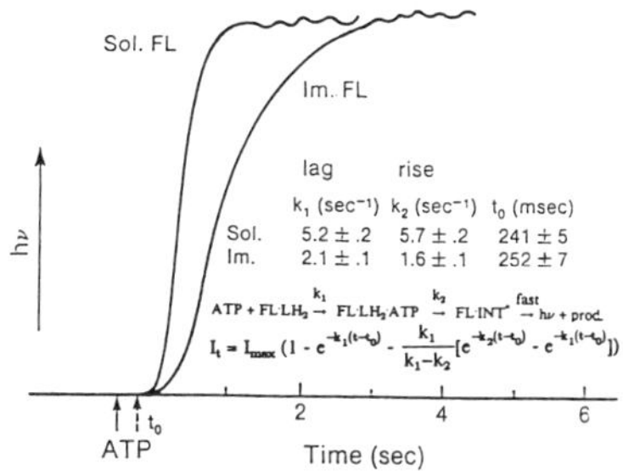

FIG.1(left): **Diffusion control in a spherical heterogenous catalytic system.** A: Distribution of substrate in the support at steady state. B: Variation of the catalytic efficiency with the effective parameters of the system (Aflalo & DeLuca, 1987).
FIG.2(right): **Time course of light output from soluble and immobilized FL.** Reaction conditions as in "EXPERIMENTAL." At the solid arrow, .1 pmol ATP was injected. The dashed arrow indicates the calculated time at which full mixing was reached. Table: calculated parameters from fitting the data to the equation shown (Aflalo & DeLuca, 1987).

RESULTS

Fig.1 represents the theoretical predictions for an enzyme immobilized within a spherical support. The effective parameters of the system determine the extent of substrate depletion (or product accumulation) in the support (a) and the macroscopic catalytic efficiency of the enzyme (b). Upon addition of ATP to immobilized FL, the onset of light production is slower than that observed with the soluble enzyme (fig.2). The reduction of the rate of both kinetic components of the light output suggests it is due to a slower diffusion of ATP to the immobilized enzyme. With immobilized PK, a relative accumulation of ATP inside the beads, as measured with coimmobilized FL, is illustrated in fig.3. The accumulation of product is enhanced when the beads suspension is not stirred

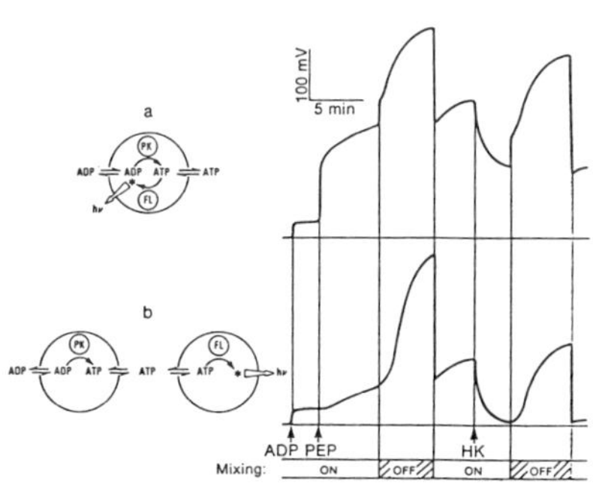

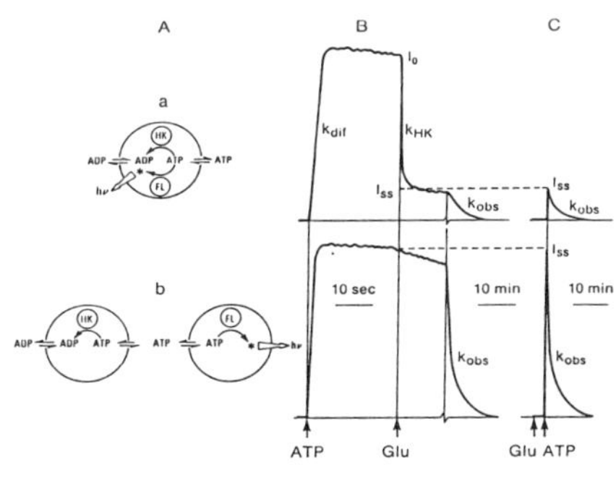

FIG.3(left): **ATP accumulation in the microenvironment of immobilized PK.** Reaction conditions as in "EXPERIMENTAL." The mixing protocol is indicated by the bar at the bottom of the figure (Aflalo & DeLuca, 1987).
FIG.4(right): **ATP depletion by HK separately or coimmobilized with FL on Sepharose beads.** Reaction conditions as in "EXPERIMENTAL." k_{dif}, k_{HK} and k_{obs} represent the rate constants for the diffusion-limited light evolution, the fast and the slow decays of the light output in b, respectively (Aflalo & DeLuca, 1987).

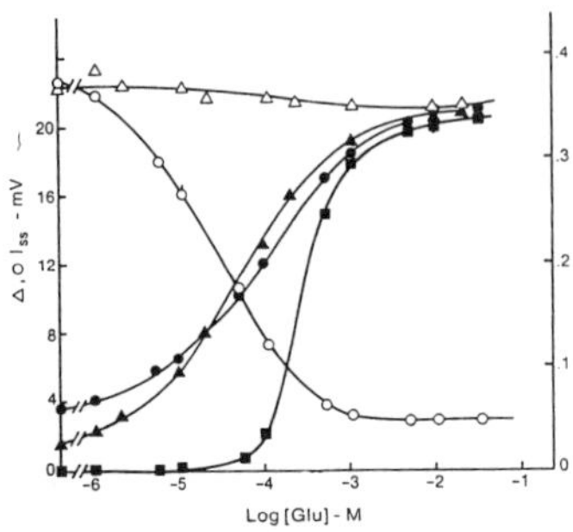

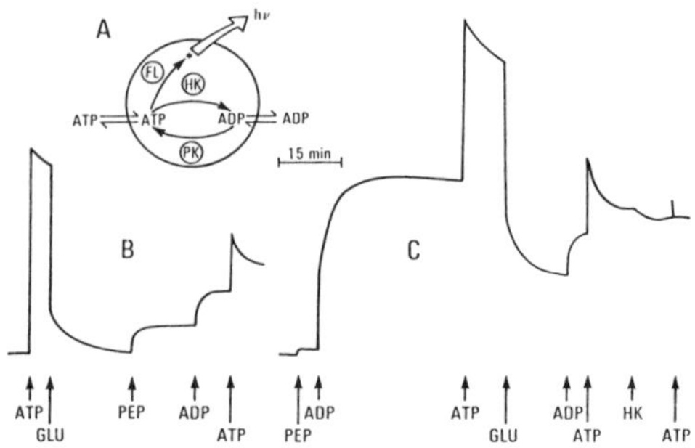

FIG.5(left): **Effect of glucose concentration on the measured parameters of coupled FL-HK systems.** Reaction conditions as in Fig. 4C. The indicated glucose concentrations were added 1 min before addition of ATP. Δ, Δ: FL and HK immobilized on separate beads. O,O: coimmobilized FL--HK. ■: calculated microscopic rate constant for HK ($k_{mic} = k_{obs}/\mu$. $\mu = I_{ss}/I_o$) (Aflalo & DeLuca, 1987).

FIG.6(right): **ATP variation in the microenvironment of a coimmobilized FL-PK-HK system.** Reaction conditions as in "EXPERIMENTAL."

(slower diffusion to the medium). This ATP in the beads is relatively inaccessible to soluble HK added to the bulk medium. Similarly, a rapid ATP depletion in the microenvironment of immobilized HK is demonstrated in fig.4. This microscopic event is kinetically distinguishable from the slower macroscopic depletion of substrate in the bulk medium. By varying glucose concentration, one can vary the local activity of HK. Fig.5 shows the dependence of the measured light output and rate constant of disappearance of ATP in the medium. Using eq.9, we were able to calculate the intrinsic activity of immobilized HK which, while reflecting a behavior similar to that of the soluble enzyme, is very different than the observed rate. Moreover, the apparent rate of depletion in the microenvironment (kHK) depends on the local activity of the immobilized enzyme but not on the total amount of enzyme in suspension, as does kobs, the macroscopic rate (not shown). This is to be expected from inherent kinetic parameter (see Aflalo & DeLuca, 1987). In figs.3-5 we used FL immobilized on separate beads to monitor [ATP] in the bulk medium as a control. The evaluation of local [ATP] can be extended to more elaborated systems like that presented in fig.6. In this case also, the light output represents ATP in the beads as an intermediate of the coupled immobilized HK-PK reactions. The virtual lack of effect of soluble HK indicates that most of the intermediate remains localized in the support at steady state.

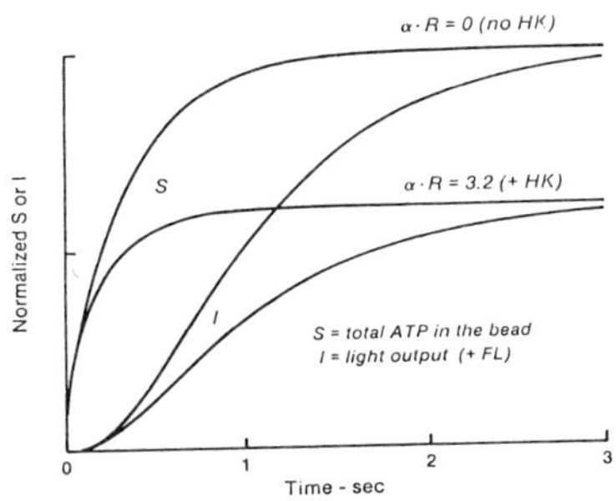

FIG.7: Computer simulation of the time course of Substrate accumulation and light production in FL beads (± HK). S was calculated by numerical integration of eq.1 in its presteady-state form according to Kernevez, 1980. I was calculated using the equation in Fig.2 by substituting I_{max}, ($=I_{ss}$) according to eq.8.

CONCLUSIONS AND PERSPECTIVES

1. The diffusion of ATP from the medium limits light production by immobilized firefly luciferase.

2. The accumulation of ATP in the microenvironment of immobilized PK is affected by the rate of diffusion in the medium and is partially insensitive to soluble HK.

3. The depletion of ATP in the microenvironment of immobilized HK can be kinetically resolved from its macroscopic disappearance in the medium.

4. The observed catalytic properties of an immobilized enzyme (probed exogeneously) do not reflect its intrinsic behavior (assessable with a localized probe only).

The approach described above permits a direct evaluation of intrinsic parameters of enzymes sequestered at high concentrations in a definite compartment simulating in vivo conditions. The physico-chemical properties of the microenvironment (buffer capacity, fixed charges density, dielectric properties, viscosity, etc.) may be varied during the immobilization procedure. However, additional investigation is needed to relate quantitatively the experimental data with theoretical considerations. At this stage, the analytic approach used above is valid for very low substrate concentrations only, where the catalytic velocity is pseudo-first order and product inhibition of FL is negligible.

The generalization of local [ATP] monitoring in more complicated cases requires the resolution of the phenomenological differential equations of the system by computer-assisted numerical techniques (Kernevez, 1980) which also permit an evaluation of the time dependence of the local substrate concentration (see fig.7, generated with physical and kinetic parameters calculated from the other data). Nevertheless, the use of FL enables the kinetic resolution of the variation of ATP in the microenvironment of the enzyme from that in the bulk phase. This tool may add a new dimension to studies on the kinetic behavior of sequestered or surface-bound enzymes, generally assessed by indirect macroscopic measurements in artificial or isolated biological systems.

The concept of a localized probe for ATP can also be extended to cellular systems. The firefly luciferase gene has been successfully cloned and expressed in bacteria, plants and cultured animal cells. However, the analysis of light output in vivo is complicated by a low permeability of luciferin through membranes and the uncertainty of the intracellular location of luciferase (Ow et al, 1986; deWet et al, 1987). It is now possible to target the enzyme into various cellular compartments using the products of fusion between the FL gene and that of various leader sequences (Colman & Robinson, 1986). With such a biologically-directed probe, a survey of the intermediary metabolism of ATP may be initiated. Such studies might provide new insights on the transfer of ATP between cellular compartments, and the energy coupling between metabolic processes.

REFERENCES

Aflalo C, DeLuca M (1987). Biochemistry 26:3913.

Colman A, Robinson C (1986). Cell 46:321.

deWet JR, Wood KV, DeLuca M, Helinski DR, Subramani S (1987). Mol Cell Biol 7:725.

Goldstein L (1976). Methods Enzymol 44:397.

Kernevez J-P (1980). "Enzyme Mathematics", Amsterdam: North Holland.

Ow DW, Wood KV, DeLuca M, deWet JR, Helinski, DR, Howell HH (1986). Science 234:856.

Wiehnhausen G, Kricka LJ, Hinkley JE, DeLuca M. (1982). Appl Biochem Biotechnol 7:463.

REGULATION

The Ion Pumps: Structure, Function, and Regulation, pages 345–347

EDITOR'S INTRODUCTION: THE REGULATION OF THE ION PUMPS

Historically, solution of problems of regulation in biology has generally followed that of the corresponding problems of function. The study of ion pumps is no exception to this generality. As the previous sections of this book show, we can say a good deal about the structure and function of the ion pumps, but we know far less about their regulation, this remaining as a major field for future advances. Repke, in his paper, emphasises the magnitude of regulation. The rate of ATP synthesis in a human subject (synthesis ultimately deriving from the activity of the proton pump) increases rapidly from a value of 0.4 g per Kg body weight per minute at rest, to 9 g/Kg.min during strenuous activity. Cellular levels of the F_1F_0 pump's substrates, ATP, ADP, and inorganic phosphate, together with the trans-membranal electrochemical gradient of protons, are all involved in the short-term regulation of this pump's activity.

Lamb's paper introduces the problem of regulation of the sodium pump. He studies cells grown in cell culture and, by treating them with inhibitors of pump function, or by depriving them of external potassium (which, of course, also inhibits sodium efflux), or by increasing trans-membrane ion leaks, brings about an alteration in the number of sodium pumps in the membrane. A change of up to four-fold in the number of pumps per cell is found, taking place over 24 hours, and requiring new protein synthesis. It is internal sodium that is the signal for this new protein synthesis, posing an exciting problem for us to solve: what is the cell's receptor for this signal, what cellular component is sensitive to internal sodium (with a Km of some 12 mM), transducing this information into the command for specific new protein synthesis? Changes in sodium pump levels can be found when potassium concentrations are manipulated in whole animals, and probably occur during adaptation to regimes of increased muscular activity. Thyroid hormones control sodium pump levels, as Famulski and Wrzosek confirm.

Regulation of pump activity is, of course, not only a result of new protein synthesis. Kellenberger and his colleagues show this in E. coli subjected to an increase in external osmolarity which results in an compensating increase in intracellular potassium. They infect the bacteria with the bacteriophage T4, before the change in external osmolarity. Now, phage infection shuts off the host's new protein synthesis. Yet osmotic regulation still takes place and the cells take up potassium just as if the phage was not present. The effects of hormones, too, can occur via short-term regulation of the activity of existing pumps. Muallem and his fellow-researchers show this for the endoplasmic reticulum of pancreas acinar cells. The isolated reticulum takes up calcium from the cytoplasm. Calcium-mobilising hormones, such as carbachol and cholecystokinin-octapeptide, markedly and rapidly increase the degree to which the reticulum pumps calcium. Indeed, at high intracellular calcium levels, the endoplasmic reticulum can become the main site of calcium uptake in these cells.

Quite another mode of regulation is found in the tubule cells of the thick ascending limb of the loop of Henle. Across this epithelium, the flux of sodium ions is precisely regulated in response to bodily needs. Yet this occurs with no change in the number of sodium pumps, nor as a result of covalent modification of the pumps. What happens, we learn from Klaerke and Jorgensen, is that an increase in the concentration of cellular calcium ions opens a potassium channel, allowing this ion to flow out of the cell. A compensating inflow of potassium, accompanied by sodium (and chloride) on their co-transporter, leads to an increase in cellular sodium levels. This, in turn, stimulates the sodium pump and results in an increase in the overall trans-epithelial sodium flux. In contrast, in the halotolerant bacterium Ba_1, studied by Avi-Dor and his colleagues, sodium and potassium fluxes are linked indirectly, each separate flux being linked to that of protons. The sodium/proton pump in these cells is a primary active transporter, the potassium/ proton system an antiporter, independent of the direct consumption of ATP.

A major theme of the regulation sessions at the Workshop was the new emphasis on the role of "slippage" in the ion pumps. We can define slippage as movement of an ion, on the pump, uncoupled to the progress of the corresponding chemical (or light) reaction, or to the progress of this reaction in the absence of ion flow on the pump. How this disjunction of reaction and pumping can occur is shown clearly in the paper by Moriyama and Nelson, who have studied the proton pump in intracellular vacuoles. This pump can build up a proton gradient that varies from 0.5 to 2 pH units, according to the biological maturity of the vacuole (as it progresses from being a primary to a secondary endosome and then to a lysosome). Vacuoles, isolated from the cell, demonstrate different rates of proton pumping depending on the anion in which they are suspended. These rates will be high or low, according to the presence or absence of a route for the trans-membrane anion flow that must compensate for the electrogenic flow of the proton. Yet under all conditions, the vacuolar membranes hydrolyse ATP at the same rate. The control of rate is due to slippage, excess protons being conducted across the membrane even though the rate of ATP hydrolysis continues unchanged. The theoretical background to the study of slippage is developed in Caplan's paper, and applied to the experimental analysis of the phenomenon in the light-driven bacteriorhodopsin. Slippage appears here as a protective device enabling the membrane to continue to provide ATP to the cell while preventing the build-up of possibly dangerous levels of proton electrochemical gradients. (Ohkawa shows how a plasma membrane (here, of the algal cell *Chara*) is, indeed, of only limited electrical tolerance, in this case, under the control of extracellular calcium.) Slippage is invoked, too, in Luvisetto's paper as a possible explanation for part of the effect of the uncouplers of proton flux and ATP production in mitochondria, while Schonfeld needs the phenomenon to complete his description of the routes that protons take across the membrane of the thylakoid. One notes that slippage is as yet confined to pumps of the F_1F_0 class (or related to them, as the vacuolar proton pump seems to be). Whether slippage will turn out to be equally significant for the E_1E_2 pumps is one of the important questions that our workshop left for future studies to resolve.

W.D. Stein
Editor

The Ion Pumps: Structure, Function, and Regulation, pages 349–354

EVIDENCE FOR ACTIVATION OF THE ER-Ca^{2+} PUMP IN INTACT AND PERMEABILIZED CELLS BY Ca^{2+} MOBILIZING HORMONES

Shmuel Muallem, Timothy G. Beeker, Stepen Pandol, and Claus J. Fimmel

Laboratory of Membrane Biology, Cedars-Sinai Medical Center, and UCLA School of Medicine, Los Angeles, California 90048.

INTRODUCTION

Calcium metabolism by the endoplasmic reticulum (ER) received much attention in recent years when it was discovered that Ca^{2+} mobilizing hormones can release Ca^{2+} from the ER. There is good evidence in many cells, including pancreatic acinar cells, that Ca^{2+} release from the ER is mediated by inositol 1, 4, 5 trisphosphate (Streb, et al 1983; Berridge, 1987) and probably also by inositol 1, 2-cyclic 4, 5 trisphosphate (Dixon and Hokin, 1987). Ca^{2+} uptake into the ER occurs through an ATP and Mg^{2+} dependent Ca^{2+} pump (Preissler and Williams, 1983). In the presence of Ca^{2+} and ATP the ER Ca^{2+} pump form a 100 kDa phosphorylated intermediate which can be blocked by $VO_4^{=}$ (Imamura and Schulz, 1985). The relative activities of the ER Ca^{2+} pump in resting and stimulated cells is not known. We show here that in intact pancreatic acinar cells at high free cytosolic Ca^{2+} concentrations ($[Ca^{2+}]_i$) the ER Ca^{2+} pump can remove Ca^{2+} from the cytosol faster than the plasma membrane Ca^{2+} pump. Further, ATP dependent Ca^{2+} uptake into the ER of permeabilized prestimulated cells is higher compared to that into the ER of unstimulated cells.

METHODS

Pancreatic acini from 75-150g rats were prepared as previously described (Pandol et al, 1985). Free cytosolic Ca^{2+} was measured with the aid of the Ca^{2+} sensitive fluorescence dye Fura 2 exactly as described before (Pandol et al, 1987).

ATP dependent Ca^{2+} uptake in permeabilized cells was measured as previously described (Muallem et al, 1987). Resting or stimulated acini cells were washed twice and resuspended in uptake medium with or without the hormones. The uptake medium also contained the following mitachondrial Ca^{2+} uptake blockers: Ruthenium red 10 uM, oligomycin 5 µg/ml and the protonophore tetrachlorosalicylanilide 1 µM. After 1 minute incubation at 37° C the cells were permeabilized with saponin and ATP dependent Ca^{2+} uptake was determined.

RESULTS AND DISCUSSION

Carbachol and cholecystokinin-octapeptide (CCK-OP) release Ca^{2+} from the ER of pancreatic acinar cells. Both hormones release the Ca^{2+} from the same stores and to similar extent (Pandol et al, 1985). When cell stimulation is terminated, the ER reload with Ca^{2+}. The rate of Ca2+ reloading of the ER at high $[Ca^{2+}]_i$ is shown in Figure 1.

Stimulation of acini suspended in Ca^{2+} free medium with carbachol increased $[Ca^{2+}]_i$ from 106 to about 897 nM. When atropine was added after 10 seconds of stimulation, the cells reduced $[Ca^{2+}]_i$ back to resting levels within about 30 seconds. Subsequent stimulation with CCK-OP resulted in an increased $[Ca^{2+}]_i$ to about 795 µm and the cells slowly reduced $[Ca^{2+}]_i$ back to near resting levels within about 2 minutes. The ability of CCK-OP to induce near maximal increase in $[Ca^{2+}]_i$ indicates that the rapid reduction in $[Ca^{2+}]_i$ after atropine addition was due to Ca^{2+} incorporation into the ER. Further, during the incubation with atropine, very little Ca^{2+} was lost from the cells by a plasma membrane mediated Ca^{2+} efflux. Thus, this experiment indicates that at high $[Ca^{2+}]_i$, the ER Ca^{2+} pump dominates Ca^{2+} removal from the cytosol. Under resting conditions, the plasma membrane Ca^{2+} pump dominates Ca^{2+} removal from the cytosol since $[Ca^{2+}]_i$ is sensitive to changes in medium Ca^{2+}.

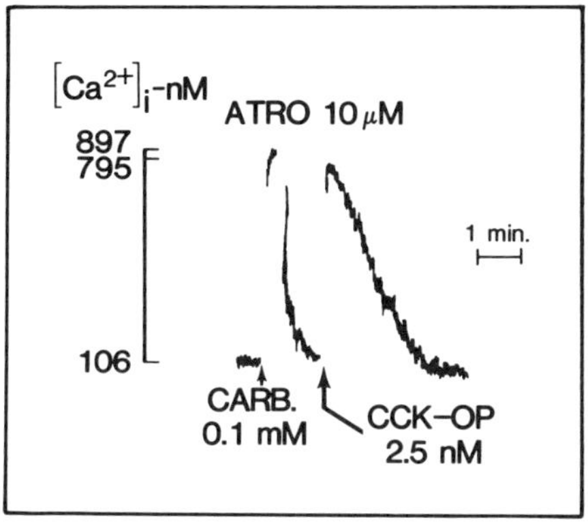

Figure 1. Effect of sequential additions of agonists and antagonist on $[Ca^{2+}]_i$ of Fura 2 loaded cells suspended in Ca^{2+} free medium.

In order to obtain direct evidence for activation of the ER Ca^{2+} pump by Ca^{2+} mobilizing hormones, we measured the rate of ATP dependent Ca^{2+} uptake into the ER of permeabilized cells (Figure 2).

Since Ca^{2+} uptake was measured in the presence of mitochondrial Ca^{2+} uptake blocker, it required ATP and it was sensitive to VO_4^-it is most likely represent Ca^{2+} uptake in the ER.

The effect of cell stimulation on Ca^{2+} uptake into the ER was then examined. The cells were prestimulated with 100 µM carbachol for 10 minutes prior to permeabilization and measurement of Ca^{2+} uptake. Figure 3 shows that prestimulation increased the rate of Ca^{2+} uptake into the ER at free Ca^{2+} concentration of about 0.13 µM and 1.25 µM. Similar results were obtained when the cells were prestimulated with 10^{-8} M CCK-OP (Muallem et al, 1987).

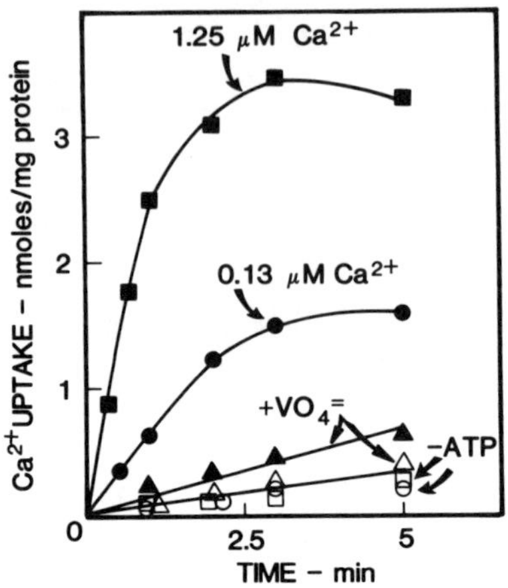

Figure 2. Rates of ATP dependent Ca^{2+} uptake into permeabilized unstimulated cells at two Ca^{2+} concentrations.

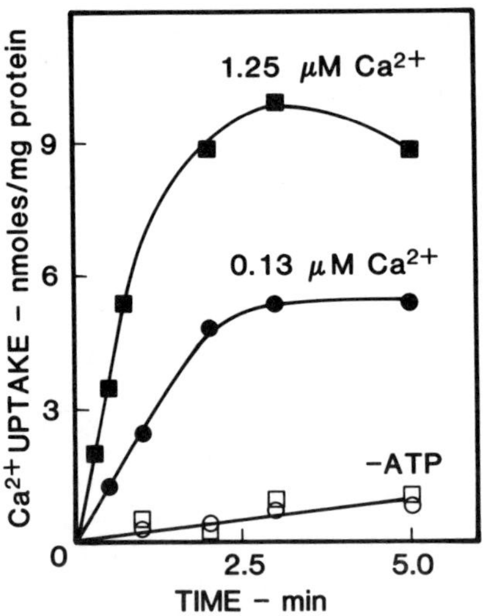

Figure 3. Rates of ATP dependent Ca^{2+} uptake into permeabilized prestimulated cells at two Ca^{2+} concentrations.

The results indicate that Ca^{2+} mobilizing hormones can stimulate the ER Ca^{2+} pump. An increased rate of Ca^{2+} uptake by the ER could be demonstrated at high and low free Ca^{2+} concentrations which corresponds to $[Ca^{2+}]_i$ measured after 10-20 seconds and 5 minutes of cell stimulation respectively. Thus, even when stimulated cells reduce Ca^{2+} back to near resting levels (Pandol et al, 1987), the rate of Ca^{2+} uptake into the ER of stimulated cells is higher compared to that of unstimulated cells.

Activation of the ER Ca^{2+} pump by Ca^{2+} mobilizing hormones can explain the finding that Ca^{2+} reloading of the ER can occur at or even below resting levels of $[Ca^{2+}]_i$ (Muallem et al, 1986; Pandol et al, 1987). Activation of the ER Ca^{2+} pump can ensure rapid reloading of the ER with Ca^{2+} upon termination of cell stimulation and allow reloading even at lower extracellular Ca^{2+}.

REFERENCES

Berridge, M. J. (1987) Inositol trisphosphate and diacylglycerol: Two interacting second messengers. Ann. Rev. Biochm. 56:159-193.

Dixon, J.F. and Hokin, L.E. (1987) Inositol 1, 2-Cyclic-4, 5-trisphosphate concentration relative to inositol1, 4, 5-trisphosphate in pancreatic minilobules on stimulation with carbamylcholine in the absence of lithium. J. Biol. Chem. 262:13892-13895.

Imamura, K. and Schulz, I. (1985) Phosphorylated intermediate of (Ca^{2+} + K^+)-stimulated Mg^{2+}-dependent transport ATPase in endoplasmic reticulum from rat pancreatic acinar cells. J. Biol. Chem. 260:11339-11347.

Muallem, S., Beeker, T. G., and Fimmel, C.J. (1987) Activation of the endoplasmic reticulum Ca^{2+} pump of pancreatic acini by Ca^{2+} mobiliizing hormones. Biochem. Biophys. Res. Comm. In press.

Muallem, S., Fimmel, C.J., Pandol, S.J. and Sachs, G. (1986) Regulation of free cytosolic Ca^{2+} in the peptic and parietal cells of the rabbit gastric gland. J. Biol. Chem. 261:2660-2667.

Pandol, S.J., Schoeffield, M.S., Fimmel, C.J. and Muallem, S. (1987) The agonist sensitive Ca^{2+} pool in the pancreatic acinar cell: Activation of plasma membrane Ca^{2+} influx mechanism. J. Biol. Chem. 262: In press.

Pandol, S. J., Schoeffield, M.S., Sachs, G. and Muallem, S. (1985) Role of free cytosolic Ca^{2+} in secretagogue-stimulated amylase release from dispersed acini from guinea pig pancreas. J. Biol. Chem. 260:10081-10086.

Preissler, M. and Williams, J.A. (1983) Localization of ATP-dependent Ca^{2+} transport activity in mouse pancreatic microsomes. J. Membrane Biol. 73:137-144.

Sterb, H., Irvine, R.F., Berridge, M.J. and Schulz, I. (1983) Release of Ca^{2+} from a nonmitochondrial intracullular store in pancreatic acinar cells by inositol-1,4,5-trisphosphate. Nature 306:67-68.

The Ion Pumps: Structure, Function, and Regulation, pages 355–360

THE EFFECT OF THYROXINE ON THE ACTIVITY OF CALMODULIN-DEPENDENT Ca-ATPase IN SKELETAL MUSCLE SARCOLEMMA

Konrad S. Famulski and Antoni Wrzosek

Department of Cellular Biochemistry and
Department of Muscle Biochemistry, Nencki
Institute of Experimental Biology,
3 Pasteur street, 02-093 Warsaw, Poland

Thyroid hormones regulate growth, development and differentiation of virtually all mammalian cells. Their profound influence is observed in the stimulation of protein turnover (Parmacek et al., 1986; Zeman et al., 1986), increased sensitivity of tissues to other hormones (Williams and Lefkowitz, 1983) and an increase of mechanical activity of heart and skeletal muscles (Nauman, 1983). The latter effect of the hormone can be partially attributed to changes of the sarcoplasmic reticulum (SR). This organelle is responsible for the release-reuptake cycle of intracellular Ca, thereby regulating the contraction-relaxation cycle. Elevation of the thyroxine level causes activation of the Ca pump in SR of heart and skeletal muscles (Ash et al., 1972; Suko, 1972; Nunes et al., 1985; Simonides and Hardeveld, 1986). However, the effect of thyroxine on Ca transport across cell membranes has been much less studied. Sarcolemma (SL) from heart and skeletal muscles contains a Ca pump which is dependent on a Ca-binding protein, calmodulin (Caroni and Carafoli, 1981; Michalak et al., 1985). This ATPase pumps Ca out of the cell with a high affinity. As result, the concentration of free Ca in the resting cell falls below 100 nM (Carafoli, 1987). This high affinity to Ca indicates that calmodulin-dependent Ca-ATPase is also involved in the maintenance of intracellular Ca homeostasis. The aim of the present investigation was to study the effect of thyroxine on SL-bound ATPase.

Hyperthyroidism was induced by injecting normal male rabbits with 0.5 mg of L-thyroxine/kg of body weight for seven consecutive days. This treatment resulted in an increased concentration of thyroxine and triiodothyronine in blood as was revealed by the radioimmunoassay technique. Then the membranes were isolated by a conventional procedure and assayed for the activity of calmodulin-dependent Ca-ATPase (Michalak et al., 1985). As it is shown in Table 1, the elevated level of thyroid hormones in the blood resulted in the activation of ATPase. On the other hand, it is known that the ATPase activity can be fully stimulated in euthyroid animals by calmodulin and a cAMP-dependent protein kinase (Michalak et al., 1985). Hence, a question arises in which way thyroxine administration causes an extra activation of the enzyme. We have recently found (Famulski et al., 1987a) that SL preparations isolated from hyperthyroid animals show the same response to calmodulin as SL from euthyroid rabbits. This opens a possibility that the activation of ATPase may be at the level of protein kinases. It has been demonstrated that thyroid hormones increase the number of β-receptors and the catecholamine production (Williams and Lefkowitz, 1983). Both events will eventually lead to an increased cAMP level in the cell, thus causing activation of protein kinases.

TABLE 1. Calmodulin-dependent Ca-ATPase activity of membranes isolated from fast skeletal muscles of rabbits having different thyreometabolic status

Hormonal status of animals	Hormone level in the blood T3* (ng/100ml)	T4* (μg/100ml)	ATPase activity (μmol/mg protein/h)
Euthyroid	68	6.2	6.9 ± 0.3**
Hyperthyroid	340	70	12.4 ± 0.1

* T3, triiodothyronine; T4, thyroxine.
** Means ± SEM for seven independent preparations.

Therefore, we attempted to elucidate the effect of cAMP-dependent protein phosphorylation on the Ca-ATPase activity tested in SL preparations isolated from euthyroid and hyperthyroid animals. Pretreatment of the membranes was performed according to Michalak et al. (1985). The following conclusions can be drawn from the data presented in Table 2: (i) thyroxine stimulates ATPase activity by increasing the Vmax value substantially above the level of the fully active enzyme; (ii) the differences in the ATPase activity between euthyroid and hyperthyroid animals sustain after dephosphorylation and subsequent rephosphorylation of membrane proteins; and (iii) thyroxine does not influence the phosphorylation of ATPase, although it affects protein kinase activities (Famulski et al., 1987a; Famulski et al., 1987b). It has to be noted that Km value of Ca-ATPase towards Ca was essentially the same in SL from euthyroid and hyperthyroid animals (40 nM) and was not affected by dephosphorylation and rephosphorylation. This excludes the possibility of modifications of the enzyme. Therefore, the increased activity at saturating Ca concentration may ensue from an elevation of the enzyme content in the membrane.

TABLE 2. The effect of protein dephosphorylation and phosphorylation on the activity of calmodulin-dependent Ca-ATPase (at saturating Ca concentration, about 1 μM) in SL preparations isolated from euthyroid an hyperthyroid animals

Membrane pretreatment	Hormonal status of animals (μmol/mg protein/h)	
	Euthyroid	Hyperthyroid
None	6.8	13.5
Dephosphorylation	3.5	6.8
Rephosphorylation	6.8	13.5

Calmodulin-dependent Ca-ATPase forms a phosphorylated intermediate in Ca-dependent manner (Michalak et al., 1985) and this property enables its identification and estimation. We followed the formation of the phosphorylated intermediate in conditions which ensure the maximal yield of phosphate incorporation (Luterbacher and Schatzmann, 1983). Fig. 1 shows that the rate of the formation and the maximal level of the Ca-induced phosphorylated intermediate are much higher in SL preparations isolated from hyperthyroid animals than from euthyroid ones. It implies that cell membranes of hormone-treated rabbits contain higher amount of ATPase molecules capable to form the intermediate as compared to SL isolated from euthyroid animals. A similar finding has been recently reported by Simonides and Hardeveld (1986) who found that thyroxine administration resulted in an increased content of Ca-ATPase in SR of slow-twitch muscles. Therefore, it can be postulated that the "extra" activation of

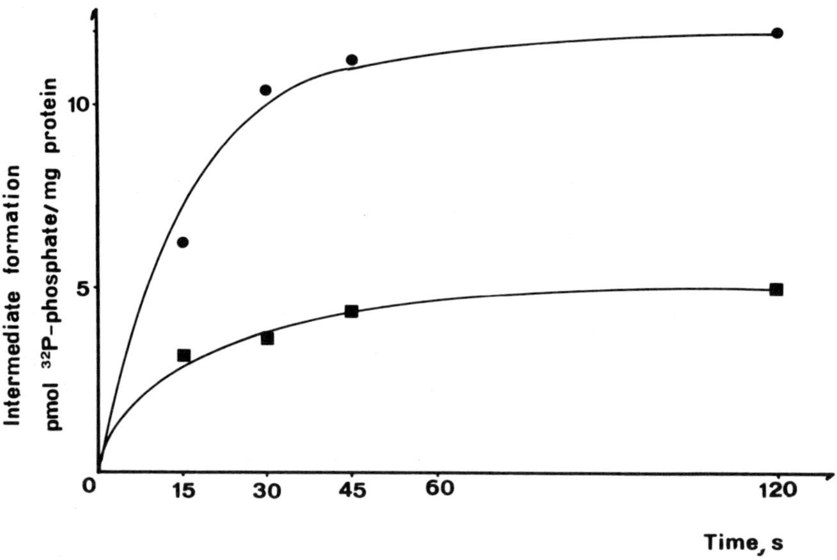

Figure 1. Time course of the Ca-dependent formation of the phosphorylated intermediate.
For experimental details see Famulski et al. (1987a). Squares indicate SL preparations from euthyroid and circles from hyperthyroid rabbits.

calmodulin-dependent Ca-ATPase evoked by thyroxine administration has its rational basis in the increased synthesis of the enzyme and its incorporation into skeletal muscle sarcolemma.

It has been reported (Davis et al., 1982) that thyroxine in vitro stimulates calmodulin-dependent Ca-ATPase. However, in our hands pretreatment of the membranes isolated from euthyroid animals with thyroxine did not alter the ATPase activity. It is therefore postulated that in some membrane systems thyroxine exerts in vitro an unspecific effect on the membrane structure rather than on the ATPase itself (Costante et al., 1986).

Acknowledgments

The authors thank Dr. Paweł Szymański for determination of thyroxine and triiodothyronine levels in blood. Valuable help of Professor Anna B. Wojtczak in preparing the manuscript is also acknowledged. This work was supported by the Polish Academy of Sciences under Project CPBP 04.01.

REFERENCES

Ash ADF, Besh HR Jr, Harigaya S, Zaimis E (1972). Change in the activity of sarcoplasmic reticulum fragments and actomyosin isolated from skeletal muscle of thyroxine-treated cats. J Physiol 244:1-19.

Carafoli E (1987). Calcium homeostasis in the cell. Ann Rev Biochem 56:395-433.

Caroni P, Carafoli E (1981). The Ca-pumping ATPase of heart sarcolemma. Characterization, calmodulin dependence and partial purification. J Biol Chem 256:3263-3270.

Costante G, Sand DC, Glinoer D (1986). In vitro effects of thyroid hormones on red blood cell Ca-dependent ATPase activity. J Endocrinol Invest 9:15-20.

Davis PJ, Davis FB, Blas SD (1982). Studies on the mechanism of thyroid hormone stimulation in vitro of human red blood cell Ca-ATPase activity. Life Sci 30:275-282.

Famulski KS, Pilarska M, Wrzosek A, Sarzała MG (1987a). The effect of thyroxine on the calmodulin-dependent (Ca-Mg)ATPase activity and protein phosphorylation in rabbit skeletal muscle sarcolemma. Eur J Biochem, in press.

Famulski KS, Szymańska G, Szymański P, Sarzała MG (1987b). Hyperthyroidism affects the activity of a cAMP-dependent protein kinase and protein phosphorylation in heart sarcolemma. Biochem Biomed Acta 46:(8/9), in press.

Luterbacher S, Schatzmann HJ (1983). The site of action of La in the reaction cycle of the human red cell membrane Ca-pump ATPase. Experientia 39:311-312.

Michalak M, Famulski KS, Carafoli E (1985). The Ca-pumping ATPase in skeletal muscle sarcolemma. Calmodulin dependence, regulation by cAMP-dependent phosphorylation, and purification. J Biol Chem 259:15540-15547.

Nauman J (1983). Thyroid Hormones. In Drake-Holland AJ, Noble MJM (eds):"Cardiac Metabolism." London: John Wiley & Sons Ltd., pp 505-525.

Nunes TM, Bianco AC, Migala A, Agostini B, Hasselbach W (1985). Thyroxine induced transformation in sarcoplasmic reticulum of rabbit soleus and psoas muscles. Z Naturforsch 40c:726-734.

Parmacek MS, Magid NM, Lesch M, Decker RS, Samarel AM (1986). Cardiac protein synthesis and degradation during thyroxine-induced left ventricular hypertrophy. Am J Physiol 251(Cell Physiol 20):C727-C736.

Simonides WS, Hardeveld C (1986). Effects of the thyroid status on the sarcoplasmic reticulum in slow skeletal muscle of the rat. Cell Calcium 7:147-160.

Suko J (1972). The calcium pump of cardiac sarcoplasmic reticulum. Functional alterations at different levels of thyroid states in rabbits. J Physiol 228:563-582.

Williams RS, Lefkowitz RJ (1983). The effect of thyroid hormone on adrenergic receptors. In Oppenheimer JH, Samuels HH (eds): "Molecular Basis of Thyroid Hormone Action." New York, London: Academic Press pp 351-385.

Zeman RJ, Berstein PL, Ludemann R, Etlinger JD (1986). Regulation of Ca-dependent protein turnover in skeletal muscle by thyroxine. Biochem J 240:269-272.

The Ion Pumps: Structure, Function, and Regulation, pages 361–368

REGULATION OF SODIUM PUMP ABUNDANCE IN ANIMAL CELLS

Joseph F. Lamb

Department of Biology & Preclinical Medicine, University of St Andrews, Scotland.

Animal cells contain fixed anions which set up a Gibbs-Donnan distribution across the cell membrane; the resulting excess of diffusible ions within the cell, together with the indiffusible anions, causes water to enter and the cell to swell. The sodium pump imposes its own distribution on the diffusible ions and so controls cell volume; it does so by keeping cell sodium low and so sets up a steep electrochemical gradient for sodium across the cell membrane. Various secondary functions are driven by this sodium gradient including ionic flow through the voltage sensitive ion channels (which produce the membrane and action potentials), symports, antiports and the ionic leaks. All of these, and other functions, are powered by the sodium pump and so must be factors in the number of sodium pumps which a cell requires to carry out its normal function. In modern cells probably only about 5% of the pumps are necessary for osmotic control (Lamb and McCall, 1972).

By the early 1970s it had become apparent that sodium pumps in various cells worked at about the same rate under equilibrium conditions, so that an increase in numbers of pumps are required to adjust the capacity of the system to various loads (Baker, 1972). In this review I shall consider regulation of pump density from three points of view:

1. The primitive cell response which occurs in cultured cells when the cell sodium changes, this is also often called the low potassium or high cell sodium effect;
2. The effect of agents from outside the cell acting on cultured cells, for example serum factors and tri-iodothyrodine (T3);
3. Integrative effects which occur when cells are organised in the body, this probably arises because of a conflict between the requirements of individual cells and that of the whole animal.

THE PRIMITIVE CELL RESPONSE

Since the primary function of the sodium pump is to keep cell sodium low then regulation is most likely to occur if cell sodium is raised chronically. This can be done in one of 3 ways; firstly by blocking a proportion of the pumps with cardiac glycosides or ethacrynate (Boardman et al, 1972), secondly by reducing the efficiency of pumping of all the membrane pumps with low external potassium (Boardman et al, 1974) or thirdly by increasing the passive leaks across the membrane.

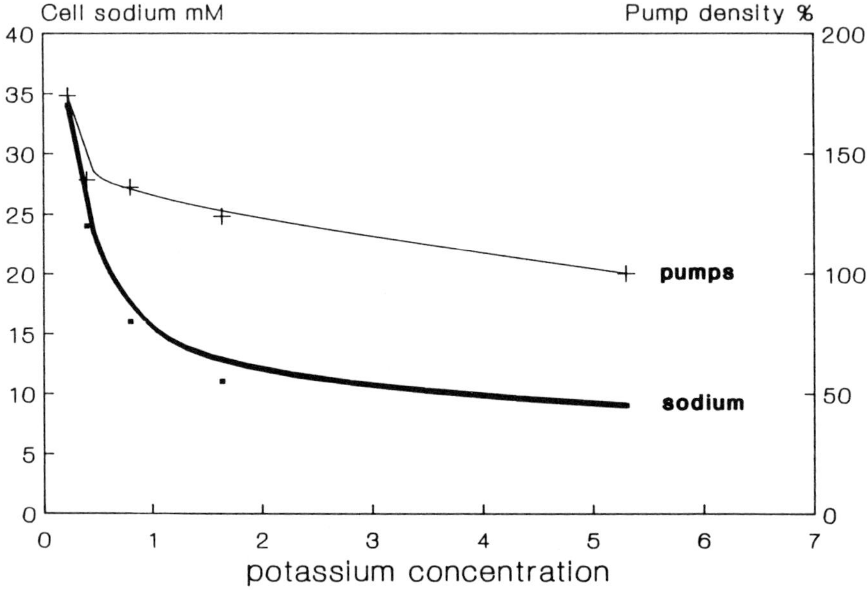

Figure 1. Upregulation of pump density in HeLa cells grown in low-potassium growth medium. More details in Boardman et al, 1974.

Figure 1 shows that when cells are grown in medium with a lower than normal external potassium concentration there is an increase in the number of sodium pumps, as well as an increase in cell sodium (and a fall in cell potassium). The rise in sodium pumps starts in about 6h, causes a reduction in cell sodium, and is largely complete in 24h. From this and other experiments the gain of the system is about 4, a normal value for a biological system.

SIGNAL

This could be due either to the rise in cell sodium, the fall in cell potassium, some secondary effect of such changes or another effect of the low external potassium treatment. A number of experiments have addressed this point:-

1. A dissociation of the sodium rise and potassium fall can be achieved by either reducing the sodium concentration in the medium, or by growing cells in an ionophore such as monensin instead of low potassium. The former experiment shows

that if cell sodium does not rise then no increase in pump density occurs (Boardman et al, 1974), and the latter experiment (Uncles et al, 1987) shows that the increase in sodium pump density is closely related to the rise in cell sodium but not to the fall in cell potassium.

2. The rate of action of the sodium pump does not affect the final density in that raising cell sodium with monensin, which allows the pump to act at a fast rate, produces the same effect as growth in low potassium, which slows the rate of pump activity.

3. The sodium effect does not appear to work by changes in intracellular calcium (Owler, 1986).

The conclusion from these and other results is that a rise in cell sodium (or lithium; Boardman et al, 1975) causes the increase in pump density, but the mechanism by which it acts on the DNA is not known. Perhaps control proteins in the nucleus bind sodium. The relationship between cell sodium and sodium pump density has a Km of around 12mM and a slope at the Km of 3% per 1mM total sodium; changes of intracellular sodium activity of 1mM should then change pump density by about 6%.

MECHANISM

Pump upregulation by high cell sodium requires protein synthesis, and consists both of a decrease in the internalisation of membrane pumps (Lewis et al, 1981) and of an increase in the rate of insertion of new pumps (Wolitzky and Fambrough, 1986; Bowen and McDonough, 1987). The former effect lasts as long as the signal does and may be the long term regulator of pump density, for the latter effect is transient and perhaps exists to increase the rate of adaptation to change in cell sodium. A puzzling feature of these results is that stopping protein synthesis abolishes the upregulation, which implies that both insertion of new pumps and the removal of old ones is abolished. A possible explanation for this latter effect is that protein synthesis is required for the removal of pumps from the membrane, as well as for the production of new pumps. This would also explain the observation that in the presence of cycloheximide or puromycin the rate of decay of pump density on the cell membrane is much slower than expected for the normal internalisation rate (Lamb and Ogden, 1982). A similar explanation has recently been given for the turnover of the Na/proton antiporter in OK cells (Miller et al, 1987). This is a sensible mechanism, for if no new pumps are being made it is an advantage to keep the old ones in the membrane.

CONTROL

When cell sodium is raised there is an increased abundance of alpha and beta mRNA (Bowen and McDonough, 1987; Pressley et al, 1987; Uncles et al, 1987; Cramb et al, 1987), a result consistent with the increased rate of synthesis of the alpha & beta subunits which occurs. As these results were obtained in different cell lines (HeLa, liver and MDCK) it is likely that this is a general finding in cultured cells. The timing of the changes are also consistent with the idea that these changes take place before the increased rate of synthesis of pump units.

It is already apparent, however, that the final picture is going to be complicated; for the following has been found:-

1. High cell sodium is mitogenic in that there is a general increase in the abundance of all mRNA, and also an increase in protein synthesis.

2. There is no simple proportionality between the increased rate of transcription of the pump DNA, the abundance of the message and the final abundance of the pumps in the cells (Edelman et al, 1987). In liver cells the increase in the mRNA for the alpha and beta subunits during high cell sodium upregulation is also not the same, whereas in MDCK cells they are similar. Presumably other factors are involved in the regulation of the final translation of the pump units.

3. The abundance of mRNA(alpha) depends on the concentration of serum in the growth medium, such that at high serum concentration there is an inhibition of mRNA(a) production. This result fits with the previous observation that high serum inhibits the cell sodium effect (Figure 2). It is not known whether this effect also occurs with plasma; if so it might be part of a mechanism by which plasma factors influenced sodium pump density

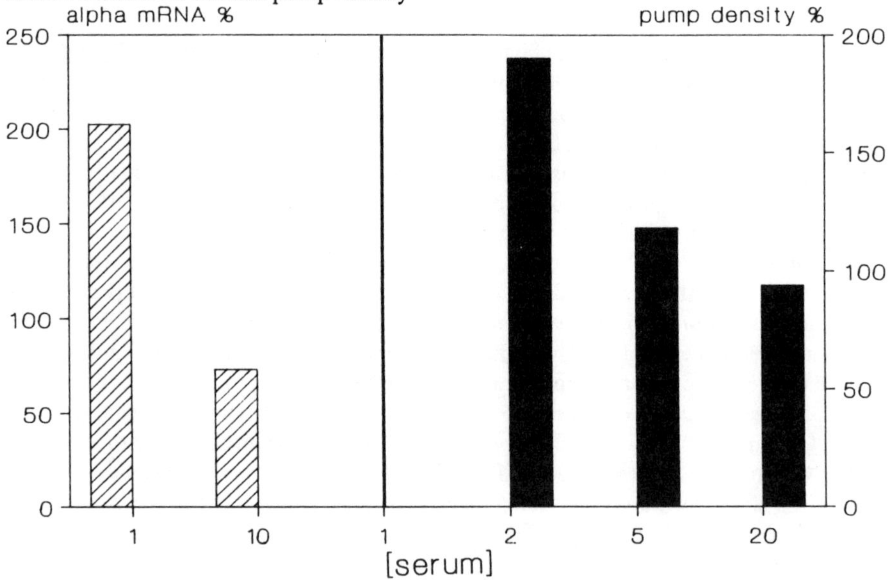

Figure 2. Interactions of serum and sodium on alpha mRNA and sodium pumps.

EXTERNAL FACTORS ON PUMP DENSITY

The serum concentration used for cell growth affects the pump density, an effect stopped by the protein synthesis inhibitor cycloheximide and the transcriptional inhibitor actinomycin D (Aiton and Lamb, 1984). This is probably due to the presence of platelet derived growth factors, though this has not been tested directly. Serum thus has complex actions on pump density, causing an increase when the cell sodium is normal, but a decrease when cell sodium is high; it is not known if this effect is caused by one or more factors.

Thyroid hormones have similar effects in increasing the density of sodium pumps on cells in vivo and in vitro (see Edelman et al, 1987 for refs). Although it would be possible for cells to control their pump density by the primitive sodium mechanism, this would be slower than the direct effect of thyroid hormone on pump density so the latter probably confers an advantage.

Recent evidence suggests that aldosterone has a direct effect on the abundance of a subfraction of the sodium pumps in cultured cells, in addition to any indirect effect through the ionic leaks into the cells (Paccolat et al, 1987) suggesting there are subpopulations of pumps in these cells. Figure 3 shows an outline of the overall schemes involved in the regulation of sodium pump density in isolated cells.

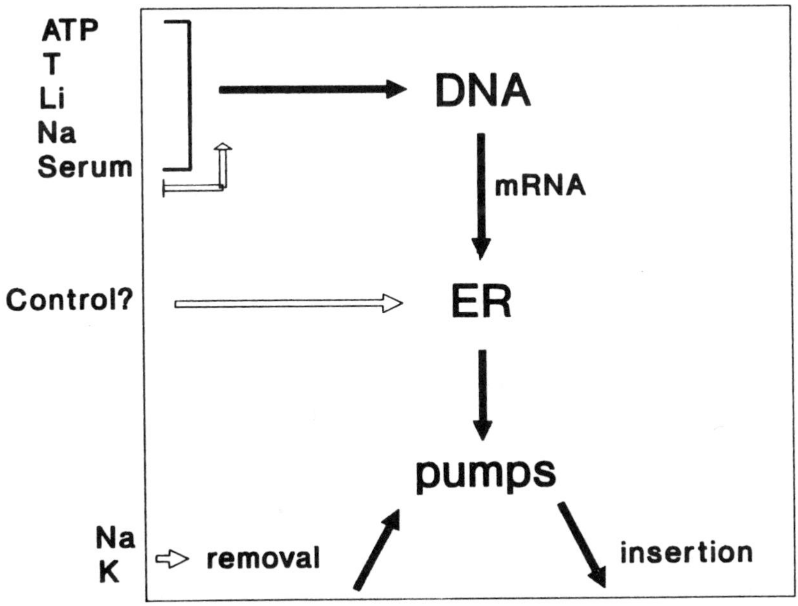

Figure 3. Factors affecting sodium pump density. Control at translation step suggested by non-linearity of mRNA and pump abundance changes. Solid line represents stimulation; hollow line inhibition.

INTEGRATED CONTROL OF SODIUM PUMP DENSITY IN ANIMALS

Cells in culture are no longer part of a whole organism, and so it might be expected that their reaction to various stimuli would not always be the same as that of similar cells within an animal. Present evidence shows that there is a spectrum of responses so that while one stimulus can cause the same response in cultured cells and in the whole animal, another can cause the opposite response. Thyroid hormone is an example of the former response, and raised cell sodium an example of the latter.

Two kinds of experiments have been done to investigate the effects of raised cell sodium on pump density in animals. In the first kind Jorgensen used adrenalectomised rats and showed that a high sodium diet caused an increase in the Na,K-ATPase of the medullary cells in the kidneys (Jorgensen, 1972). His explanation for this was that raised cell sodium caused an increase in the sodium pumps in the cells. The second kind of experiment was first done by Chan and Sanslone who raised cell sodium in the whole animal by lowering the serum K with special diets. (This in itself is somewhat surprising in that the plasma potassium concentration of about 2mM in these experiments rarely raises cell sodium in cultured cells). Using this technique they showed that newly synthesised erythrocytes had both a higher cell sodium and an increased density of pumps in their membranes (Chan and Sanslone, 1969), whereas others showed in rat skeletal muscle that there is a large decrease of pump density in such experiments (Norgaard et al, 1981). Clausen has argued that this is a protective mechanism to keep the plasma potassium from falling too far, with consequent effects on excitability of heart and other tissues. Such a mechanism would work because skeletal muscle is the largest tissue in the body and so has a big store of potassium; thus in conditions of low plasma potassium reducing pump density will allow potassium to leave the muscle and tend to elevate the plasma potassium.

ISOFORMS OF THE SODIUM PUMP

For many years it has been suggested that cells, particularily in rat heart, contain sodium pumps with differing sensitivities to the cardiac glycoside ouabain (Greef, 1981). Several recent papers on the molecular biology of the pump have shown that there are different isoforms of the pump with differing sensitivities to the cardiac glycosides (see references in Shull et al, 1987). Two reports have demonstrated both the existence of two isoforms of the pump and receptors of two different sensitivities in rat heart (Charlemagne et al, 1987; Maixent et al, 1987), and so it may be possible that these differing isoforms are responsible for the differing sensitivities seen; although it is of course difficult to assign a particular isoform to a particular receptor.

How are these different isoforms regulated by the factors so far discussed? If one of the isoforms performed a housekeeping function then high cell sodium might only upregulate it, leaving the other isoforms to adapt to other functions; this is perhaps most likely as the pump DNA for the different isoforms are found on different genomes. There is, as far as I am aware no direct evidence on this, but experiments in Erdman's laboratory show that in rat heart a high cell sodium stimulus causes an increase in the abundance of both the high and low affinity pumps (Erdman et al, 1987). This would suggest that in this tissue both isoforms are upregulated.

This work was supported by grants from the British Heart Foundation and the Wellcome Trust. Recent experiments were carried out with G Cramb, T McDevitt, P Ogden, S Uncles & C Voy. Much help was given by J Aiton in the preparation of the manuscript.

REFERENCES

Aiton JF, Lamb JF, (1984). Effect of the serum concentration of the growth medium on the sodium pump site density of cultured HeLa cells. Q J Exp Physiol 69: 97-115.

Baker PF, (1972). The sodium pump in animal tissues and its role in the control of cellular metabolism and function. In Hokin LE (ed): "Metabolic pathways in Metabolic transport 4," New York: Academic Press, pp 243-268.

Boardman LJ, Huett M, Lamb JF, Newton JP, Polson JM (1974). Evidence for the genetic control of the sodium pump density in HeLa cells. J Physiol 241:771-794.

Boardman LJ, Hume SP, Lamb JF, Polson J (1975). Effect of growth in Lithium on ouabain binding, Na-K-ATPase and Na and K transport in HeLa cells. J Physiol 244: 677-682.

Boardman LJ, Lamb JF, McCall D (1972). Uptake of ^{3}H-ouabain and Na pump turnover rates in cells cultured in ouabain. J Physiol 225:619-635.

Bowen JW, McDonough A (1987). Pretranslational regulation of Na-K-ATPase in cultured canine kidney cells by low K. Am J Physiol 252 (c 21): c179-c189.

Chan PC, Sanslone WR (1969). The influence of a low potassium diet on rat erythrocyte membrane adenosine triphosphatase. Arch Biochem Biophys 134: 48-52.

Charlemagne D, Mayoux E, Poyard M, Oliviero P, Geering K (1987). Identification of Two isoforms of the catalytic subunit of Na,K-ATPase in myocytes from adult rat heart. J Biol Chem 262 (19): 8941-8943.

Cramb G, Lamb JF, Kinghorn JF, Uncles S, Voy C (1987). High internal sodium on sodium pump density and mRNA for the alpha subunit of the sodium pump in human cultured cells. J Physiol 390: 96p.

Edelman IS, Gick G, Ismail-Beigi F (1987). Hormonal regulation of Na,K-ATPase. In Maunsbach AB (ed): "Na K-ATPase (sodium potassium ATPase): Proceedings of the fifth International Conference on Na K-ATPase," New York: Alan R. Liss.

Erdman E, Brown L, Werdan K, Berger HJ (1987). Multiple forms of the cardiac glycoside receptor with different affinities for cardiac glycosides In Beamish RE, Panagia V, Dhalla NS (eds): "Pharmacological Aspects of Heart Disease," Boston: Martinus Nijhoff Publishing, pp 261-271.

Greeff K (1981). "Cardiac Glycosides. Handbook of Experimental Pharmacology 56." Berlin: Springer-Verlag.

Jorgensen PL (1972). Acute and chronic adaptation of (Na + K)-ATPase in rat kidney in response to aldosterone stimulation. In "Proc of the 4th International Congress of Endocrinology," Excerpta medica 273: 433-438.

Lamb JF, McCall D (1972). Effect of prolonged ouabain treatment on Na K Cl and Ca concentration and fluxes in cultured human cells. J Physiol 225: 599-617.

Lamb JF, Ogden Pat (1982). Internalization of Ouabain and Replacement of Sodium Pumps in the Plasma Membranes of HeLa cells following block with Cardiac Glycosides. Q J Exp Physiol 67: 105-109.

Lewis R, Tate Emily, Cook JS (1981). Turnover and regulation of Na-K-ATPase in HeLa cells. Am J Physiol 241: C173-C183.

Maixent JM, Charlemagne D, de la Chapelle B, Lelievre LG (1987). Two Na K-ATPase isoenzymes in canine cardiac myocytes. J Biol Chem 262: 6842-6848.

Miller RT, Pollock AS, Warnock DG (1987). The effects of cycloheximide on Na/H antiporter activity in cultured opossum kidney cells. BBA 900: 157-159.

Norgaard A, Kjeldsen K, Clausen T (1981). Potassium depletion decreases the number of ^{3}H-ouabain binding sites and the active Na-K transport in skeletal muscle. Nature 293: 739-741.

Owler D (1986). Uptake of cardiac glycosides by HeLa cells. PhD Thesis University of St Andrews.

Paccollat MP, Geering K, Gaeggeler HP, Rossier BC (1987). Am J Physiol 252 (c21):c468-c476.

Pressley TA, Ismail-Beigi F, Gick GG, Edelman IS (1987). Inreased abundance of Na K-pump mRNAs by low external K in a rat liver cell line.J Biol Chem (in press).

Shull GE, Young RM, Greeb J, Lingrel JB (1987). Amino acid sequence of the Alpha and Beta subunits of the Na,K-ATPase. In Maunsbach AB (ed): "Na K-ATPase (sodium potassium ATPase): Proceedings of the fifth International Conference on Na K-ATPase," New York: Alan R. Liss, pp3–18.

Uncles S, Cramb G, Kinghorn JR, Lamb JF, McDevitt T, Voy C (1987). The effect of growth in monensin or low potassium on internal sodium, alpha sub-unit mRNA and sodium pump density in human cultured cells. In Maunsbach AB (ed): "Na K-ATPase (sodium potassium ATPase): Proceedings of the fifth International Conference on Na K-ATPase," New York: Alan R. Liss.

Wolitzky BA, Fambrough DM (1986). Regulation of the (Na,K)-ATPase in cultured chick skeletal muscle J Biol Chem 261 (21): 9990-9999.

The Ion Pumps: Structure, Function, and Regulation, pages 369–372

OSMOREGULATION IN THE HALOTOLERANT BACTERIUM, Ba_1

Ronit Regev, Shifra Ken-Dror and Yoram Avi-Dor

Department of Biology, Technion-Israel Institute of Technology, Haifa 32000, Israel

INTRODUCTION

The moderately halophilic, halotolerant eubacterium referred to as Ba_1 can grow on a wide range of salt concentrations (0.2-4.0 M). Either Na and K salts can serve as osmoticum (Rafaeli-Eshkol, 1968). Like other microorganisms which belong to this group also in Ba_1 the intracellular salt concentration is regulated and is significantly lower than in the extracellular fluid (Kushner, 1978). Previous studies conducted in our laboratory revealed that both K^+ (Shnaiderman and Avi-Dor, 1982) and Na^+ (Ken-Dror et al., 1984; Ken-Dror et al., 1986) ions can be extruded against the concentration gradient from the cytoplasm and that Na^+ in catalytic amounts stimulates strongly net extrusion of K salts in the alkaline pH range. The underlying reason for this latter observation was not clear. A primary Na pump, which was first described by Tokuda and Unemoto (1982) in the marine *Vibrio alginolyticus*, was also characterized in Ba_1 (Ken-Dror et al., 1984). This pump is most active in the same pH range in which the Na^+ was found to affect K^+ movements. The present work was undertaken to test whether the stimulatory effect of Na^+ described above is connected with the Na pump activity.

RESULTS

In the experiment shown in Fig. 1, as well as in all the following experiments, cells of Ba_1 were grown on a "low-salt" medium (Rafaeli-Eshkol, 1968) and equilibrated with 0.2 M KCl by repeated washing. The cells were then

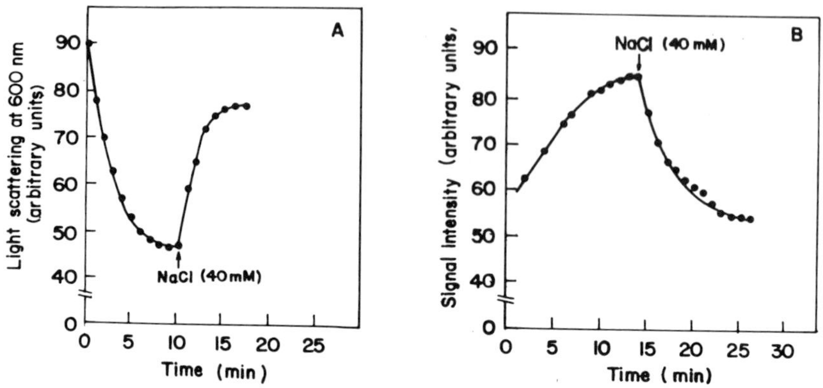

Figure 1. Volume changes induced by net flux of KCl, monitored by light-scattering (A) or by ESR (B).

injected into buffered (pH 8.5 in Figs. 1 and 2), hypertonic (0.8 M) KCl solution which contained ethanol as the respiratory substrate. The subsequent volume changes due to influx or efflux of the K salt were followed either by the conventional light-scattering method (Fig. 1A and Fig. 2) or as in Fig. 1B by ESR (electron spin probe, Tempone; quencher of extracellular signals, $K_3Fe(CN)_6$), developed by Blumenwald et al. (1983). The declining trace in Fig. 1A and the rising in Fig. 1B indicate volume increases and the changes in the opposite direction induced by Na^+ ions correspond to volume contractions. In the experiment shown in Table 1, when the phase of swelling conducted at pH 8.3 terminated, HCl ± NaCl was injected into the assay medium in order to induce variations in the pH at which the recontraction took place. It was found that at lower pH values recontraction also occurred in the absence of Na^+ ions and the rate of the latter increased with increasing acidity between pH 8.3 and 6.4. Na^+ ions raised the respiratory rate and equalised the latter between pH 8.3 and pH 7.0. In contrast, the stimulatory effect of Na^+ on the rate of recontraction declined sharply with the downshift in the pH. Thus, no obvious correlation could be demonstrated between the Na^+-induced respiratory stimulation and its effect on the rate of contraction. The contraction was suppressed by uncouplers irrespectively whether Na^+ was added or not (not documented).

TABLE 1. Effect of Na^+ on rate of respiration and on salt extrusion

pH	Rate of oxygen uptake n atom (O)/min/mg/protein		Changes in light-scattering at 600 nm/min	
	-	NaCl (40 mM)	-	NaCl (40 mM)
8.3	2.5	6.0	0	12
7.5	4.0	6.0	1	9
7.0	4.0	6.0	3	6
6.4	3.5	4.5	6	6

At pH 8.5, at which the efflux of the K salt showed an obligatory requirement for Na^+, the ΔpH (pH_{in}-pH_{out}) component of the electrochemical potential difference is known to drop significantly (Padan et al., 1981). On the other hand, it was shown (Ken-Dror et al., 1984) that cells of Ba_1 respiring in 0.8 M NaCl as the osmoticum, concomitantly with the ejection of Na^+ produced an inwardly directed, uncoupler insensitive proton current. As seen in Fig. 2 such inward proton current also was generated when at the end of the swelling phases NaCl was injected into the suspension of the cells in 0.8 M KCl. Moreover, the uncoupler, FCCP (carbonylcyanide-p-trifluoromethoxyphenylhydrozone) increased the

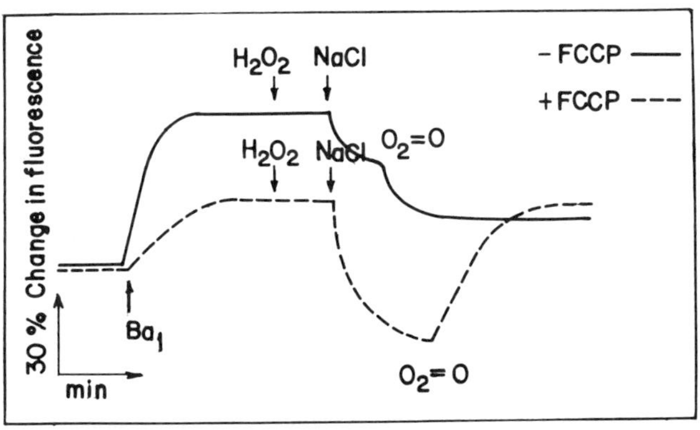

Figure 2. Na^+-induced proton fluxes (± 2μM FCCP) recorded with the fluorescent dye, 9-amino acridine (see for ref. Ken-Dror et al., 1984).

inward proton current and produced in the steady state a ΔpH of negative sign.

Based on the present findings and on previous circumstantial evidence (Shnaiderman and Avi-Dor, 1982) for the involvement of a K^+/H^+ exchanger in the extrusion of the K salt, we suggest that Na^+ shifted the balance between K^+ influx and efflux in favour of the latter by activating the primary Na pump. The inward proton current produced by the Na pump served apparently not only for charge compensation (Tokuda and Unemoto, 1982) but also for driving the K^+/H^+ exchanger. The recontraction induced by acidification is plausibly due to perturbation of the aforementioned balance by diminishing the permeability of the membrane to K^+ ions.

REFERENCES

Blumwald E, Melhorn RJ and Packer L (1983). Studies of osmoregulation in salt adaptation of cyanobacteria with ESR spin-probe techniques. Proc Natl Acad Sci USA 80: 2599-2602.

Ken-Dror S, Shnaiderman R and Avi-Dor Y (1984). Uncoupler-stimulated Na^+ pump and its possible role in the halotolerant bacterium, Ba_1. Arch Biochem Biophys 229: 640-649.

Ken-Dror S, Preger R and Avi-Dor Y (1986). Functional characterization of the uncoupler-insensitive Na^+ pump of the halotolerant bacterium, Ba_1. Arch Biochem Biophys 244: 122-127.

Kushner DJ (1978). Life in high salt and solute concentrations. In Kushner DJ (ed): "Microbial life in extreme environments", New York, Academic Press pp 318-368.

Padan E, Zilberstein D and Shuldiner S (1981). pH homeostatis in bacteria. Biochim Biophys Acta 650: 151-166.

Rafaeli-Eshkol D (1968). Studies on halotolerance in a halophilic bacterium. Biochem J 109: 679-685.

Shnaiderman R and Avi-Dor Y (1982). The uptake and extrusion of salts by the halotolerant bacterium, Ba_1. Arch Biochem Biophys 213: 177-185.

Tokuda H and Unemoto T (1982). Characterization of the respiration-dependent Na^+ pump in marine Vibrio alginolyticus. J Biol Chem 257: 10007-10014.

The Ion Pumps: Structure, Function, and Regulation, pages 373–376

IS THE CONSTITUTIVE Kfp-K^+ PUMPING SYSTEM OF E. COLI ABLE TO ADAPT TO OSMOTIC SHIFTS WITHOUT NEW PROTEIN SYNTHESIZED?

Eduard Kellenberger, Tapani Alatossava, Esther Stauffer, Helga Jütte, Hans Seiler*

Departments of Microbiology and of Inorganic Chemistry*, University of Basel
Klingelbergstrasse 70
CH-4056 Basel /Switzerland

INTRODUCTION

The situation of K^+ pumping in bacteria is best described by Hamann et al. (1987) whom we quote textually: "Bacterial cells accumulate K^+-ions to high concentrations in the cytoplasm. This high K^+-content plays a role in the maintenance of cell turgor pressure, in the activation of enzymes, and possibly in cytoplasmic pH homeostasis (for ref. see Hamann et al. 1987). The high internal K^+-concentration is maintained by the balance of activities of K^+ uptake and K^+ exit systems. Escherichia coli is able to express two K^+ uptake systems (For ref. see Epstein 1984, 1986). The inducible system Kdp functions as a K^+-translocating ATPase and has been characterized in detail. Much less is known about the second K^+ uptake system, Trk, which is expressed constitutively. Trk transports K^+ with a 10^3-fold lower affinity but severalfold larger capacity than Kdp does and is therefore responsible for K^+ uptake under most conditions of growth. The activity of the Trk system is regulated by the cell turgor pressure and Trk also requires both a high cytoplasmic ATP concentration and a high transmembrane proton motive force for activity."

We were interested in the adaptation of the intracellular K^+-concentration after osmolarity shifts in the growth medium. Is it mainly achieved by an activation of the constitutive Trk system or is an induction of the KdP system also involved? We choose to study the osmotic adaptation after T4 bacteriophage infection. For T4 it is known that the transcription of the host-genes ceases rapidly after infection, and shortly afterwards, the DNA of the host is digested. Functions of host genes can therefore no longer be induced. After T4 infection, the number of Kdp pumps can no longer be increased. Any osmotic adaptation after phage infection must therefore be due to an activation of existing pumps, or, possibly, to the additional expression of a putative phage coded K^+-pumping system.

In the present paper we summarize results which show that the osmotic adaptation after phage infection is as efficient and rapid as with uninfected bacteria (Alatossava et al., 1987). To exclude a phage-coded new potassium pump we decided - following the suggestion of a reviewer - to study the osmotic adaptation in presence of chloramphenicol (CM) where an induction of the Kdp system cannot be expressed in the form of an increased number of Kdp pumps. It is well known that many antibiotics induce K^+-leakages (or export) - as do other inhibitors of cellular metabolism - and we therefore had to demonstrate that CM does not do so. (Experiment of Fig. 2).

We then compared the potassium content of CM treated cells with and without an osmotic shift. (Exper. of Fig. 3)

RESULTS

Fig. 1 shows osmotic adaptation of <u>E. coli</u> with and without T4-phage infection. A shift from 100 mM to 350 mM NaCl was made at time 0. The K^+-content of the cells was determined by harvesting the cells by filtration on a Millipore filter. During filtration the O_2 supply was maintained as well as the temperature. The deposit was then washed with an isotonic solution containing Li and sucrose. The filter deposit was dissolved in 1 M HNO_3 and analysed by atomic absorption. The potassium content is expressed relative to the dry weight, which was determined by o.d. and a calibration curve relating dry weight to o.d.

Phage T4 infection was made so as to achieve lysis inhibition, i.e. continued phage synthesis during more than an hour. To do this, the cells were infected and twice superinfected at multiplicities of 5 phages per cell (indicated by ↑T4). By the osmotic shift, the inside osmolality

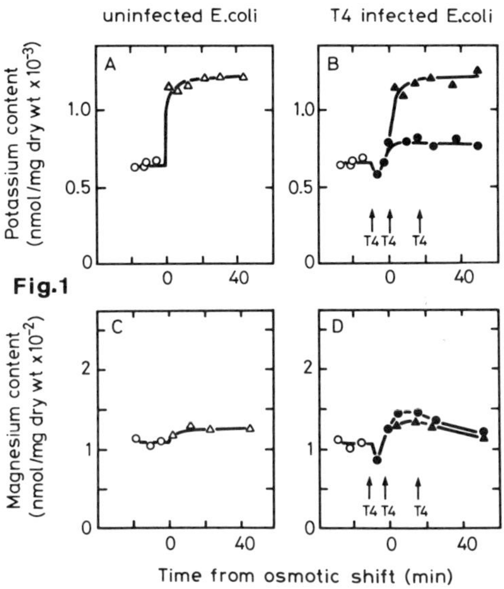

Fig.1

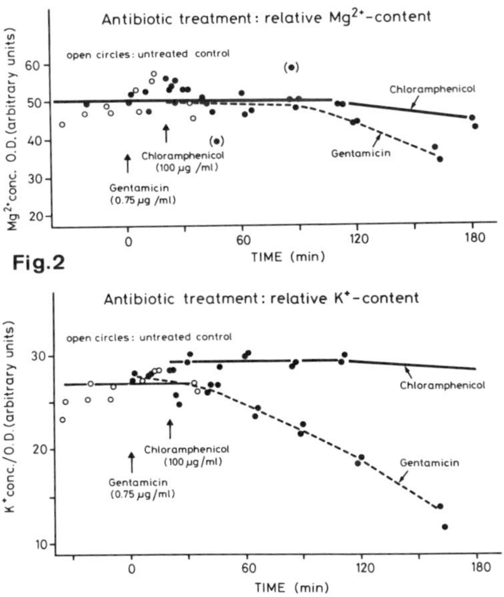

Fig.2

was changed from 220 to 690 m osmol. It is known that phage infection often leads first to a transient leakage which is rapidly repaired (Kuhn & Kellenberger, 1985). With bacteriophage ghost (phages without DNA) the leakage continues without repair.

Fig. 2 shows tests for possible leaks of K^+ and Mg^{2+} induced by chloramphenicol and gentamicin. The ion content was determined as described in fig.1. It is evident that CM (100μgr/ml) induces no K^+ export but possibly some Mg^{2+} leakage. Both ions leak strongly after gentamicin treatment.

Fig. 3. shows that osmotic adaptation is possible when protein synthesis is inhibited by CM. The curve depicts the ratios of the K^+-contents of aliquots of *E. coli* cultures which were or were not osmotically shifted. It is visible that within one minute after the shift (100 to 350 mM NaCl) the K-content rises sharply. Afterwads only a slow increase is observed. A detailed quantitative interpretation is hampered by the fact that we know nothing about a possible volume increase after CM.

DISCUSSION AND CONCLUSIONS

The reported experiments show conclusively that osmotic adaptation after T4 phage infection is possible and makes it feasible to manipulate the intracellular K^+-concentration and to study its influence on phage development. After phage infection the inducibility of the Kdp system is abolished and only the already existing host pumps continue to function. We cannot exclude yet, that the phage codes for its own pump system, although nothing suggests it. After the action of CM, the uninfected E. coli is still able to adapt as shown by our experiments, but in a way which is not easy to interprete in respect of activations of the (constutive) Kfp and (inducible) Kdp systems when we know that no new proteins are made. Further experiments that include phage infected cells are needed to disentangle the activities of the different pumps.

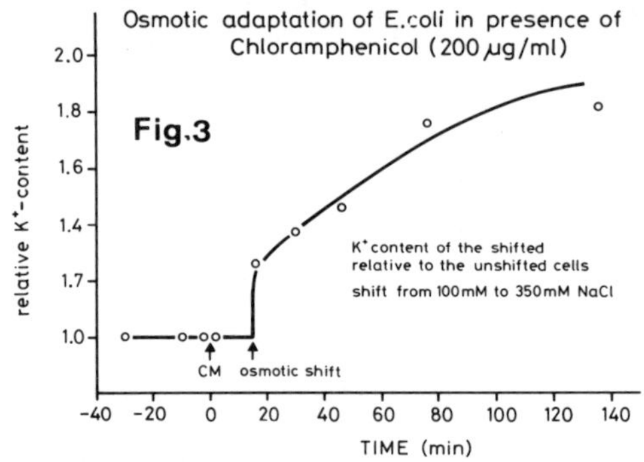

The experiments had been subsidized by the Swiss National Science Foundation (EK) and by the Finnish Academy of Science (TA).

REFERENCES

Alatossava T, Stauffer E, Seiler H (1987). Osmotic adaptation of T4 infected E. coli. FEMS Microbiol. Lett. 43:177-181.

Epstein W (1986). Osmoregulation by potassium transport in E. coli FEMS Microbiol. Rev. 39:73-78.

Epstein W (1984). The Kdp system, a bacterial K^+-transport ATPase. Curr. Top. Membr. Transp. 23:153-175.

Hamann A, Bossemeyer D, Bakker E (1987) Physical mapping of the K^+-transport trkA gene of E. coli and overproduction of the TrkA protein. J. Bact. 169:3138-3145.

Kuhn A, Kellenberger E (1985). Productive phage infection in E. coli with reduced internal levels of the major cations. J. Bacteriol. 163:905-912.

The Ion Pumps: Structure, Function,
and Regulation, pages 377–386

THE CONTRIBUTION OF INTRINSIC UNCOUPLING ("SLIP") TO THE REGULATION OF ION PUMPS, IN PARTICULAR BACTERIORHODOPSIN

S. Roy Caplan

Department of Membrane Research
Weizmann Institute of Science
Rehovot 76100, Israel

INTRODUCTION

It is generally assumed that ion pumps are completely coupled and that all uncoupling may be attributed to external leaks. However, there is no thermodynamic or kinetic justification for this assumption. Recently several reports have appeared showing that a certain degree of intrinsic uncoupling is indeed present in ion pumps and transport systems. Analysis of the six-state model of a redox proton pump studied by Pietrobon and Caplan (1985) shows the regulatory influence of the thermodynamic forces on the extent and relative contributions of redox slip and proton slip. Comparison between simulated behavior and experimental results leads to the conclusion that the typical relation between rate of electron transfer and proton potential difference $\Delta\tilde{\mu}_H$ found in mitochondria at static head is a manifestation of intrinsic uncoupling in the redox pumps, which may well have physiological significance for the efficiency of oxidative phosphorylation (Pietrobon et al., 1986).

In bacteriorhodopsin the photoreaction proceeds at an essentially fixed rate, and hence a $\Delta\tilde{\mu}_H$-controlled slip regulation would provide a "safety valve" as suggested by the author some years ago (Caplan, 1982; see also Eisenbach and Caplan, 1979). This has been observed by a number of workers (e.g. Westerhoff and Dancshazy, 1984). A detailed compartmental analysis of light-induced proton movement in bacteriorhodopsin vesicles has been given by Klausner et al. (1982). The fitting of a compartmental model to a large number of experimental data required the introduction of an arbitrary "control function". It is now shown by means of the Hill diagram method (Hill, 1977) that this control function is a straight-forward consequence of intrinsic uncoupling. Reaction slip increases as the proton gradient increases. As Westerhoff and Dancshazy have pointed out, this property in pumps which constitute the first step in a free-energy transducing pathway permits those pumps to be at least partially controlled by the processes constituting the later steps.

INTRINSIC UNCOUPLING

The assumption of complete coupling in an ion pump is essentially equivalent to the postulate that single-cycle kinetics holds for the system. However, in many instances branching cannot be excluded, especially between two states which differ only in conformation. The addition to a pump cycle of a single branching transition of this kind between two non-adjacent states immediately adds two further cycles to the kinetic scheme. If ligand binding occurs in each of these they may be regarded as "slip" cycles, and they contribute together with the parent cycle to the overall process. A six-state model of a pump in which such a slip transition can occur has recently been used by Pietrobon and Caplan (1985) to study the mitochondrial proton pumps. In this case the slip cycles correspond to proton slip and reaction slip, and it was shown in particular that the latter increases as the force against which the proton pump operates increases.

Intrinsic uncoupling of the mitochondrial redox pump was adduced by Zoratti et al. (1986) to account for the following experimental observation. They found that in rat liver mitochondria oxidizing succinate, about 70% of the rate of electron transfer seen when the redox pumps are at static head (without inhibitor) cannot be accounted for by the influx of protons through leaks. Pietrobon et al. (1986) presented simulations of these findings with the six-state model indicating that redox slip is the main uncoupled cycle at high reaction affinity (redox potential). Fig. 1 shows some of these simulations. The algorithm used is based on the diagram method of Hill (1977).

To exemplify this approach, the contribution of the fully-coupled cycle to the electron flow is given by

$$J_e = N(\Pi_+ - \Pi_-)/\Sigma = (N\Pi_+/\Sigma)\left(1 - \frac{\Pi_-}{\Pi_+}\right) = \kappa\left[1 - \frac{1}{\exp\{(A - n\Delta\tilde{\mu}_H)/RT\}}\right]$$

where N refers to the number of units in the ensemble of pumps, Π_+ and Π_- refer to the product of pseudo-first order rate constants around the cycle in the forward and backward directions of pumping, respectively, and Σ is a sum of terms corresponding to the "directional diagrams" of the complete kinetic scheme (Hill, 1977), in this case 95 in all. The thermodynamic forces operating in the cycle appear explicitly in the exponential term, where A is the affinity of the "driving" reaction (i.e., negative Gibbs free energy change) and $\Delta\tilde{\mu}_H$ is the proton electrochemical potential difference <u>against</u> which the pump works. The stoichiometry, n, is $4H^+/e$. Thus the proton flow resulting from the fully coupled cycle is

$$J_H = nJ_e$$

The complex kinetic coefficient $\kappa \equiv (N\Pi_+/\Sigma)$ is directly proportional to $[H^+]^n{}_i$, where $[H^+]_i$ is the proton concentration on the inside of the mitochondrion, i.e. the uptake side. It is also sensitive to the concentrations of succinate, fumarate, and O_2 (all of which may reasonably be expected to be constant). Furthermore some of the terms in the denominator Σ depend on proton concentrations on either side of the membrane and on the electrical potential difference $\Delta\psi$. If we write $\kappa = k[H^+]_i^n$, it turns out that for the redox pump, which is essentially irreversible, the "rate constant" k is indeed constant when $A >> n\Delta\tilde{\mu}_H$ but tends to a very low value when $A < n\Delta\tilde{\mu}_H$. This accounts for the flow-force relationships seen in curves a of Fig. 1, which represent the behavior of a fully-coupled pump. The pump operates between an upper saturated region when $A >> n\Delta\tilde{\mu}_H$ and a lower saturated region in which it is virtually shut down. The control region in which the forces determine the flows is limited to small positive values of $(A - n\Delta\tilde{\mu}_H)$.

The addition of a slip transition gives rise to two slip cycles, one for the reaction flow and one for the proton flow, whose contributions must be added to that of the fully-coupled cycle. Curves b-e in Fig. 1 correspond to increasing slip with e at the limit of complete uncoupling (i.e. the rate constants of the slip transition have here become so high as to thermodynamically decouple the process entirely). It is seen that at static head, i.e. when $J_H = 0$, $J_e > 0$ in all the uncoupled cases and it

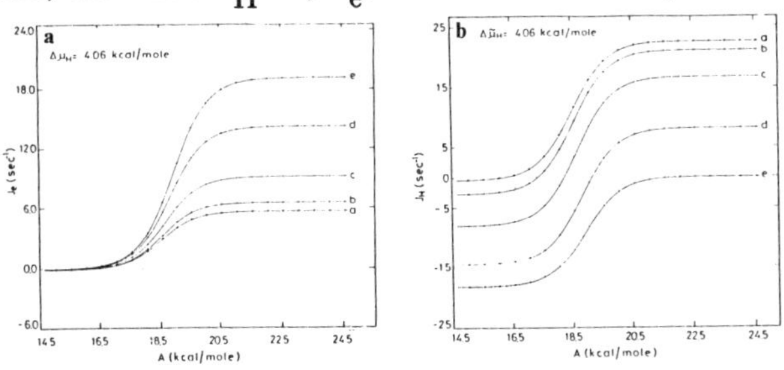

Figure 1. Simulated behavior of a mitochondrial redox pump. Dependence of the rate of electron transfer J_e (a) and of proton translocation J_H (b) on the affinity A of the redox reaction at constant $\Delta\tilde{\mu}_H$. The curves correspond to degrees of coupling varying from 1 (no slip--curves a) to 0 (complete slip--curves e). A is varied by varying substrate concentration. (From Pietrobon et al. (1986), reprinted with permission. Copyright American Chemical Society.)

increases as the degree of coupling increases.

BACTERIORHODOPSIN - THE "BACK PRESSURE" EFFECT

The light-driven proton pump bacteriohodopsin (bR) may be treated similarly, but with some modification to the thermodynamic aspect of the approach stemming from the problem of incorporating a beam of photons into the formalism (Hill, 1977). Figure 2 shows, in the central hexagonal kinetic scheme, a conventional representation of the bR photocycle. At the top is seen the "light-adaptation" reaction (bR_{548}--->bR_{570}) which has a relaxation half-time of about 20 minutes and is usually accounted for by pre-illumination prior to experimentation. Proceeding clockwise, the photoreaction occurs in a time scale of ps, the thermal reactions from K to M in a time scale of μs, and the remainder in ms. The parentheses around the O intermediate merely indicate some uncertainly regarding its position relative to proton uptake and release (the words "high" and "low" refer to the electrochemical potential of the proton in the corresponding medium). Uptake is most generally thought to take place in the O--->bR transition as shown in the inset to the left, with release occurring in the L--->M transition, but some authors consider release to occur between M and O (see, e.g., Stoeckenius, 1985). The relaxation of M has been shown to be $\Delta\psi$-dependent (Dancshazy et al., 1983). The branch pathway between L and bR may or may not include the M′ intermediate (Kalisky et al., 1981; Stoeckenius, 1985; Alshuth and Stockburger, 1986), with a possible associated release and

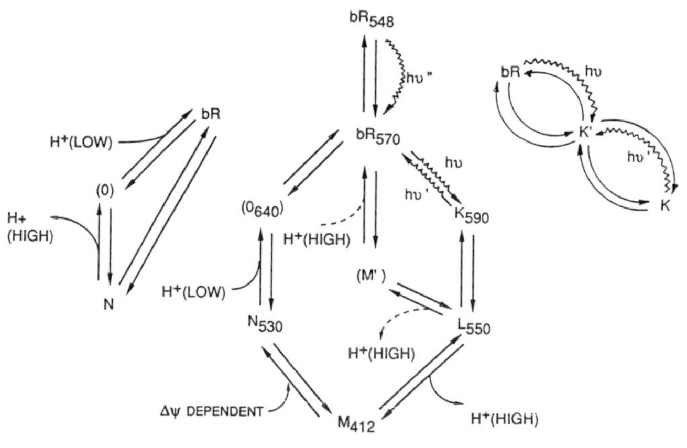

Figure 2. The bacteriorhodopsin photocycle, including possible slip transitions (see text). Inset to left: slip cycle proposed by Westerhoff and Dancshazy (1984). Inset to right: probable mechanism of photoreaction.

uptake of protons. The photoreaction, which is light-driven in both directions, probably involves a transient intermediate K′ as shown in the inset to the right, indeed an additional transient intermediate J has been postulated (Stoeckenius, 1985). Nevertheless the representation of the photoreaction as a single transition is a permissible kinetic "reduction" (Hill, 1977), and indeed Goldschmidt et al. (1976) have shown that the ratio of the forward to backward quantum yields is 0.4. The rate constants of the forward and backward photoreactions depend, of course, on the product of the light intensity (I), the quantum yield, and the molecular cross section for photon absorption.

Since bR is an essentially irreversible pump as can be deduced from its extremely weak fluorescence (Alfano et al., 1976), it will behave very much like the redox pump discussed above. Although we do not at present have sufficient information for a detailed simulation, the rate of photocycling, considering the main (fully-coupled) six-state cycle only, will be given by

$$J_p = k(I)[H^+]^n{}_{LOW}\left[1 - \frac{1}{r\ \exp\{(h\nu - n\Delta\tilde{\mu}_H)/RT\}}\right]$$

Here the photon energy $h\nu$ approximates the <u>basic</u> free energy change for the photoreaction bR--->K (Hill, 1977). Assuming narrow band absorption, if the numbers of substates (vibrational, rotational, etc.) are not very different in bR and K, as is probably the case, the approximation will be close (Parson, 1978; Juretic, 1984). The kinetic coefficient κ is in this case directly proportional not only to $[H^+]^n{}_{LOW}$ but also to I, and the "rate constant" k has an indirect dependence on I (since I appears in some of the terms of the denominator Σ). The pre-exponential factor r, which represents the forward-to-backward ratio of the rate constants of the photoreaction, arises because the radiation field excludes the usual detailed balance consideration (Hill, 1977). In general r may be a function of I, but here it is evidently essentially independent of I with a value not differing greatly from the quantum yield ratio 0.4 mentioned above. The number of protons pumped per photocycle, n, is not yet well established; it may be taken to be 1 at low ionic strength but has been claimed to rise to 2 at higher ionic strengths or even to be 2 at all salt concentrations (Ort and Parson, 1979; Govindjee et al., 1980). These results are based on single turnover flashes, which may not take into account the membrane Bohr effect (Eisenbach et al., 1978). Under steady illumination no kinetic evidence has been seen either in subbacterial particles or in liposomes to suggest a value for n greater than 1 (Eisenbach et al., 1978; Klausner et al., 1982).

At 570nm the photon energy $h\nu$ has a value of 50 kcal/einstein (or 210 kjoule/einstein). In proteoliposomes reconstituted with bR the direction of pumping is inward and hence $[H^+]_{LOW}$ varies very little. It is seen that if n=2 the value of $\Delta\tilde{\mu}_H$ must approach some 25 kcal/mole before a so-called "back-pressure" effect (Hellingwerf et al., 1978; Seigneuret and Rigaud, 1986) could appreciably reduce the rate of photocycling in the main cycle, i.e. move the system off an upper saturated region as shown in Fig. 1; if n=1 $\Delta\tilde{\mu}_H$ must rise to nearly 50 kcal/mole. These values are much higher than those experimentally achievable, and hence it is not surprising that such effects, if they occur, are very small (Westerhoff and Dancshazy, 1984; Dancshazy et al., 1984). On the other hand, the addition of slip cycles would alter the picture. These would also result in an effective stoichiometry that decreases with increasing $\Delta\tilde{\mu}_H$ as has in fact been observed (Westerhoff and Dancshazy, 1984; Dancshazy et al., 1984) in single turnover experiments.

EVIDENCE FOR SLIP IN bR-CONTAINING VESICLES UNDER STEADY ILLUMINATION

Klausner et al. (1982) used the compartmental model shown in Fig. 3, which takes into account conformational or Bohr protons, to analyze proton movements under steady illumination in reconstituted bR vesicles. Two preparations were studied: in one bR was present in monomeric form, in the other in trimeric form. The following three-compartmental representation (cf. Fig. 3) gave a completely faithful simulation of all their kinetic data, for both preparations, at several light intensities and on switching off the light after reaching stationarity (see Fig. 4), providing the empirical "control function" f was introduced as indicated:

$$d[H^+]_1/dt=-k_{12}(I)[H^+]_1-k_{13}(I)[H^+]_1f+k_{21}[H^+]_2+k_{31}[H^+]_3$$

$$d[H^+]_2/dt=k_{12}(I)[H^+]_1-k_{21}[H^+]_2$$

$$d[H^+]_3/dt=k_{13}(I)[H^+]_1f-k_{31}[H^+]_3$$

where f is given by

$$f=1/(1+k_c[H^+]_3)$$

Here $k_{ij}(I)$ indicates a functional dependence on I, but since this dependence is direct proportionality it may be read as $k_{ij}\cdot(I)$, k_{ij} being constant. It is seen that the rate constant k'_{31} appearing in Fig. 3 plays no part in this representation; instead an empirical light-independent rate constant k_c governs the control function. This kinetic scheme takes no account of the electrical potential difference, which is assumed to be constant since it develops within a few milliseconds while the concentration changes to be simulated occur on a time scale of seconds to minutes. As Klausner et al. point out, only a fraction of the protons pumped are electrogenic, the remainder being coupled to the movement through the membrane of counterions or coions.

Klausner et al. give no physical interpretation of their control function, except to suggest that it reflects an inhibition of the pump by $\Delta\tilde{\mu}_H$. No explicit consideration was given to slip. However, as shown schematically in Fig. 5, the two possible slip transitions indicated in Fig. 2 (omitting the M′ intermediate) are readily accounted for by the

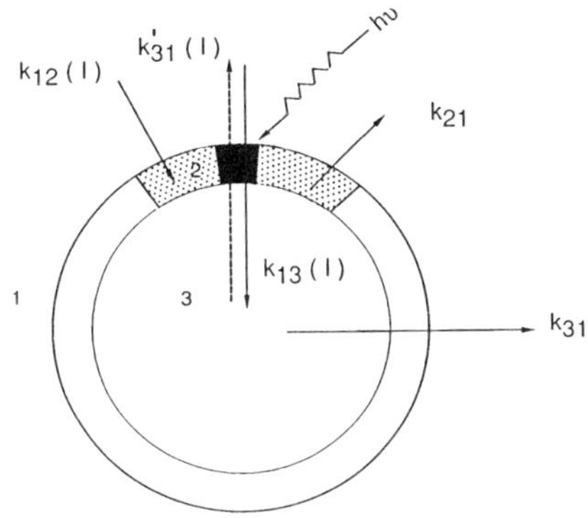

Figure 3. Compartmental model for proton uptake in vesicles reconstituted with bacteriorhodopsin. Compartment 2 represents proton binding due to conformational changes of the protein under steady illumination.

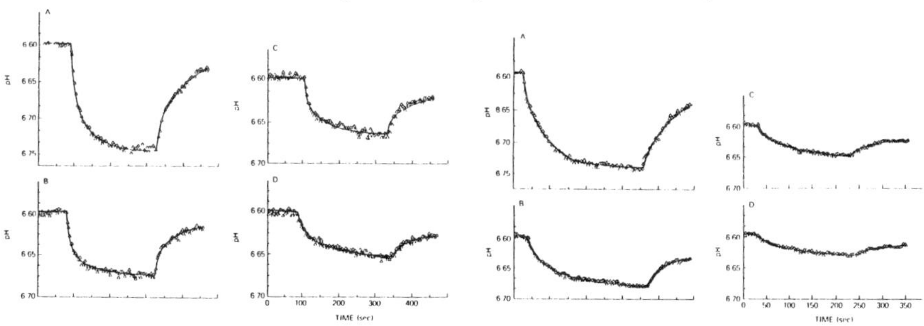

Figure 4. Observed data and simulated curves (see text) for multimeric (left panel) and monomeric (right panel) bR vesicles. Illumination is at the following intensities (A) 425, (B) 325, (C) 175, and (D) $100 W/m^2$. (From Klausner et al. (1982), reprinted with permission. Copyright American Chemical Society.)

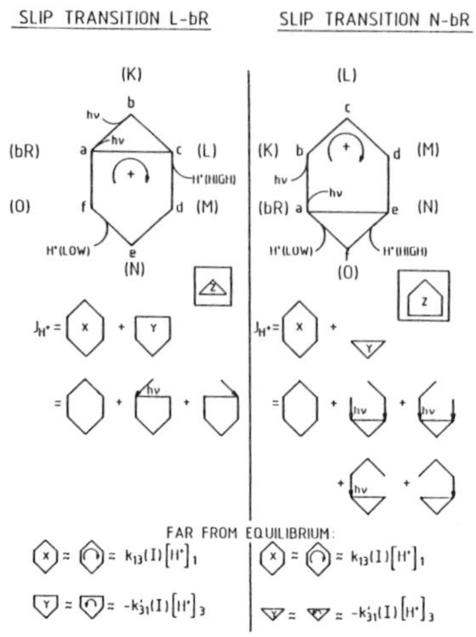

Figure 5. Cycle and flux diagrams for bR (cf. Hill, 1977). Cycles X and Y represent the main (fully-coupled) cycle and the proton slip cycle, respectively. Cycle Z (insets) represents the photoreaction slip cycle. Far from equilibrium only the unidirectional cycles need be considered.

diagram method (Hill, 1977). In both cases the net proton flux through the pump is given by a kinetic expression of the form

$$J_{H^+} = k_{13}(I)[H^+]_1 - k'_{31}(I)[H^+]_3 = k_{13}(I)[H^+]_1\{1 - \frac{k'_{31}[H^+]_3}{k_{13}[H^+]_1}\}$$

Since the variation of $[H^+]_1$ is very small, the term in parentheses may be denoted f where, using k_c as an abbreviation,

$$f = 1 - k_c[H^+]_3 \approx 1/(1 + k_c[H^+]_3)$$

for sufficiently small values of k'_{31}/k_{13}.

CONCLUSIONS

It has been shown that the control function of Klausner et al. (1982) may be formally identified with the contribution of slip to the bR photocycle kinetics under steady illumination. The possibility of a direct "back pressure" effect may be discounted, since according to their measurements the maximum value of $\Delta\tilde{\mu}_H$ could not have exceeded 5 kcal/mole in their experiments.

Although we have dealt only with the proton slip cycles, as pointed out by Pietrobon et al. (1986) the reaction cycles (indicated as insets in Fig. 5) probably play a major role. These undoubtedly provide the "safety valve" alluded to earlier, since reaction slip increases as $\Delta\tilde{\mu}_H$ increases. Thus regulation of slip cycles by $\Delta\tilde{\mu}_H$, and hence by processes which "consume" $\Delta\tilde{\mu}_H$, may well have profound physiological significance in such systems. The effective degree of coupling of primary energy transducers may be subject to control by their "customers", rising and falling as needed, and preventing dangerously high values of $\Delta\tilde{\mu}_H$ being reached which might threaten the integrity of the membrane.

REFERENCES

Alfano RR, Yu W, Govindjee R, Becher B, Ebrey TG (1976). Picosecond kinetics of the fluorescence from the chromophore of the purple membrane protein of Halobacterium halobium. Biophys J 16:541-545.

Alshuth R, Stockburger M (1986). Time-resolved resonance Raman studies on the photochemical cycle of bacteriorhodopsin. Photochem Photobiol 43:55-66.

Caplan SR (1982). Efficiency of energy transduction by bacteriohodopsin: an approach using the Hill diagram method. In Oosawa F (ed): "Dynamic Aspects of Biopolyelectrolytes and Biomembranes", Tokyo: Kodansha, pp 431-441.

Dancshazy Zs, Groma GI, Oesterhelt D (1984). The stoichiometry of the bacteriorhodopsin photocycle and "protoncycle" is not constant: it is regulated by $\Delta\tilde{\mu}_{H^+}$. EBEC Reports 3:229-230.

Dancshazy Zs, Helgerson SL, Stoeckenius W (1983). Coupling between the bacteriorhodopsin photocycle kinetics and the protonmotive force. I. Single flash measurements in Halobacterium halobrium cells. Photobiochem and Photobiophys 5:347-357.

Eisenbach M, Caplan SR (1979). The light-driven proton pump of Halobacterium halobrium: mechanism and function. In Bronner F, Kleinzeller A (eds): "Current Topics in Membranes and Transport" Vol 12, New York: Academic Press, pp 165-248.

Eisenbach M, Garty H, Bakker EP, Klemperer G, Rottenberg H, Caplan SR (1978). Kinetic analysis of light-induced pH changes in bacteriorhodopsin-containing particles from Halobacterium halobium. Biochemistry 17:4691-4698.

Goldschmidt CR, Ottolenghi M, Korenstein R (1976). On the primary quantum yields in the bacteriorhodopsin photocycle. Biophys J 16:839-843.

Govindjee R, Ebrey TG, Crofts AR (1980). The quantum efficiency of proton pumping by the purple membrane of Halobacterium halobium. Biophys J 30:231-242.

Hellingwerf KJ, Schuurmans JJ, Westerhoff HV (1978). Demonstration of coupling between the protonmotive force across bacteriohodopsin and the flow through its photochemical cycle. FEBS Lett 92:181-186.

Hill TL (1977). "Free Energy Transduction in Biology". New York: Academic Press.

Juretic D (1984). Efficiency of free energy transfer and entropy production in photosynthetic systems. J Theor Biol 106:315-327.

Kalisky O, Ottolenghi M, Honig B, Korenstein R (1981). Environmental effects on formation and photoreaction of the M_{412} photoproduct of bacteriorhodopsin: implications for the mechanism of proton pumping. Biochemistry 20:649-655.

Klausner RD, Berman M, Blumenthal R, Weinstein JN, Caplan SR (1982). Compartmental analysis of light-induced proton movement in reconstituted bacteriorhodopsin vesicles. Biochemistry 21:3643-3650.

Ort DR, Parson WW (1979). The quantum yield of flash-induced proton release by bacteriorhodopsin-containing membrane fragments. Biophys J 25:341-354.

Parson WW (1978). Thermodynamics of the primary reactions of photosynthesis. Photochem Photobiol 28:389-393.

Pietrobon D, Caplan SR (1985). Flow-force relationships for a six-state proton pump model: intrinsic uncoupling, kinetic equivalence of input and output forces, and domain of approximate linearity. Biochemistry 24:5764-5776.

Pietrobon D, Zoratti M, Azzone GF, Caplan SR (1986). Intrinsic uncoupling of mitochondrial proton pumps. 2. Modeling studies. Biochemistry 25:767-775.

Seigneuret M, Rigaud J-L (1986). Analysis of passive and light-driven ion movements in large bacteriorhodopsin liposomes reconstituted by reverse-phase evaporation. 2. Influence of passive permeability and back-pressure effects upon light-induced proton uptake. Biochemistry 25:6723-6730.

Stoeckenius W (1985). The rhodopsin-like pigments of halobacteria: light-energy and signal transducers in an archaebacterium. TIBS 10:483-486.

Westerhoff HV, Dancshazy Zs (1984). Keeping a light-driven proton pump under control. TIBS 9:112-117.

Zoratti M, Favaron M, Pietrobon D, Azzone GF (1986). Intrinsic uncoupling of mitochondrial proton pumps. 1. Non-ohmic conductance cannot account for the nonlinear dependence of static head respiration on $\Delta\tilde{\mu}_H$. Biochemistry 25:760-767.

The Ion Pumps: Structure, Function, and Regulation, pages 387–394

THE VACUOLAR H^+-ATPase, A PROTON PUMP CONTROLLED BY A SLIP

Yoshinori Moriyama and Nathan Nelson

Department of Biochemistry, Roche Institute of Molecular Biology, Roche Research Center, Nutley, New Jersey 07110

INTRODUCTION

The main function of vacuolar H^+-ATPase is a controlled acidification of internal organelles of eukaryotic cells. These organelles are constituents of the secretory pathway, which is characteristic of those cells. The vacuolar H^+-ATPase has been detected in lysosomes, plant and fungal vacuoles, clathrin-coated vesicles, synaptic vesicles, chromaffin granules and several other storage vesicles (Cidon and Nelson, 1983; Bennett and Spanswick, 1984; Uchida et al., 1985; Manolson et al., 1985; Xie and Stone, 1986; Bowman et al., 1986; Mandala and Taiz, 1986; Randall and Sze, 1987). It is also present in the Golgi-complex and may function in the sorting of post Golgi vesicles, which are destined to various locations in the cell (Loh et al., 1984; Chanson and Taiz, 1985; Mellman et al, 1986). Although we do not know whether the enzymes that function in these organelles are identical, the similarity of their properties clearly distinguishes them as members of one family of proton-ATPases. A unique feature of this class of enzymes is ability to maintain a precise ΔpH, the value of which can be varied in different organelles or at different times. Thus, in a primary endosome the ΔpH may be maintained at 0.5 units, while in a secondary endosome it increases to about 1 unit, and finally in the lysosomes a ΔpH of about 2 is maintained (Mellman et al., 1986). This communication addresses the question of how this pH regulation is achieved.

PROPERTIES OF THE VACUOLAR H^+-ATPases

The subunit structure of the vacuolar H^+-ATPases clearly distinguishes this family of proton pumps from the other two families, the eubacterial-type and the plasma membrane-type (Nelson, 1987a). Table 1 describes the possible subunit compositions of vacuolar H^+-ATPases from various sources. Although the subunit structure of these

TABLE 1.

Source	Membrane	Subunits molecular weight (kDa)					Reference
Mammalian	Chromaffin granules	115	72	57	39	17	Cidon and Nelson, 1986
Plant	Vacuole		72	60		16	Mandala and Taiz, 1986
Fungi	Yeast vacuole		70	62		15	Bowman and Bowman, 1986
Archaebacteria	Cell membrane		86	64			Nanba and Mukohata, 1987

enzymes is not absolutely defined, the tendency of complicating the subunit composition during evolution prevailed (Nelson and Cidon, 1984). Several evidences, including nucleotide binding and immunological cross reactivity, suggest that the subunit of about 70 kDa is the active site of the enzyme (Mandala and Taiz, ,1986; Moriyama and Nelson, 1987b; Randall and Sze, 1987). Even though electron microscopy of chromaffin granule membranes (Winkler and Westhead, 1980) and of the reconstituted purified H^+-ATPase (Boublik, Moriyama and Nelson, unpublished) revealed structural similarities to the eubacterial proton ATPase, the two families of enzymes are not homologous. If indeed they have evolved from a common ancestor, the diversion occurred before the eubacteria existed (Nelson, 1987b). The lack of a phosphoenzyme intermediate and the presence of the N,N'-dicyclohexylcarbodiimide (DCCD)-binding proteolipid in both systems, however, indicate some similarities in the mechanism of catalysis and proton uptake of these two different proton pumps (Nelson, 1987a). The most striking differences between the vacuolar H^+-ATPase and the eubacterial H^+-ATPase lie in the regulation of H^+-pumping and the extent of ΔpH formed in the two systems.

REGULATION OF PROTON PUMPING

The proton pumping activity of vacuolar H$^+$-ATPases is affected by the presence of anions in the medium (Bennett et al., 1984; Chanson and Taiz, 1985; Bowman and Bowman, 1986). However, direct anion uptake by the enzyme has never been demonstrated. It is accepted that this is an electrogenic proton pump, which forms a positive potential inside the organelles as a consequence of the proton movement into the organelles. An anion channel or carrier is also present in most of the acidified organelles. Its activity results in a collapse of large part of the membrane potential, while anions are driven inside by the protonmotive force generated by the H$^+$-ATPase. The effect of anions, inside and outside the vesicles, could be studied best following isolation and reconstitution of the H$^+$-ATPase in the absence of the chloride channel (Moriyama and Nelson, 1987a; Moriyama and Nelson, 1987c). When the enzyme was reconstituted in the presence of limited ion concentrations or in the presence of potassium acetate, the proton uptake activity was absolutely dependent on the presence of chloride or bromide outside the vesicles. Sulfate inhibited H$^+$ pumping when present outside the vesicles and several organic anions had no effect. On the other hand, sulfate induced the proton uptake activity when present inside the vesicles (Moriyama and Nelson, 1987c). The proton uptake is also very sensitive to membrane potential. Positive membrane potential inside the vesicles strongly inhibited the proton uptake activity. However, negative membrane potential had a positive effect and it overcame some of the inhibitory effects of the anions (Moriyama and Nelson, 1987c).

Inhibition of proton uptake and ATPase activities by nitrate is one characteristic of the vacuolar H$^+$-ATPases (Mellman et al., 1986). The effect of nitrate with respect to time revealed that when added from outside, it initially promoted proton uptake and later inhibited the reaction (Moriyama and Nelson, 1987a). It was concluded that nitrate outside the vesicles induced proton uptake and inhibited the reaction when present inside. Nitrate appears to have a differential effect on H$^+$ pumping and ATPase activity, the former being more sensitive. Moreover, the proton pumping activity of these enzymes stopped prior to reaching thermodynamic equilibrium (Bennett and Spanswick, 1984). These effects led us to propose that the proton uptake activity of the vacuolar H^+-ATPases is controlled by a "slip" (for thermodynamics details see chapter by R. Kaplan in this volume).

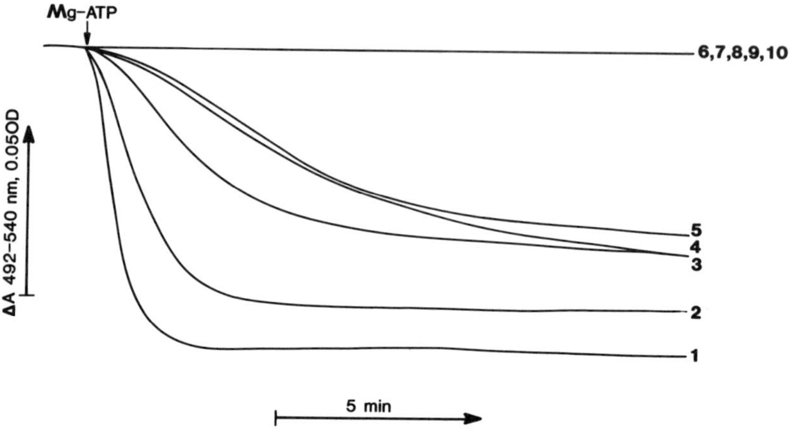

Figure 1. Effect of ionophores and anions on the H^+-pump and ATPase activities of reconstituted H^+-ATPase from chromaffin granules. Proteoliposomes were formed from the purified ATPase in the presence of 0.1 KCl (traces No. 1,2,3,6,7) or K-acetate (traces No. 4,5,8,9,10) and the H^+ pump activity was measured at room temperature as previously described (Moriyama and Nelson, 1987c) in the presence of 0.1 M KCl (traces No. 1,3,4,5,6,7) or K-acetate (traces No. 2,8,9,10). Valinomycin (1 μg) (traces No. 1,2,4,6,8,9) and/or FCCP (1 μM) (traces No. 6,7,9) also included in the assay mixture. [γ-^{32}P] Mg-ATP (1 mM) was added to initiate the H^+ transport and then the liberated ^{32}P was extracted, immediately after the recording, from the reaction mixture in the cuvette. Mg-ATPase activities (μ mole/mg prot./min at the specified traces No.) in each reaction cuvette were as follows: 1.6(1), 1.8(2), 1.6(3), 1.7(4), 1.6(5), 3.5(6), 3.2(7), 1.8(8), 2.6(9), 1.6(10).

Figure 1 shows the effect of anions inside and outside the reconstituted H^+-ATPase from chromaffin granules, as well as valinomycin and carbonyl cyanide p-trifluoromethoxyphenylhydrazone (FCCP), on the proton uptake and ATPase activities of the preparation. Proton uptake activity was not detected in the presence of potassium acetate inside and outside the vesicles, yet, substantial ATPase activity was observed under the same conditions. When chloride was present outside the vesicles, substantial proton uptake activity was recorded while the ATPase activity, measured in the same tube, was marginally enhanced. Inclusion of valinomycin in the presence of potassium chloride markedly increased the proton uptake activity with a very small effect on the ATPase activity. FCCP by itself had little effect on the ATPase activity, and only when present with valinomycin was about two-fold increase of the ATPase activity observed. It is apparent that after reconstitution of the purified H^+-ATPase, effectors which markedly affect the proton uptake activity, hardly affect the ATPase activity of the enzyme.

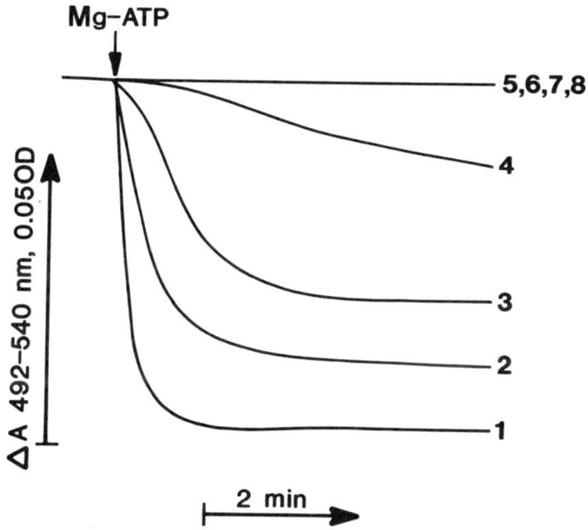

Figure 2. Effect of ionophores and anions on the H^+ pump and ATPase activities of chromaffin granule membrane vesicles. H^+ pump and ATPase activities were measured as in Fig. 1 in the presence of 0.1 M KCl (traces No. 1,2,5,6) or K acetate (traces No. 3,4,7,8). The following reagents were also included during the assay: vanadate (100 µM), Na-azide (500 µM), FCCP (1 µM) (traces No. 5,6,7,8), and valinomycin (1 µg) (traces No. 1,3,5,7). ATPase activities (µ mole/mg prot./min at the specified traces No.) were: 0.08(1,2,3), 0.07(4,8), 0.1(5,7), 0.09(6).

It can be argued that part of the differences in the effects observed with the reconstituted enzyme are due to incomplete incorporation of the enzyme into the lipid vesicles. Figure 2 depicts similar experiments carried out with chromaffin membranes. In order to prevent involvement of contaminating ATPases, vanadate and azide were included in the reaction mixture. The differences between the effect of anions and uncouplers on proton uptake and ATPase activities of the preparation are even more striking than those obtained with the reconstituted enzyme. While a negligible effect by the various factors on the ATPase activity was observed, remarkable effects on the proton uptake were recorded. Therefore, the proton uptake activity of vacuolar H^+-ATPases can be completely blocked while the ATPase activity of the same preparation is hardly affected.

A stoichiometry of 2 H^+ to ATP was measured in vacuoles from Red Beet and ghosts of chromaffin granules (Johnson et al., 1982; Bennett and Spanswick, 1984). Theoretically, at thermodynamic equilibrium, a protonmotive force of about 200 mV should be formed, yet, the most common protonmotive force of the vacuolar organelles does not exceed 120 mV (Van Dyke, 1987). It is apparent that a thermodynamic equilibrium is never reached and the proton pumping activity of the system is inhibited in order to prevent over acidification of the organelles. In mitochondria and especially chloroplasts, the protonmotive force produced by ATPase activity is controlled by a specific ATPase inhibitor (Nelson and Cidon, 1984). This mechanism is especially valuable for inhibiting ATP hydrolysis in the darkness. This kind of control is not very precise and it is very difficult to maintain a defined pH at the other face of the membrane from where the inhibitor resides. The ΔpH in the vacuolar system is achieved by modulation of anion concentrations and the sensitivity of the proton pumping activity to membrane potential and ΔpH (Nelson, 1987a). As was noticed before and stressed in this work, modulators have only slight effect on the ATPase activity of the vacuolar ATPase, while H^+ pumping activity is strongly affected. This kind of control can be achieved by two distinct mechanisms: A spill in which the excess protons leak outside the membrane by a carrier evolved specifically for this function, or a slip in which the excess protons are not conducted across the membrane even though the system is still utilizing energy. The best example of the first mechanism is brown fat mitochondria, whereas the H^+-ATPase of the vacuolar system is an example of a slip, which controls ΔpH.

REFERENCES

Bennett AB, O'Neill SD, Spanswick RM (1984). H^+-ATPase activity from storage tissue of *Beta vulvagis* I. Identification and characterization of an anion-sensitive H^+-ATPase. Plant Physiol 74:538-544.

Bennett AB, Spanswick RM (1984). H^+-ATPase activity from storage tissue of *Beta vulgaris*. II. H^+-ATPase stoichiometry of an anion-sensitive H^+-ATPase. Plant Physiol 74:545-548.

Bowman BJ, Bowman EJ (1986). H^+-ATPases from mitochondria, plasma membranes, and vacuoles of fungal cells. J Membr Biol 94:83-97.

Bowman EJ, Mandala S, Taiz L, Bowman BJ (1986). Structural studies of the vacuolar membrane ATPase from *Neurospora crassa* and comparison with the tonoplast membrane ATPase from *Zea mays*. Proc Natl Acad Sci USA 83:48-52.

Cidon S, Nelson N (1983). A novel ATPase in the chromaffin granule membrane. J Biol Chem 258:2892-2898.

Cidon S, Nelson N (1986). Purification of N-ethylmaleimide-sensitive ATPase from chromaffin granule membranes. J Biol Chem 261:9222-9227.

Chanson A, Taiz L (1985). Evidence for an ATP-dependent proton pump on the Golgi of *Corn coleoptiles*. Plant Physiol 78:232-240.

Johnson RG, Beers MF, Scarpa A (1982). H^+-ATPase of chromaffin granules: Kinetics, regulation, and stoichiometry. J Biol Chem 257:10701-10707.

Loh YP, Brownstein MJ, Gainer H (1984). Proteolysis in neuropeptide processing and other neural functions. Ann Rev Neurosci 7:189-222.

Mandala S, Taiz L (1986). Characterization of the subunit structure of the Maize tonoplast ATPase, immunological and inhibitor binding studies. J Biol Chem 261:12850-12855.

Manolson MF, Rea PA, Poole RJ (1985). Identification of 3-O-(4-benzoyl)benzoyladenosine 5'-triphosphate and N,N'-dicyclohexylcarbodiimide-binding subunits of a higher plant H^+-translocating tonoplast ATPase. J Biol Chem 260:12273-12279.

Mellman I, Fuchs R, Helenius A (1986). Acidification of the endocytic and exocytic pathways. Ann Rev Biochem 55:663-700.

Moriyama Y, Nelson N (1987a). The purified ATPase from chromaffin granule membranes is an anion-dependent proton pump. J Biol Chem 262:9175-9180.

Moriyama Y, Nelson N (1987b). Nucleotide binding sites and chemical modification of the chromaffin granule proton ATPase. J Biol Chem 262:14723-14729.

Moriyama Y, Nelson N (1987c) Internal anion binding site and membrane potential dominate the regulation of proton pumping by the chromaffin granule ATPase. Biochem Biophys Res Commun, in press.

Nanba T, Mukohata Y (1987). A membrane-bound ATPase from *Halobacterium halobium*: Purification and characterization. J Biochem 102:591-598.

Nelson N (1987a). The vacuolar proton-ATPase of eukaryotic cells. BioEssays, in press.

Nelson N (1987b). Structure, function and evolution of proton-ATPases. Plant Physiol, in press.

Nelson N, Cidon S (1984). New molecular aspects of energy-transducing protein complexes. J Bioenerg Biomembr 16:11-36.

Randall SK, Sze H (1987). Probing the catalytic subunit of the tonoplast H^+-ATPase from oat roots, binding of 7-chloro-4-nitrobenzo-2-oxa-1,3-diazole to the 72-kilodalton polypeptide. J Biol Chem 262:7135-7141.

Uchida E, Ohsumi Y, Anraku Y (1985). Purification and properties of H^+-transducing, Mg^{2+}-Adenosine triphosphatase from vacuolar membranes of *Saccharomyces cerevisiae*. J Biol Chem 260:1090-1095.

Van Dyke RW (1987). Proton pump-generated electrochemical gradients in rat liver multivesicular bodies: Quantitation and effect of chloride. J Biol Chem, in press.

Winkler H, Westhead EW (1980). The molecular organization of adrenal chromaffin granules. Neuroscience 5:1803-1823.

Xie X-S, Stone DK (1986). Isolation and reconstitution of the clathrin-coated vesicle proton translocating complex. J Biol Chem 261:2492-2495.

The Ion Pumps: Structure, Function,
and Regulation, pages 395–400

UNCOUPLING OF OXIDATIVE PHOSPHORYLATION INDUCED BY FCCP, OLEIC ACID AND CHLOROFORM IN RAT LIVER MITOCHONDRIA.

Siro Luvisetto, Daniela Pietrobon and Giovanni Felice Azzone.

CNR Unit for the study of the Physiology of Mitochondria and Institute of General Pathology, University of Padova, 35100 Padova, Italy.

INTRODUCTION

The mechanism by which a number of substances uncouples oxidative phosphorylation has been the subject of intense research, and in particular, much work has been devoted to establishing if the protonophoric action of these substances could quantitatively account for the uncoupling of oxidative phosphorylation (McLaughlin and Dilger, 1980). From these studies it can be generally concluded that certainly there exists a correlation between protonophoric action and uncoupling action; however, no clear-cut evidence that the uncoupling is exclusively due to the increase in membrane conductance has been ever reported.

We have further investigated the mechanism of uncoupling of a typical weak acid protonophore (FCCP), a fatty acid (oleic acid), and a general anesthetic (chloroform) by measuring their effect on the inner mitochondrial membrane conductance in addition to their effects on the rates of electron transfer and ATP synthesis and on the transmembrane electrical potential difference. We have also simulated the results by using a kinetic model of chemiosmotic free energy coupling described by Pietrobon and Caplan (1986).

RESULTS

The results reported in fig. 1 show that the three agents decrease the rate of ATP synthesis and increase the rate of respiration in static head, thus justifying their classification as uncouplers of oxidative phosphory lation. However, a given stimulation of the rate of respiration (or of the rate of ATP hydrolysis, not shown) in static head is accompanied by different depressions of $\Delta\psi$ for the different uncouplers. In particular, chloroform decrease less the static head $\Delta\psi$ than FCCP and oleic acid.

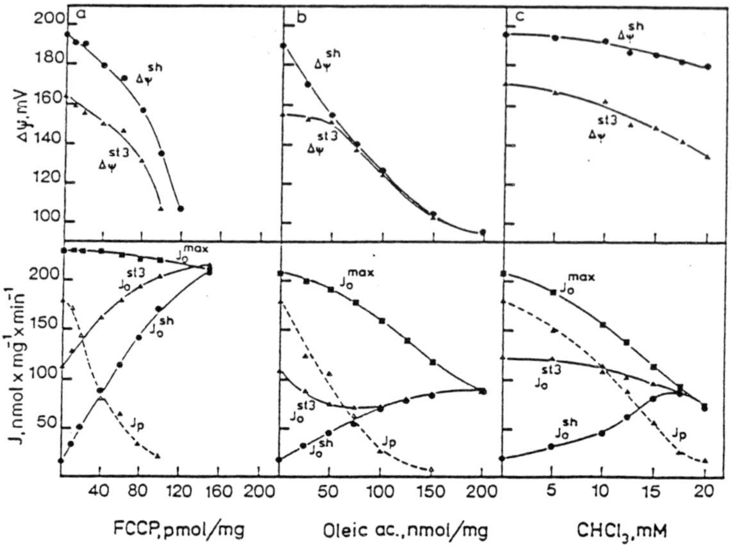

Figure 1. (lower panels) Mitochondrial rate of respiration in static head, Josh (●), and in state 3, Jost3 (▲) and maximal rate of respiration, Jomax (■), and rate of ATP synthesis, Jp (Δ), and (upper panels) difference of electrical potential across the inner mitochondrial membrane in static head, $\Delta\psi$sh (●), and in state 3, $\Delta\psi$st3 (▲), as a function of the concentrations of FCCP (a), oleic acid (b), and chloroform (c) (see Luvisetto et al., 1987, for more details).

Fig. 2 show that chloroform increases only slightly the dissipative ionic current. The small increase in dissipative ionic current does not account for the increase in the proton efflux by either the redox pumps or the ATPases (as calculated from the rate of electron transfer and a stoichiometry H+/e-=3 or from the rate of ATP hydrolysis and a stoichiometry H+/ATP=4, assuming fully coupled pumps). In agreement with the decrease of $\Delta\psi$ shown in fig. 1, FCCP and oleic acid increase considerably the passive conductance of the membrane as indicated by the increase of the ionic dissipative current in fig. 2. However, this increase does not completely account for the stimulation of the rate of electron transfer in static head.

Fig. 3 shows how the P/O ratio changes upon redox inhibition in control mitochondria and in mitochondria incubated with a certain amount of uncoupler. As expected for an increased membrane conductance the P/O ratio is more markedly depressed upon redox inhibition in the presence of either FCCP or oleic acid than in control mitochondria. In contrast, in the presence of chloroform the P/O ratio increases in a certain range instead of decreasing.

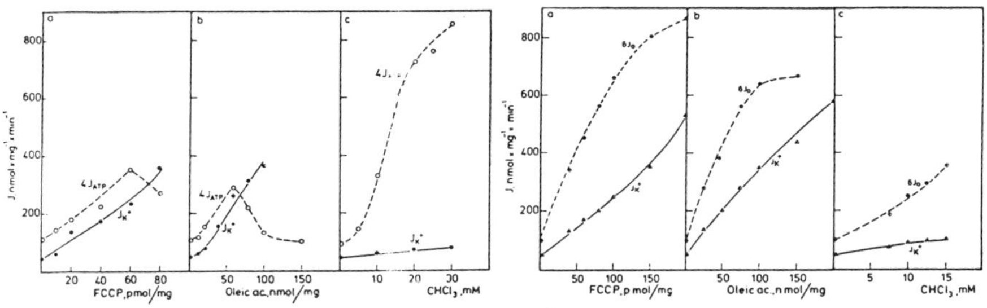

Figure 2. (right) Rate of respiration in static head multiplied by the H+/O stoichiometry and (left) rate of ATP hydrolysis multiplied by the H+/ATP stoichiometry, dashed lines, and rate of K+ efflux after addition of valinomycin to respiration-inhibited mitochondria, continuous lines, as a function of the concentrations of FCCP (a), oleic acid (b), and chloroform (c) (see Luvisetto et al., 1987, for more details).

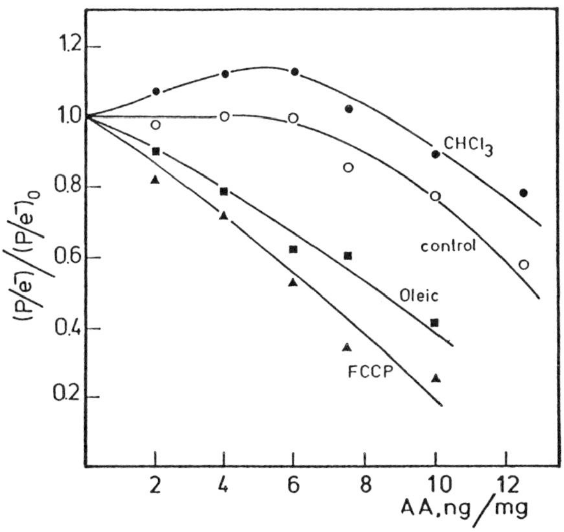

Figure 3. Normalized P/O ratio as a function of antimycin in the presence of a constant amount of chloroform (15 mM, ●), oleic acid (40 nmol/mg, ■) ,and FCCP (20 pmol/mg, ▲) and in the absence of uncoupler (○). (P/O)o is the value in absence of antimycin (see Pietrobon et al., 1987, for more details).

DISCUSSION

The uncoupling of oxidative phosphorylation by the three different types of uncouplers is not quantitatively accounted for by the increase in membrane conductance. All the three agents appear to cause the transfer of a certain number of electrons without translocation of a stoichiometric number of protons across the membrane from the mitochondrial matrix to the external medium. In addition, chloroform gives rise also to hydrolysis of ATP not accompanied by the translocation of a stoichiometric number of protons. At least two different mechanisms of uncoupling can give rise to electron transfer and ATP hydrolysis without apparent translocation of proton across the membrane: i) intrinsic uncoupling (slip) of the proton pumps (Pietrobon et al., 1981; Walz, 1983; Pietrobon et al., 1986); ii) decoupling of a localized protonic coupling mechanism (Rottenberg, 1983; Hitchens and Kell, 1983; Westerhoff et al., 1984; Slater et al., 1985).

In the kinetic model of chemiosmotic free energy coupling described by

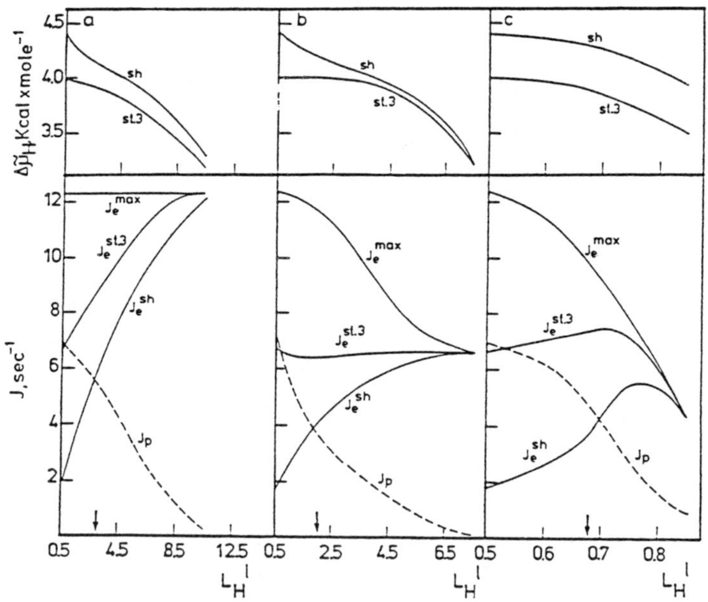

Figure 4. Simulations of the rates of ATP synthesis (dashed line) and of electron transfer and of $\Delta\psi$ in static head and state 3 as a function of leak conductance and of extent of intrinsic uncoupling of the redox pumps (a,b) and of both the redox and ATPase pumps (c) (see Pietrobon et al., 1987, for more details).

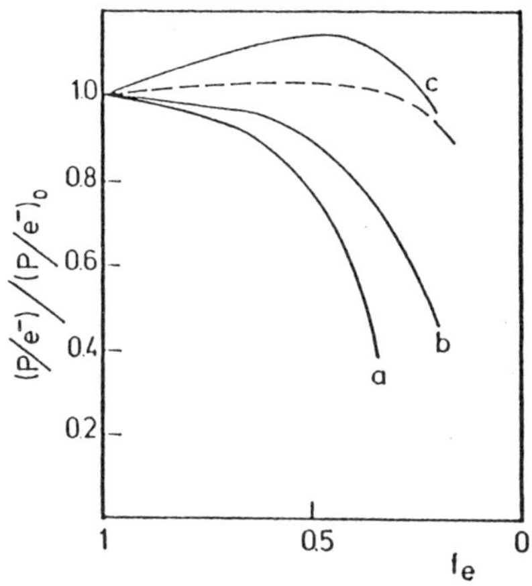

Figure 5. Simulations of the normalized P/e ratio as a function of the fraction of active redox pumps, fe; dashed line is obtained without uncoupler (see Pietrobon et al., 1987, for more details).

Pietrobon and Caplan (1986; Pietrobon, 1986) it is possible to introduce intrinsic uncoupling in the proton pump models (Pietrobon and Caplan, 1985), thus allowing the simulation of the effects of an increased slip at the level of either pump on the process of oxidative phosphorylation.

The simulations a, b, c, in Figs. 4 and 5 reproduce the behaviour of mitochondria in the presence of FCCP, oleic acid and chloroform respectively. Panels a and b in Fig. 4 have been obtained by increasing, together with the leak conductance, LHl, the rate constants of the slip transition in the redox pump model (and in addition in panel b by decreasing the fraction of active redox pumps to account for the additional secondary inhibitory effect of oleic acid shown in fig. 1). Panel c has been obtained by increasing the rate constants of the slip transition in both the redox and ATPase pump models together with a very little increase of LHl (and a decrease of the fraction of active redox pumps). The parameters corresponding to the arrows in Fig. 4 have been used in Fig. 5 to obtain the simulations of the P/O ratio upon redox inhibition.

The comparison between simulations and experiments leads to the conclusions that i) the results are consistent with a mixed action of FCCP and oleic acid as protonophores and intrinsic uncouplers of the redox proton pumps and with an action of chloroform as intrinsic uncouplers of the redox proton pumps and ATPase pumps and ii) the results do not necessarily require the rejection of the delocalized chemiosmotic model of energy coupling and do not appear to be able to unequivocally discriminate between different energy coupling mechanisms.

REFERENCES

Hitchens CD, Kell DB (1983). On the functional unit of energy coupling in photophosphorylation by bacterial chromatophores. Biochim Biophys Acta 723:308-316.

Luvisetto S, Pietrobon D, Azzone GF (1987). Uncoupling of oxidative phosphorylation. I. Protonophoric effects account only partially for uncoupling. Biochemistry 26:7332-7338.

McLaughlin SGA, Dilger JP (1980). Transport of protons across membranes by weak acids. Physiol Rev 60:825-863.

Pietrobon D (1986). A non-linear kinetic model of chemiosmotic energy coupling. Bioelectrochem Bioenerg 15:193-209.

Pietrobon D, Caplan SR (1985). Flow-force relationships for a six-state proton pump model. Intrinsic uncoupling, kinetic equivalence of input and output forces, and domain of approximate linearity. Biochemistry 24:5764-5776.

Pietrobon D, Caplan SR (1986). Double-inhibitor and uncoupler-inhibitor titrations. II. Analysis with a nonlinear model of chemiosmotic energy coupling. Biochemistry 25:7690-7696.

Pietrobon D, Azzone GF, Walz D (1981). Effect of funiculosin and antimycin A on the redox-driven H+-pumps in mitochondria: on the nature of 'leaks'. Eur J Biochem 117:389-394.

Pietrobon D, Luvisetto S, Azzone GF (1987). Uncoupling of oxidative phosphorylation. II. Alternative mechanisms: intrinsic uncoupling or decoupling ? Biochemistry 26:7339-7347.

Pietrobon D, Zoratti M, Azzone GF, Caplan SR (1986). Intrinsic uncoupling of mitochondrial proton pumps. II. Modeling studies. Biochemistry 25:767-775.

Rottenberg H (1983). Uncoupling of oxidative phosphorylation in rat liver mitochondria by general anesthetics. Proc Natl Acad Sci USA 80:3313-3317.

Slater EC, Berden JA, Herweijer MA (1985). A hypothesis for the mechanism of respiratory-chain phosphorylation not involving the electrochemical gradient of protons as obligatory intermediate. Biochim Biophys Acta 811:217-231.

Walz D (1983). Thermodynamic assessment of process in isolated energy converting organelles. In Optalaka A, Balaban M (eds): "Biological Structure and Coupled Flows", New York: Academic, pp 45-60.

Westerhoff HV, Melandri BA, Venturoli G, Azzone GF, Kell DB (1984). A minimal hypothesis for membrane-linked free energy transduction. The role of independent, small coupling units. Biochim Biophys Acta 768:257-292.

The Ion Pumps: Structure, Function, and Regulation, pages 401–407

ALLOSTERIC SHORT-TERM REGULATION OF ATP SYNTHASE: A PLAUSIBLE MODEL

Kurt R.H. Repke

Energy Conversion Unit in Central Institute of Molecular Biology, Academy of Sciences of G.D.R., DDR 1115 Berlin

THE NEED FOR SHORT-TERM REGULATION

In man, the rate of ATP synthesis varies from 0.4 g ATP at rest to 9.0 g ATP per min and kg body weight during strenuous exercise. This indicates that there has to be a correspondingly large reserve capacity in oxidative phosphorylation, and that the enzymes in the ATP synthetic pathway should not be rate-controlling. Within the required large dynamic range, the rate of energy production essentially equals the rate of energy utilization. This requires a finely tuned mechanism for the short-term regulation of oxidative phosphorylation to meet the actual ATP demands. The molecular basis of the respiratory control has been much disputed (see Erecińska and Wilson, 1982), but has remained unknown in detail (R.A. Mitchell, 1984) and on the whole enigmatic (Vignais and Satre, 1984). As critically reviewed by Erecińska and Wilson (1982) the long-standing dispute has centered on the question whether the [ATP]/[ADP] ratio or the phosphate potential, [ATP]/[ADP] [P_i], would be the parameter which determines the immediate rates of oxygen consumption.

The present brief account intends to show that, exploiting the topical data of current literature, one is able to construct a coherent model for the allosteric short-term regulation of ATP synthase/H^+ ATPase (F_1 enzyme) in mitochondrial and chloroplast membrane. The various pieces of evidence apt to puzzle it out may be found in the papers referred to in connection with the conclusions drawn here by me, but clearly often not directly by their authors. To meet the allotted space limit, preferentially review papers will be cited.

THE STARTING SIGNAL AND ITS SENSOR SYSTEM

As reviewed by McCormack and Denton (1979, 1986), decreases of the [ATP]/[ADP] ratio increase several-fold the activity of three exclusively intramitochondrial dehydrogenases: the pyruvate dehydrogenase complex (through kinase inhibition), the NAD^+-isocitrate dehydrogenase (through decrease in K_m for isocitrate) and the oxoglutarate dehydrogenase (through decrease in K_m for oxoglutarate). This leads to stimulated rates of pyruvate oxidation and citrate cycle flux as well as stimulation of NADH production, and hence to the rise of the protonic potential difference for that increase in ATP supply which is required to meet the enhanced demand. Any transient fall of the [NADH]/[NAD^+] ratio resulting from raised phosphorylating proton fluxes will further activate the above [ATP]/[ADP]- and [NADH]/[NAD^+]-dependent dehydrogenases. These properties of the three dehydrogenases will ensure that their turnover rate may be closely matched to cell ATP requirements.

THE ACTIVATION-DEACTIVATION TRANSITIONS OF F_1 ENZYME

Gräber et al. (1984) have demonstrated that the rate of ATP synthesis by the F_1 enzyme depends in a sigmoidal way on the transmembrane electro-chemical potential difference of protons, $\Delta\tilde{\mu}H^+$, and that the functional dependence of the ATP synthesis rate on $\Delta\tilde{\mu}H^+$ does not reflect the catalytic events at ATP synthase, but the transformation of F_1 enzyme from a deactivated to a catalytically active state. Thus, proton-elicited activation and catalytic activity come off consecutively so that the rate of ATP synthesis at saturating [ADP] and [P_i] is controlled by the fraction of activated enzyme molecules. The transition from the deactivated to the activated state appears to be provoked through binding of H^+ and OH^- to F_1 on inside or outside sites, respectively (Schlodder et al., 1982). The activation is correlated with profound conformational changes of F_1 protein which loosen the binding of the inhibitor peptide and thus render it ineffective (see Pedersen, 1982; Schlodder et al., 1982; Vignais and Satre, 1984; Matsuno-Yagi and Hatefi, 1986). In this way, the proper alignment of the peptide chain segments forming the catalytic center is restored so as to accept ADP and P_i as substrates for ATP synthase activity or ATP as substrate for H^+ ATPase activity, i.e., the activated F_1 enzyme can operate in the one or the other way, depending on the phosphate potential (see next paragraph).

In terms of the concept of allosterically regulated enzymes (Monod et al., 1965), H^+ and OH^- can be viewed as heterotropic allosteric activators and the inhibitor peptide as heterotropic allosteric inhibitor, and the activation $\rightleftarrows$ deactivation transitions as transitions between R and T states of F_1 enzyme, for which activator, inhibitor and substrate use to have very differential affinities. Thus, the F_1 enzyme behaves like a K system. The sigmoidal relationship between the rate of ATP synthesis and the $\Delta\tilde{\mu}H^+$ value appears to reflect homotropic cooperative interactions between nH^+ and nOH^-, respectively, in binding to their sites on the oligomeric enzyme ($\alpha_3\beta_3\gamma\delta\varepsilon$: see Senior and Wise, 1983).

The bound protons, after having worked as allosteric activator, transform F_1 protein into a high-energy conformation which, through proton translocation across the membrane, relaxes into a low-energy conformation. This delivers the energy to form the bond between ADP and P_i, and to release product ATP (cf. Schlodder et al., 1982).

THE DIRECTIONAL SIGNAL AND ITS DETECTOR SYSTEM

The reaction: $ADP + P_i + \Delta\tilde{\mu}H^+ \rightleftharpoons ATP + H_2O$, catalyzed by F_1, is potentially reversible in situ. ATP synthesis and ATP hydrolysis occur at the same catalytic sites on F_1 (Senior and Wise, 1983; Vignais and Satre, 1984). The K_m values for ADP and P_i, and the K_i value for ATP measured on ATP synthesis, as well as the K_m value for ATP and the K_i value for ADP measured on ATP hydrolysis are so low that the catalytic centers could be all occupied by the fueling substrates or inhibitory products if in situ would be no competition between them for occupancy of the centers (Tiedge et al., 1982; Schlodder et al., 1982; Junesch and Gräber, 1987). Apparently the [ATP]/[ADP] [P_i] ratio, i.e. the phosphate potential, directs the catalytic potentiality of F_1 in the synthetic or hydrolytic operation mode. Actually, the phosphate potential may fall, e.g., as much as 50-fold when a physiological work load is imposed on skeletal muscle.

In greater detail, ADP and P_i are known to promote ATP synthesis (Minkov et al., 1980; Schlodder et al., 1982, LaNoue et al., 1986; Komatsu-Takaki, 1987) and to inhibit ATP hydrolysis (Moyle and P. Mitchell, 1975; Minkov et al., 1980; Feldman and Boyer, 1985; Urbauer et al., 1987). The back reaction may be suppressed also by free Mg^{2+} (Moyle and Mitchell, 1975; Minkov et al., 1980; Gómez-Puyou et al., 1983; Feldman and Boyer, 1985) which might thus have an additional regulatory action.

The early suggestion that the catalytic mechanism of ATP synthase involves flip-flop (alternating site) cooperativity between the catalytic subunits of F_1 (Repke and Schön, 1974; Kayalar et al., 1977) was shown right by Stroop and Boyer (1985) as well as by Matsuno-Yagi and Hatefi (1986). Although the maximal rate of ADP phosphorylation may increase about 30-fold within a low ADP concentration range (Stroop and Boyer, 1985), the catalytic cooperativity seems to have in situ a minor, if any, regulatory relevance. According to Matsuno-Yagi and Hatefi (1986), the $\Delta\tilde{\mu}H^+$ value (the 'energy pressure'), which determines the fraction of active F_1 enzyme molecules (see above), appears to be the regulatory determinant of greater physiological significance.

In principle, the phenomenon of cooperativity could be considered as cooperativity through time via consecutive substrate bindings to the same catalytic center. In the present case, however, the observed cooperativity can be classified as cooperativity through space via consecutive substrate bindings to the catalytic centers on the three subunits, i.e., as interprotomeric positive homotropic interactions. In line with this interpretation is the finding of Lunardi et al. (1986) that the binding of ADP or ATP to native F_1 enzyme yields curvilinear Scatchard plots.

TWO ASPECTS OF THE INTERPLAY BETWEEN THE REGULATED PROCESSES

The long-lasting dispute whether the response of oxidative phosphorylation to changes in the rate of ATP utilization is controlled by the $[ATP]/[ADP]$ ratio or by the $[ATP]/[ADP]\ [P_i]$ ratio might be viewed as being settled in the framework of the model developed here that postulates differential control functions for both parameters.

The thermodynamic significance of the allosterically regulated R ⇌ T transitions, derived here for the activation/deactivation of ATP synthase/H^+ ATPase, appears to be that the associated quasi-unidirectionality of the ATP synthase operation mode circumvents thermodynamic equilibration.

CONCLUDING REMARKS

The concept I have puzzled out from literature data may be designated, I believe, as a model that is, according to H.A. Krebs (1966) "a simplified system which presents some of the essential features of reality and provides explanations of experimental observations, insights, and starting points for fuller explanation of the real thing".

As might have been expected, the model presented here for the allosteric regulation of ATP synthase shows some similarities to the model for the allosteric regulation of Na^+/K^+-transporting ATPase (Repke, 1986).

REFERENCES

Erecińska M, Wilson DF (1982). Regulation of cellular energy metabolism. J Membrane Biol 70: 1-14.

Feldman RI, Boyer PD (1985). The role of tightly bound ADP on chloroplast ATPase. J Biol Chem 260: 13088-13094.

Gómez-Puyou A, Ayala G, Muller U, de Gómez-Puyou MT (1983). Regulation of the synthesis and hydrolysis of ATP by mitochondrial ATPase. Role of Mg^{2+}. J Biol Chem 258: 13673-13679.

Gräber P, Junesch U, Schatz GH (1984). Kinetics of proton-transport-coupled ATP synthesis in Chloroplasts. Activation of the ATPase by an artificially generated ΔpH and $\Delta\psi$. Ber Bunsenges Phys Chem 88: 599-608.

Junesch U, Gräber P (1987). Influence of the redox state and the activation of the chloroplast ATP synthase on proton-transport-coupled ATP synthesis/hydrolysis. Biochim Biophys Acta 893: 275-288.

Kayalar C, Rosing J, Boyer PD (1977). An alternating site sequence for oxidative phosphorylation suggested by measurement of substrate binding patterns and exchange reaction inhibitions. J Biol Chem 252: 2486-2491.

Komatsu-Takaki M (1987). Correlation between the ATP synthetic active state and the ATP hydrolytic active state in chloroplast ATP synthase-ATPase complex $CF_o \cdot FC_1$. J Biol Chem 262: 8202-8205.

Krebs HA (1966). Theoretical concepts in biological sciences. In Kaplan NO, Kennedy EP (eds): "Current Aspects of Biochemical Energetics", New York, London: Academic Press, pp 83-95.

LaNoue KF, Jeffries FMH, Radda GK (1986). Kinetic control of mitochondrial ATP synthesis. Biochemistry 25: 7667-7675.

Lunardi J, Klein G, Vignais PV (1986), Interaction between aurovertin and adenine nucleotide binding sites on mitochondrial F_1-ATPase and the isolated β subunit. J Biol Chem 261: 5350-5354.

Matsuno-Yagi A, Hatefi Y (1986). Kinetic modalities of ATP synthesis. Regulation by the mitochondrial respiratory chain. J Biol Chem 261: 14031-14038.

McCormack JG, Denton RM (1979). The effects of calcium ions and adenine nucleotides on the activity of pig heart 2-oxoglutarate dehydrogenase complex. Biochem J 180: 533-544.

McCormack JG, Denton RM (1986). Ca^{2+} as a second messenger within mitochondria. Trends Biochem Sci 11: 258-262.

Minkov IB, Vasilyeva EA, Fitin AF, Vinogradov AD (1980). Differential effects of ADP on ATPase and oxidative phosphorylation in submitochondrial particles. Biochem Int 1: 478-485.

Mitchell RA (1984). Control of forward and reverse electron flow in mitochondria: thermodynamic versus kinetic considerations. Current Topics in Cellular Regulation 24: 387-395.

Monod J, Wyman J. Changeux J-P (1965). On the nature of allosteric transitions: a plausible model. J Mol Biol 12: 88-118.

Moyle J, Mitchell P (1975). Active/inactive state transitions of mitochondrial ATPase molecules influenced by Mg^{2+}, anions and aurovertin. FEBS Lett 56: 55-61.

Pedersen PL (1982). H^+-ATPase in biological systems: an overview of their function, structure, mechanism, and regulatory properties. Ann NY Acad Sci 402:1-20.

Repke KRH and Schön R (1974). Flip-flop model of energy interconversion by ATP synthetase. Acta biol med germ 33: K27-K38.

Repke KRH (1986). A model for allosteric regulation of Na^+/K^+-transporting ATPase. Biochim Biophys Acta 864: 195-212.

Schlodder E, Gräber P, Witt HT (1982). Mechanism of phosphorylation in chloroplasts. In Barber J (ed): "Electron Transport and Photophosphorylation", Amsterdam: Elsevier Biomedial Press, pp. 105-175.

Senior AE and Wise JG (1983). The proton-ATPase of bacteria and mitochondria. J Membrane Biol 73: 105-124.

Stroop SD, Boyer PD (1985). Characteristics of the chloroplast ATP synthase as revealed by reaction at low ADP concentrations. Biochemistry 24: 2304-2310.

Tiedge H, Lücken U, Weber J, Schäfer G (1982). High-affinity binding of ADP and of ADP analogues to mitochondrial F_1-ATPase. Eur J Biochem 127: 291-299.

Urbauer JL, Dorgan LJ, Tomaszek Jr TA, Schuster SM (1987). Dependence of the activity of beef heart mitochondrial adenosinetriphosphatase on the properties of the catalytic metal ion. Biochemistry 26: 2914-2924.

Vignais PV, Satre M (1984). Recent developments on structural and functional aspects of the F_1 sector of H^+-linked ATPase. Mol Cell Biochem 60: 33-70.

The Ion Pumps: Structure, Function, and Regulation, pages 409–416

MECHANISMS OF PROTON EFFLUX FROM CHLOROPLAST THYLAKOIDS.

Mordechay Schonfeld, Hedva Schickler and Bente S. Kopeliovitch.

The Hebrew University, Faculty of Agriculture, Department of Agricultural Botany, Rehovot, Israel.

INTRODUCTION

Photosynthetic electron transport functions in the thylakoid membrane as a proton pump, generating a proton gradient which drives ATP synthesis. A question currently under consideration is whether the driving force for ATP synthesis is a delocalized transmembrane proton gradient (Mitchell, 1966), or whether redox reactions are directly linked to ATP synthesis via membrane-localized proton fluxes (Dilley et al ,1987). Better characterization of proton fluxes in thylakoids and their interactions with electron transport and ATP synthesis, should help to resolve this issue. Here we present a kinetic analysis of passive proton fluxes in the thylakoid membrane. We compare the characteristics of proton efflux from thylakoids during steady state illumination, with those of dark proton efflux measured on cessation of illumination. We also evaluate the effects of internal buffering groups on the kinetics.

MATERIALS AND METHODS

Envelope-free chloroplasts were isolated from lettuce leaves and assayed as previously described (Schonfeld and Neumann, 1977). A magnetically stirred glass cuvette, equipped with a glass combination electrode and an oxygen electrode, was installed in the sample compartment of a Jasco FP-550 spectrofluorometer. A projector-lamp equipped with a heat filter and red cut-off filters provided 1000 µE m^{-2} s^{-1} of red actinic light ($\lambda >$ 600 nm) at the position of the cuvette. This arrangement permitted simultaneous measurements of the internal pH via fluorescence changes of 9-aminoacridine (Schuldiner et al, 1972), of electron transport, and of the pH in the medium. Changes in pH were calibrated by addition of standard HCl solutions. The reaction mixture contained in a final volume of 3 ml: 50 mM KCl, 0.2mM methyl viologen, 1 mM Sodium azide, 2μM 9-aminoacridine and chloroplasts equivalent to 24 μg Chl/ml.

RESULTS AND DISCUSSION

Light vs Dark Proton Efflux from Thylakoids.

Proton efflux from chloroplast thylakoids can be measured by the pH changes in the chloroplast incubation medium, during the dark decay of the light induced proton gradient. This process exhibits first order kinetics (Karlish and Avron, 1968), so that the initial rate can be readily calculated. Alternatively one can obtain the steady state rate of proton efflux, indirectly, from the measured rate of electron transport, assuming a fixed stoichiometry between the two processes. A value of $2H^+/e^-$ was assumed in the present study (Davenport and McCarty, 1984), and in addition we used a value of 2.2 nm^2 for the membrane area occupied by one Chl molecule (Wolken and Schwertz 1953), in order to obtain proton fluxes per unit area, rather than per mg Chl.

A comparison of proton efflux in the light ($J_H{}^L$) and in the dark ($J_H{}^D$), measured by these two methods respectively, showed close correlation at low pH levels (Fig 1). At pH > 7 however, $J_H{}^L$ increased with pH in a manner typical of light induced electron transport (Avron, 1971), while $J_H{}^D$ exhibited a slight decrease as the pH was increased above 7.

In a previous study (Schonfeld and Schickler, 1984) we used simultaneous measurements of electron transport and pH to resolve proton efflux in the light into three components.

$$J_H{}^L = P_H \times H_i + K_1 \frac{H_i}{H_o} + K_2 \left(\frac{H_i}{H_o}\right)^b \qquad (1)$$

where P_H is the proton permeability of the membrane, H_i and H_o are the internal and external proton concentrations respectively, and K_1, K_2 and b are constants. The first term represents proton efflux by diffusion and is characterized by a linear dependence on H_i. At low pH, this is practically the only efflux pathway. In the absence of membrane potential, efflux by diffusion should be given by $P_H (H_i - H_o)$ but this is reduced to $P_H \times H_i$, where $H_i >> H_o$. The second term seems to be due to an additional leakage pathway, whose nature is not clear as yet, which is characterized by a linear dependence on H_i/H_o. Efflux by this pathway can be observed above pH 7 in the presence of inhibitors of proton efflux associated with the CF (the ATP synthase of the thylakoids), such as ATP. The third term was found to be due to proton efflux associated with the CF complex. It is characterized by an exponential dependence on H_i/H_o and is the dominant proton flux above pH 7. Similar characteristics of proton efflux associated with the CF-complex were previously described (Graber and Witt, 1976).

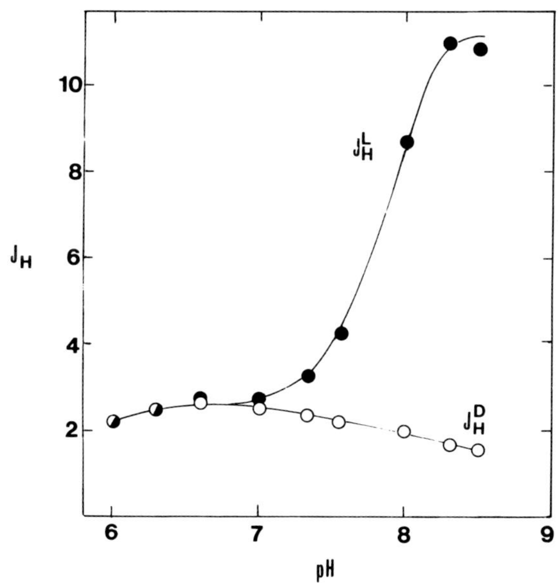

Figure 1: Dependence on pH of proton efflux in the light (J_H^L) and in the dark (J_H^D). Proton fluxes are in $pmol \cdot cm^{-2} \cdot s^{-1}$.

A similar analysis carried out for dark proton efflux (Schonfeld and Kopeliovitch, 1985), by parallel measurements of pH changes inside and outside the thylakoids, indicates that proton efflux in the dark is given by

$$J_H^D = P_H \times H_i + K_1 \frac{H_i}{H_o} \qquad (2)$$

A comparison of Eqn 2 and Eqn 1 shows that proton efflux via the two leakage pathways is evident both in the light and in the dark. However, proton efflux associated with the CF-complex is evident only in the light. This is conspicuous at high pH (Fig 1), where proton efflux in the light takes place largely via the CF-complex (Schonfeld and Schickler, 1984).

This conclusion is further supported by the finding that specific inhibitors of proton transport associated with the CF-complex, such as DCCD, do not inhibit proton efflux in the dark, and inhibit the efflux in the light to about the level of the dark process (Schonfeld et al, 1987). DCCD was previously shown to suppress the stimulated rates of electron transport obtained at high pH (Avron, 1971).

Under these experimental conditions, proton efflux via the ATPase is apparently deactivated within a short time after cessation of illumination. Therefore, the initial rate of dark proton efflux, measured on cessation of illumination, does not portray the steady state in the light, as commonly assumed.

Kinetics of Dark Proton Efflux

More than 99% of the protons accumulated during steady state illumination, are bound in the thylakoids during the buildup of the gradient, and are gradually released on darkening the chloroplast suspension. The apparent first order rate constant for dark proton efflux (K_d), is proportional to the permeability of thylakoid membrane for protons (P_H) and is inversely proportional to C_H, the proton binding capacity of the thylakoids (Schonfeld and Kopeliovitch, 1985).

$$K_d = P_H / C_H \qquad (3)$$

C_H is measured by the ratio between Hx, the total amount of protons accumulated in the thylakoids at any moment, and H_i, the internal proton concentration, (i.e $C_H = H_x / H_i$). Parallel measurements of H_x and H_i at different light intensities showed indeed a linear relationship between the two parameters, and C_H was also found to remain constant during the buildup and decay of the proton gradient. C_H was found however to vary with the pH in the medium (Schonfeld and Kopeliovitch, 1985).

A recent theoretical analysis (Whitmarsh, 1987) supports our concept of internal buffering groups affecting the proton gradient decay kinetics. For a single internal proton-binding group under the conditions of $A_t >> Ka >> H_i$; (A_t being the group concentration and Ka its dissociation constant,) this analysis yields

$$K_d = \frac{S}{V} \times \frac{P_H}{A_t/Ka} \qquad (4)$$

where S/V is the surface to volume ratio of the thylakoids. A comparison of Eqn 3 and Eqn 4 shows that for these conditions the proton binding capacity is given by

$$C_H = \frac{A_t}{Ka} \times \frac{V}{S} \qquad (5)$$

Under the specified conditions, C_H is thus expected to be proportional to the concentration of the proton-binding group and inversely proportional to its dissociation constant. It is also expected to increase with the volume / surface ratio of the thylakoids, e.g. increase during swelling. These predictions were largely confirmed by experiments with permeant buffers (unpublished observations).

Effects of Protonophores on Dark Proton Efflux.

Gramicidin stimulated the rate of proton efflux very effectively at pH 6.5, producing a significant increase in K_d even at 1 nM. Its potency, however, decreased sharply at higher pH values: the concentrations required to induce at pH 8 the same stimulation of K_d as at pH 6.5, were two to three orders of magnitude higher (Fig 2).

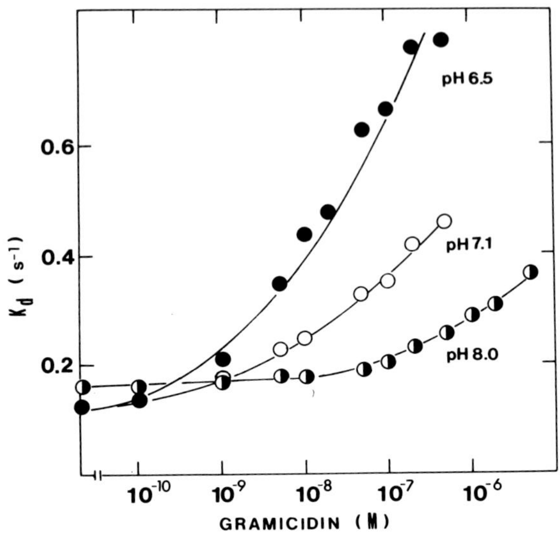

Figure 2: Dependence of the rate constant for dark proton efflux (K_d) on gramicidin concentration, at three pH levels.

This decrease in potency was found to be due to two separate factors: 1. An increase in the pH of the medium resulted in a gradual shift from proton efflux by diffusion to efflux via the second endogenous leakage pathway (Schonfeld et al, 1987), thus also decreasing the relative contribution of gramicidin channels to the total proton flux. 2. At pH 6.5 gramicidin had no effect on C_H, as expected. The stimulation of P_H by gramicidin was therefore fully reflected in K_d (Eqn 3 & Fig 3A). At pH 8, however, the effects on P_H were partially offset by an unexpected stimulation of C_H by gramicidin, so that its effects on K_d were rather limited (Fig 3B).

The effects of nigericin on dark proton efflux at pH 6.5 (Fig 3C), were similar to those of gramicidin. It stimulated P_H and K_d parallely, without affecting C_H. At pH 8 however, P_H and C_H were stimulated by nigericin in a parallel manner, so that K_d was hardly affected (Fig 4D)

These experiments, and those carried out with other protonophores such as FCCP, NH_4^+ and several alkyl amines, indicate that different protonophores affect differently not only the proton permeability, but also the proton binding capacity of the thylakoids. Effects of protonophores on the rate of dark proton efflux, cannot therefore be used directly as a measure for changes in the proton-permeability. They have first to be resolved from effects on the proton binding capacity (Eqn 2).

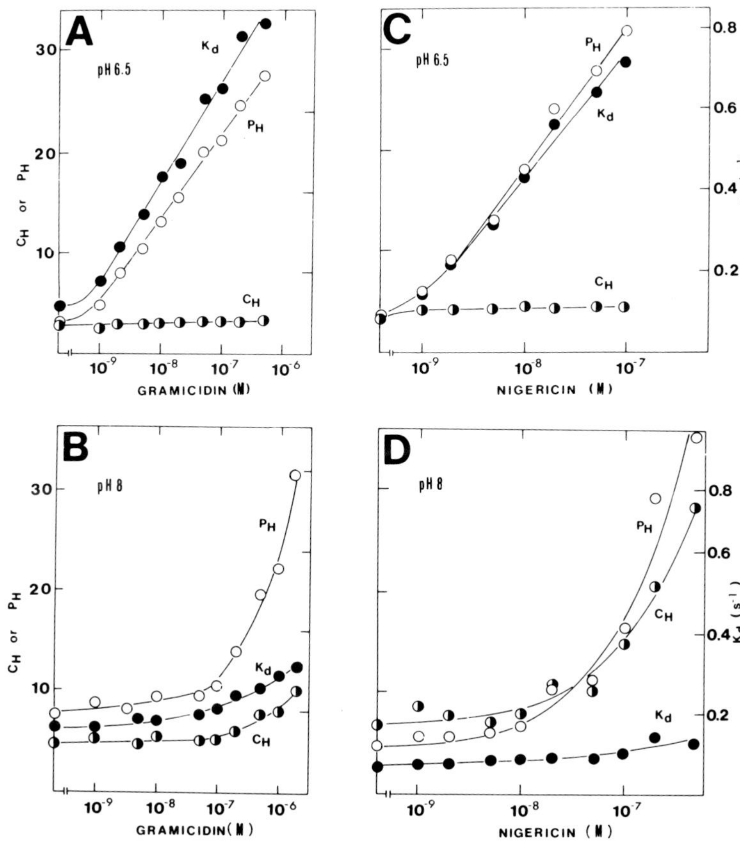

Figure 3: Effects of gramicidin and nigericin on K_d, P_H and C_H at pH 6.5 and 8.0. C_H is in cm x 10^{-4} and P_H in cm/s x 0^{-5}.

REFERENCES

Avron M (1971). The relation of light induced reactions of chloroplasts to proton concentration. In Forti G, Avron M, Melandri BA (eds): "Proc 2nd int Cong Photosynth," The Hague: Junk, vol 2, pp 861-871.

Davenport JW and RE McCarty 1984. An analysis of proton fluxes coupled to electron transport and ATP synthesis in chloroplast thylakoids. Biochim Biophys Acta 766: 363-374.

Dilley RA, SM Theg, and WA Beard (1987). Membrane proton interactions in chloroplast bioenergetics: localized proton domains. Ann Rev Plant Physiol 38:347-389.

Graber P and HT Witt (1976). Relations between the electrical potential, pH gradient, proton flux and phosphorylation in the photosynthetic membrane. Biochim Biophys Acta 423: 141-163.

Karlish SJD and Avron M (1968). Analysis of light induced proton uptake in isolated chloroplasts. Biochim Biophys Acta 153:878-888.

Mitchell P (1966). Chemiosmotic coupling in oxidative and photosynthetic phosphorylation. Biol Rev Cambridge Phil Soc 41:445-540.

Schonfeld M and Neumann J (1977). Proton conductance of the thylakoid membrane: modulation by light. FEBS Letters 73:51-54.

Schonfeld M and Schickler H (1984). The permeability of the thylakoid membrane for protons. FEBS Letters 167: 231-234

Schonfeld M and Kopeliovitch BS (1985). Kinetics of dark proton efflux in chloroplasts. FEBS Letters 193:79-82.

Schonfeld M, Kopeliovitch BS and Schickler H (1987). Pathways for dark proton efflux from chloroplasts. in Biggins J (ed): "Progress in Photosynthesis Research," Dordrecht: Martinus Nijhoff, vol III pp 153-156.

Schuldiner S, Rottenberg H and Avron M (1972). Determination of ΔpH in chloroplasts. 2. Fluorescent amines as a probe for the determination of ΔpH in chloroplasts. Eur. J. Biochem 25:64-70.

Wolken JJ and Schwertz FA (1953). Chlorophyll monolayers in chloroplasts. J Gen Physiol 37:111-126.

Whitmarsh J (1987). Proton decay kinetics for vesicles containing buffers - an analytical solution. Photosynth Res 12:43-62

The Ion Pumps: Structure, Function, and Regulation, pages 417–422

PURIFICATION OF Ca^{2+}-ACTIVATED K^+ CHANNEL AND ITS ROLE IN REGULATION OF ACTIVE NaCl TRANSPORT

Dan A. Klaerke and Peter L. Jørgensen

Institute of Physiology, University of Aarhus, DK-8000 Aarhus C, Denmark

INTRODUCTION

The tubule cells in the thick ascending limb of Henle's loop (TAL) are specialized for active reabsorption of NaCl with a highly efficient transport system for which the sum of paracellular and transcellular transport across the epithelium amounts to 5-6 NaCl per ATP split in the cell cytoplasm (Fig. 1). The driving force for this transport is the Na,K-pump in the basolateral membrane, but the maximum stoichiometry of the Na,K-pump is 3 Na^+ per ATP split. The high efficiency is therefore due to the coupling of fluxes through the Na,K-pump with fluxes through conductances in luminal and basolateral cell membranes and in the paracellular pathway. The luminal membrane of the tubule cells contains the diuretic sensitive Na,K,Cl-cotransport system and the K^+ channel. The basolateral membrane has a specific Cl^- conductance and specific conductances are found in the tight junctions of the paracellular pathway (Greger and Schlatter, 1983; Hebert et al., 1981; Jørgensen, 1986).

The Ca^{2+}-activated K^+ channel has been identified in luminal membrane vesicles from TAL and characterized after reconstitution into phospholipid vesicles (Klaerke et al. 1987a). The K^+ channel is stimulated by Ca^{2+} in physiological concentrations and its activity depends on calmodulin and phosphorylation from cAMP dependent protein kinase (Klaerke et al., 1987b). These observations suggest that the K^+ channel may be of primary importance for regulation of transcellular NaCl transport in TAL, e.g. in response to stimulation of the cells with antidiuretic hormone (ADH).

It has been possible to purify the K^+ channel protein from

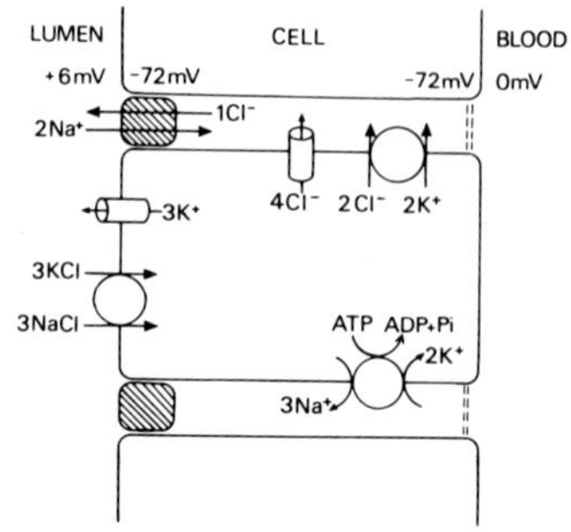

Figure 1. Model for tubulus cell from thick ascending limb of Henle's loop. Modified after Greger and Schaltter (1983).

TAL by affinity chromatography after solubilization of the membrane proteins in CHAPS. The purified protein can be reconstituted into phospholipid vesicles and its properties have been compared with those of the channel in native plasma membrane vesicles. (Klaerke et al., 1987a+b).

ASSAY OF Ca^{2+}-ACTIVATED K^+ CHANNEL IN LUMINAL MEMBRANE VESICLES

The K^+ channel from the luminal plasma membrane vesicles from TAL can be stimulated 4 fold with Ca^{2+} after initial depletion for Ca^{2+} with EGTA (Klaerke et al., 1987a; Burnham et al., 1986). The $K_{0.5}$ for the Ca^{2+} stimulation of the K^+ channel is 200 nM at pH 7.2 ranging from 1 μM at pH 6.4 to 20 nM at pH 8.0. This affinity for Ca^{2+} is among the highest reported (Petersen and Maruyama, 1984). Inhibition of the K^+ channel in the luminal plasma membrane vesicles and subsequent stimulation with Ca^{2+} only takes place in the presence of the Ca^{2+} ionophore A23187. This implies that the binding site for Ca^{2+} activation is located to the interior of the plasma membrane vesicles.

Patch-clamping and reconstitution of channels have shown that the transmembranal potential alters the Ca^{2+} activation; hyperpolarization of the membrane reduces the sensitivity to Ca^{2+} whereas depolarization increases the sensitivity (Findlay, 1984). This relationship may form a feed back mechanism in physiological conditions: a rise in intracellular Ca^{2+} opens the K^+ channels making a hyperpolarization of the cell membrane. This then reduces the sensitivity of the K^+ channel to Ca^{2+} and thus counteract the effect of the initial rise in intracellular Ca^{2+}.

EFFECT OF Ca^{2+} AND CALMODULIN ON RECONSTITUTED K^+ CHANNEL IN PHOSPHOLIPID VESICLES

Assay of K^+ channel activity with access to the luminal cytoplasmic face is possible after solubilization and reconstitution of the K^+ channel into phospholipid vesicles (Burnham et al., 1985; Klaerke et al., 1987a). In reconstituted vesicles the K^+ channel is activated by Ca^{2+} in the same range of concentration as in the luminal plasma membrane vesicles indicating that the K^+ channel has not been denatured by solubilization and reconstitution. However there is one important difference between the two systems. In reconstituted vesicles Ca^{2+} activation of the K^+ channel takes place without or with ionophore present, whereas in plasma membrane vesicles the ionophore is required for Ca^{2+} activation (Klaerke et al., 1987a). This means that in the reconstituted vesicles the cytoplasmic Ca^{2+} activation site of the K^+ channel is mainly exposed to the exterior of the vesicles. This allows for direct examination of the effect of compounds normally present in cell cytoplasm.

Addition of calmodulin to the reconstituted vesicles doubles the Ca^{2+} activation of the K^+ channel, whereas there is no effect of addition of calmodulin in the absence of Ca^{2+} (Klaerke et al., 1987b). The $K_{0.5}$ for the calmodulin effect is 0.1 nM. Regulation by calmodulin is often mediated through phosphorylation. However, calmodulin stimulation of the Ca^{2+}-activated K^+ channel takes place under conditions where phosphorylation is unlikely (Klaerke et al., 1987b). Calmodulin may thus bind directly to cytoplasmic aspects of the Ca^{2+}-activated K^+ channel in presence of Ca^{2+}, like it is the case for the plasma membrane Ca^{2+}-ATPase.

PURIFICATION OF K^+ CHANNEL PROTEINS

High affinity for calmodulin in presence of Ca^{2+} suggests that the K^+ channel protein from TAL can be purified by calmodulin affinity chromatography. The K^+ channel in luminal membranes is effectively solubilized by CHAPS, but only 0.7 per cent of the soluble protein binds to the calmodulin affinity column in presence of Ca^{2+} and is subsequently eluted from the column with EGTA (Fig. 2A). SDS gel electrophoresis of the purified

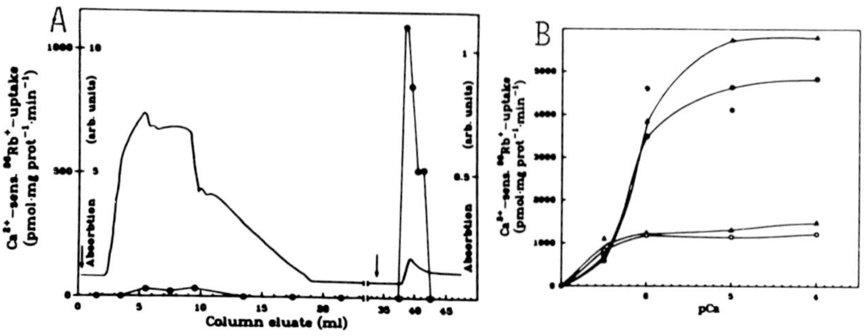

Figure 2. A: Purification of K^+ channel protein by calmodulin affinity chromatography showing UV-absorption (unbroken line) and Ca^{2+}-activated K^+ channel activity ( ). B: $^{86}Rb^+$ uptake after phosphorylation of purified K^+ channel protein in reconstituted vesicles with (▲,●) and without phosphorylation (△,○) compared to uptake in presence (▲,△) and absence (●,○) calmodulin. For details see Klaerke et al. (1987b).

protein show only 2 major proteins with molecular masses of 36.000 Da and 51.000 Da respectively (Klaerke et al., 1987b). Data in Fig. 2A+B shows that these proteins have a high Ca^{2+}-activated K^+ channel activity after reconstitution. However the purified protein seems to have lost some of its sensitivity to calmodulin during affinity chromatography (Fig. 2B).

Gel electrophoresis after tryptic digestion of the purified protein have shown that the 36.000 Da band was cleaved much faster than the 51.000 Da band (Klaerke et al., 1987b). From the cytoplasmic face this tryptic digestion activated the K^+ channel in the absence of Ca^{2+} to the level of activity seen with saturating concentrations of Ca^{2+} (Klaerke et al., 1987a). This provides indirect evidence for involvement of the 36.000 Da protein in opening and closing of the K^+ channel in response to Ca^{2+} binding.

Phosphorylation catalyzed by a cAMP-dependent protein kinase of the purified protein after reconstitution into phospholipid vesicles stimulates the maximal Ca^{2+}-activated K^+ channel activity several fold (Fig. 2B). Autoradiography after cAMP dependent protein kinase phosphorylation of the purified soluble protein in presence of [γ-^{32}P]-ATP showed ^{32}P labelling of the 51.000 Da band (Klaerke et al., 1987b). Since the K^+

channel activity is stimulated by this phosphorylation our data suggest that the 51.000 Da band forms a part of the K^+ channel that is involved in regulation of the K^+ channel after phosphorylation. Other laboratories have earlier proposed that proteins with molecular masses ranging from 14.000 to 88.000 Da could be part of K^+ channels (Wen et al., 1984, Seagar et al., 1985).

ROLE OF K^+ CHANNEL IN REGULATION OF NaCl TRANSPORT

Our study thus shows that K^+ channel activity depends on phosphorylation and activation by Ca^{2+} and calmodulin and that H^+ effects Ca^{2+} activation and transport through the channel. This raises the question of how modulation of channel activity can be important for regulation of transcellular NaCl transport in the intact epithelium.

Activation of the K^+ channel causes a loss of intracellular K^+, which stimulates the reabsorption of NaCl through the luminal Na,K,Cl-cotransport system. At saturating concentrations of ATP (2-3 mM) the rate limiting step of the Na,K-pump is the E_1P-E_2P transition, and the cytoplasmic $[Na^+]$ activity is the important regulating factor determining Na,K-pump turnover rates. Small changes in intracellular $[Na^+]$ activity causes large changes in pump rates (Jørgensen, 1986). A rise in intracellular Na^+ due to stimulation of transport through the Na,K,Cl-cotransport system thus activated the Na,K-pump. An opening of the luminal K^+ channel together with a rise in intracellular Cl^- concentration and a opening of the basolateral Cl^- channel also means a rise in the transepithelial potential that drives the paracellular transport across the tight junctions. In this way a activation of the luminal K^+ channel means a stimulation of the total NaCl reabsorption.

Stimulation of the K^+ channel by phosphorylation by cAMP dependent protein kinase and Ca^{2+}/calmodulin raises the possibility that the channel may mediate hormone responses. In mouse kidney the reabsorption of NaCl in TAL is stimulated by ADH (Hebert et al., 1981). The properties of the K^+ channel as discussed above makes it an obvious candidate as one of the targets in the cell for this hormonal regulation.

An interesting aspect of using the K^+ channel as mediator of regulatory influences on transport is conservation of energy. The ratio of the number per cell of K^+ channels to Na,K,Cl-cotransport systems or Na,K-pumps is only 400/4 x 10^6 or 400/40 x 10^6. It is energetically very favorable to use the K^+ channel as target for hormone action as the phosphorylation or dephosphorylation of only a few hundred channels may alter the turnover rate of millions of cotransport systems or Na,K-pump units.

REFERENCES

Burnham C, Karlish SJD and Jørgensen PL (1985). Identification and reconstitution of a Na/K/Cl cotransporter and K^+ channel form luminal membranes of renal red outer medulla. Biochim. Biophys. Acta 821: 461-469.

Burnham C, Braw R and Karlish SJD (1986). A Ca^{2+}-dependent K^+ channel in luminal membranes from the renal outer medulla. J. Membrane Biol. 93: 177-186.

Findlay I (1984). A patch-clamp study of potassium channels and whole-cell currents in acinar cells of the mouse lacrimal gland. J. Physiol. 350: 179-195.

Greger R and Schlatter E (1983). Proporties of the basolateral membrane of the cortical thick ascending limb of Henles loop of rabbit kidney. Pfluegers Arch. 396: 325-334.

Hebert SC, Culpepper RM and Andreoli TE (1981). NaCl transport in mouse medullary thick ascending limbs. I. Functional nephron heterogenity and ADH-stimulated NaCl cotransport. Am. J. Physiol. 241: F412-F431.

Jørgensen PL (1986). Structure, function and regulation of Na,K-ATPase in the kidney. Kidney Int. 29: 10-20.

Klaerke DA, Karlish SJD and Jørgensen PL (1987a). Reconstitution in phospholipid vesicles of calcium-activated potassium channel from outer renal medulla. J. Membrane Biol. 95: 105-112.

Klaerke DA, Petersen J and Jørgensen PL (1987b). Purification of Ca^{2+}-activated K^+ channel protein on calmodulin affinity columns after detergent solubilization of luminal membranes form outer renal medulla. FEBS Letters 216: 211-216.

Petersen OH and Maruyama Y (1984). Calcium-activated potassium channels and their role in secretion. Nature 307: 693-696.

Seagar MJ, Labee-Jullie C, Granier C, Van Rietschoten J and Couraud F (1985). Photoaffinity labelling of components of the apamin-sensitive K^+-channel in neuronal membranes. J. Biol. Chem. 260: 3895-3898.

Wen Y, Famulski KS and Carafoli E (1984). Ca^{2+}-dependent K^+ permeability of heart sarcolemma vesicles. Modulation by cAMP-dependent protein kinaseactivity and by calmodulin. Biochem. Biophys. Res. Commun. 122: 237-243.

The Ion Pumps: Structure, Function,
and Regulation, pages 423–426

ELECTRICAL TOLERANCE OF THE *CHARA CORALLINA* PLASMALEMMA

Taka-aki Ohkawa and Izuo Tsutsui

Department of Biology,
College of General Education,
Osaka University, Toyonaka 560, Osaka, Japan.

Introduction

When the membrane potential Vm of the *Chara corallina* plasmalemma is beyond about -400 mV under voltage clamp, the so-called breakdown phenomenon is observed. To examine this phenomenon is, in other words, to examine the membrane electrical tolerance. This communication reports that the external Ca^{2+} concentration $[Ca^{2+}]o$ is the major factor of the control of the electrical tolerance and that $[Ca^{2+}]o$ regulates both the electrogenic H^+-pump and the diffusion channels, competing with $[Mg^{2+}]o$. These channels are the main components of the origin of RP (resting potential) of the *Chara* plasmalemma (Kishimoto et al.,1984).

Various electrical properties such as the voltage dependent K^+ channel and the Ca^{2+} activated Cl^- channel have been analyzed with an equivalent circuit shown in Fig.1**A** which is electrically equal to that in **B**. Then, the following relations are satisfied:

$$I=G(Vm-Em)=Ip+Id \quad (1)$$
$$G=Gp+Gd \quad (2)$$
$$Em=(GpEp+GdEd)/G \quad (3)$$

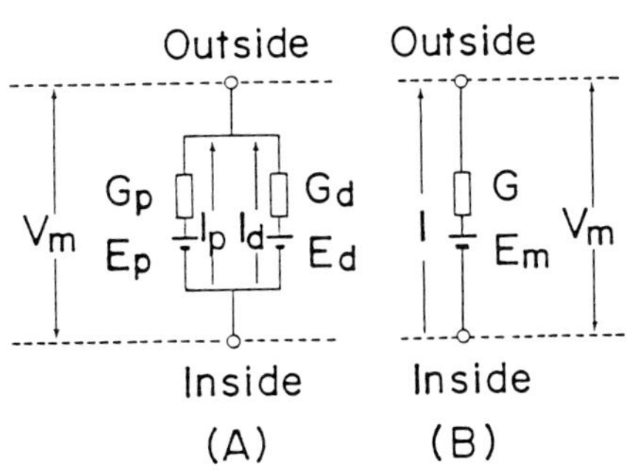

Fig.1. Equivalent circuits. Vm: membrane potential. A). Suffix p means the pump and d the diffusion. B). Em: membrane electromotive force. G: membrane conductance. I: membrane current(=Ip+Id).

Material and Methods

Material: Internodal cell of *Chara corallina*. *Solutions*: 0.5 mM KCl, 0.2 mM NaCl, 0.1 mM $CaCl_2$, and 0.1 mM $MgCl_2$. Otherwise, $[Ca^{2+}]o$ and $[Mg^{2+}]o$ are shown with $[Ca^{2+}]o/[Mg^{2+}]o$ in mM ratio, R. pH=7. *Voltage clamp*: Vm was shifted with a rate of 100mV/60sec from the resting level to a more negative level, which yielded almost a steady state I-Vm curve. A train of small and short square pulses was superimposed on the slow ramp voltage change to evaluate G as a function of Vm during the voltage span. Thus determined G corresponds to the instantaneous chord conductance. Since I, Vm and G were known, Em could be calculated from Eq.1.

Results and Discussion

In Fig.2 a series of I-Vm curves, which was obtained when R changed, is shown in the uppermost, and the voltage dependencies of Em and G in the middle and bottom respectively. The increase of R causes the decrease of G around the resting level without a marked change in the resting potential, which suggests that both Gp and Gd decrease with the increase of R without a marked change in the Gp/Gd ratio. On the other hand, the break-down region where the I-Vm curve descends shifts toward more negative with the increase of R, which means the increase of the membrane elec-

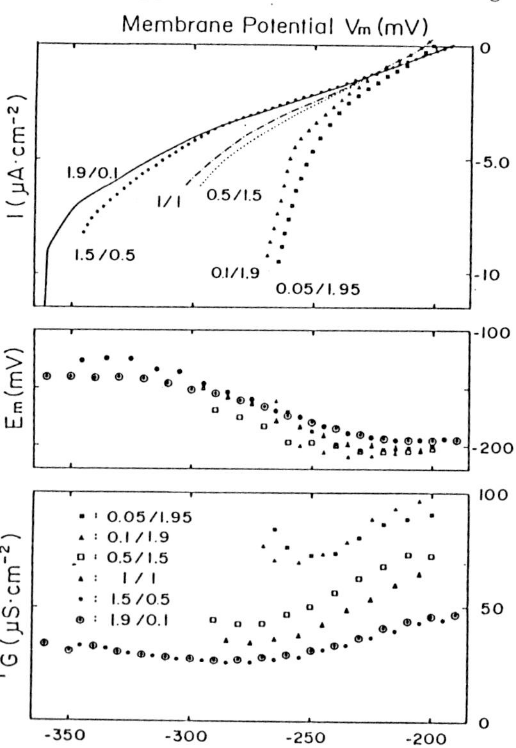

Fig.2. Effects of varying the $[Ca^{2+}]o/[Mg^{2+}]o$ ratio, R, keeping the total concentration at 2.0 mM. Uppermost: I-Vm curves. Middle: Em-Vm curves. Bottom: G-Vm curves.

trical tolerance. In the breakdown region, G increases and Em shifts toward less negative. The increase in G is mainly due to the increase in the efflux of Cl^- (Coster and Hope, 1968), i.e., G_{Cl} (also cf.Fig.4).

Such effects of varying R could be mainly due to varying $[Ca^{2+}]o$. However, it should be also kept in mind that $[Mg^{2+}]o$ competes with $[Ca^{2+}]o$ at the Ca^{2+} binding sites in the membrane and thereby Mg^{2+} at the Ca^{2+} binding sites will mask the actions of Ca^{2+} (Ohkawa and Tsutsui,in preparation).

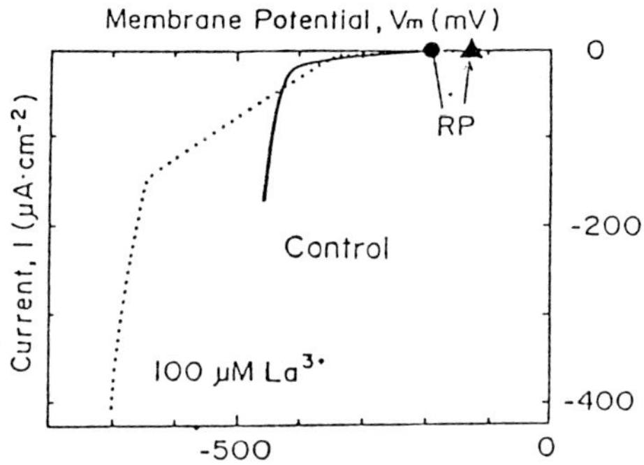

Fig.3. Effects of addition of La^{3+} on I-Vm curve. 100 µM $LaCl_3$ was added to the medium containing 0.1/0.1 of $[Ca^{2+}]o/[Mg^{2+}]o$ in mM ratio.

Since La^{3+} causes a large increase in the membrane electrical tolerance (Fig. 3), it binds with the Ca^{2+} binding sites more strongly than Ca^{2+}. In contrast, 0.1 mM EGTA causes decrease of the electrical tolerance (data not shown), because it removes Ca^{2+} from the membrane. However, it should be noticed that both La^{3+} and EGTA cause both membrane depolarization and decrease of G near the resting level. Such a depolarization could mainly be due to the inhibition of the H^+-pump, because pump inhibitors, DCCD or DES, cause both membrane depolarization and decrease of G (Beilby,1984, Kishimoto et al.,1984). In fact, low concentrations of La^{3+} not only block the Ca^{2+} activated Cl^- channel specifically in about 10 min (Tsutsui et al.,1986), but also inhibit the H^+-pump later (Beilby, 1984, Tsutsui et al.,1987a,b).

Summary

$[Ca^{2+}]o$ controls at least three distinct electrical phenomena in the *Chara* plasmalemma. 1). Increase of the electrical tolerance. 2). Decrease of the G without a marked change of the Gp/Gd ratio. 3).Elevation of the threshold for excitation (data not shown). On the other hand, $[Mg^{2+}]o$ shows quite opposite effects. Thus, optimum $[Ca^{2+}]o$ and $[Mg^{2+}]o$ are expected in vivo.

Voltage dependencies of various conductances in the

Chara plasmalemma are schematically shown in Fig.4.

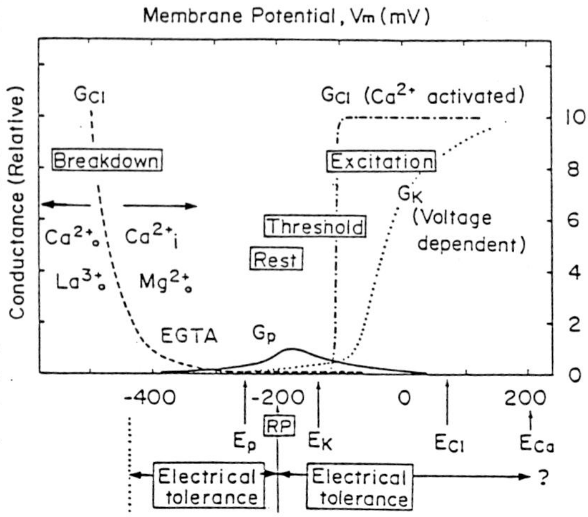

Fig.4. Voltage dependencies of various conductances in the *Chara* plasmalemma.

Acknowledgment:

We thank emeritus Prof. U. Kishimoto, Dr. N. Kami-ike and Mr. S.A. Salehi for their critical reading and discussions.

References

Beilby MJ (1984). Current-voltage characteristics of the proton pump at *Chara* plasmalemma:I.pH dependence. J Memb Biol 81:113-125.

Coster HGL, Hope AB (1968). Ionic relations of cells of *Chara australis*. XI.Chloride fluxes. Aust J Biol Sci 21: 243-254.

Kishimoto U, Kami-ike N, Takeuchi Y, Ohkawa T (1984). A kinetic analysis of the electrogenic pump of *Chara corallina*. I.Inhibition of the pump by DCCD. J Memb Biol 80:175-183.

Ohkawa T, Tsutsui I, Kishimoto U (1986). K^+ channel in the *Chara* plasmalemma: Estimation of K^+ channels density and single K^+ channel conductance. Plant Cell Physiol 27:1429-1438.

Tsutsui I, Ohkawa T, Kishimoto U (1986). Inhibition of Cl^- channel activation in *Chara corallina* membrane by lanthananum ion. Plant Cell Physiol. 27:1197-1200.

Tsutsui I, Ohkawa T, Nagai R, Kishimoto U (1987). Role of calcium ion in the excitability and electrogenic pump activity of the *Chara corallina* membrane. I.Effects of La^{3+}, Verapamil, W-7, and TFP on the action potential. J Memb Biol 96:65-73

Tsutsui I, Ohkawa T, Nagai R, Kishimoto U (1987). Role of calcium ion in the excitability and electrogenic pump activity of the *Chara corallina* membrane. II.Effects of La^{3+}, EGTA, and calmodulin antagonists on the current-voltage relation. J Memb Biol 96:75-84.

The Ion Pumps: Structure, Function, and Regulation, pages 427–434

DOPAMINE INHIBITION OF NaKATPase ACTIVITY IN PROXIMAL CONVOLUTED TUBULE CELLS

Anita C. Aperia

Department of Pediatrics, Karolinska Institute, St. Göran's Children's Hospital, 112 81 STOCKHOLM, Sweden

The understanding of the short-term in vivo regulation of NaKATPase has been progressing slowly. Among the major reactants (Na, K and ATP) of the enzymatic reaction, K and ATP have generally been considered to be present in optimal concentrations required for maximal NaKATPase activity. The physiological intracellular Na concentration (Na_i) which is approximately 15 mM, is definitely lower than the Na concentration required for halfmaximum activity of the pure enzyme (approximately 35 mM). Thus variations in Na_i will be an important determinant of NaKATPase activity in vivo. Quite often it has been assumed that variations in Na_i is the only in vivo regulator of the activity of NaKATPase and that hormone regulation of transcellular Na transport occurs only at the site of Na entry.

During recent years evidence has been accumulating that there is also a direct hormonal control of NaKATPase. Norepinephrine (Beach et al. 1987) and insulin (Clausen & Kohn 1977; Lytton 1985) appear to stimulate the enzyme. Dopamine (DA) inhibits the enzyme. Studies of the in vivo hormonal regulation of NaKATPase activity have required intact cells where the Na entry step can be controlled. It has been an advantage to use cells that have a high concentration of the enzyme.

In our studies of the short-term in vivo regulation of NaKATPase we have used freshly dissected rat proximal convoluted (S2) segments that have been permeabilized to clamp the Na entry step. All assays have been performed under microscopical supervision. Only segments with intact cells without membrane blebs have been used. NaKATPase has been determined as ouabain sensitive ^{32}P-ATP hydrolysis in single 0.5 - 1 mm tubule segments. In each protocol studies have been performed in 3 - 6 rats. Total and ouabain insensitive ATPase was determined in 14 - 20 segments in each rat.

Dopamine (DA) inhibition of NaKATPase activity in PCT segments

In looking for a hormone that controls of NaKATPase activity in proximal convoluted tubule (PCT) we chose to study the effect of DA because 1) DA is synthesized in PCT cells (Hagege & Richet 1985). PCT cells have a high concentration of the enzyme dopadecarboxylase that converts L-dopa to DA (Goldstein et al. 1972). 2) DA has a natriuretic effect that is blocked by dopadecarboxylase inhibitors (Alexander et al. 1974; Ball & Lee 1977; Sowers et al. 1984). 3) Studies of brain tissue have suggested that DA might interact with NaKATPase (Desaiah & Ho 1976; Kometiani et al. 1984; Rodriguez & Antonelli 1981; Sawas & Gilbert 1982). Most of those studies were carried out on cell homogenates and they have not been conclusive.

We started those studies by incubating single permeabilized segments with DA (Aperia et al. 1987a) and we found that DA caused a rapid, reversible and dose-dependent inhibition of NaKATPase activity. The DA precursor L-dopa also inhibited NaKATPase. Inhibition was not observed in the presence of a dopadecarboxylase inhibitor, and we therefore concluded that PCT NaKATPase could be inhibited by locally produced DA. The DA metabolites dihydroxyphenyl acetic acid and homovanillic acid were much less potent inhibitors of NaKATPase activity than DA (Bertorello 1987a). In studies performed in media where K concentration ranged between 0.5 and 40 mM it was shown that DA deceased both Vmax and K affinity (Km was 0.9 mMK in DA free medium and 3.5 mMK in medium containing DA 10^{-5}M (Aperia et al. 1987a). We have also demonstrated that in suspensions of proximal tubule cells, L-dopa will, in a dose-dependent manner, inhibit ouabain sensitive O_2 consumption. The inhibition is not observed in the presence of a dopadecarboxylase inhibitor (Seri et al. 1987).

There are at least two types of DA receptors, DA_1 and DA_2. NaKATPase activity was inhibited in PCT segments that were incubated with both DA_1 and DA_2 agonists but not with either agonist alone (Bertorello & Aperia 1988). Similar synergistic effects between DA_1 and DA_2 have previously been observed in the globus pallidum (Walters et al. 1987). DA has no effect on the purified enzyme (Bertorello & Aperia 1987a).

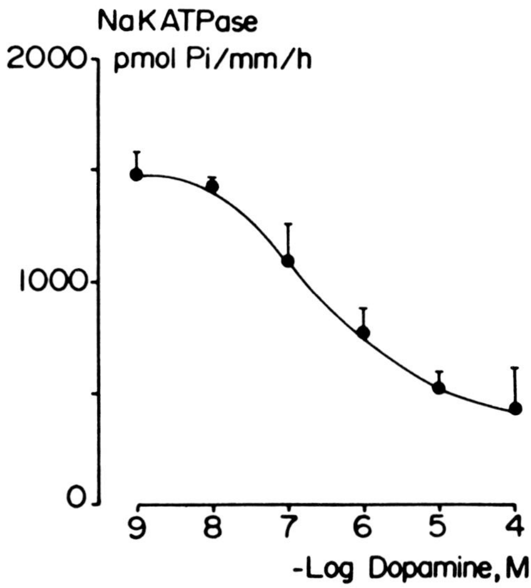

Figure. Proximal convoluted tubule (PCT) segments have been incubated in modified Hank's medium with various concentrations of dopamine. Filled circles indicate mean values obtained in 3 - 6 experiments. Bars, SE. Dopamine (DA) inhibited Na^+-K^+-ATPase activity in a dose-dependent manner.

TABLE 1. DA inhibition of PCT Na^+-K^+-ATPase activity is reversible

Condition of incubation	n	Na^+-K^+-ATPase activity, pmol $P_i \cdot mm^{-1} \cdot h^{-1}$
MHM + DA, 10^{-5} M	6	528 ± 85
MHM + DA, 10^{5} M, tubules washed in DA-free MHM	4	1.305 ± 43*

Values are means ± SE. DA, dopamine; PCT, proximal convoluted tubule; MHM, modified Hank's medium. Incubation time with DA = 30 min. * Significantly different from MHM + 10^{-5} M DA, $P < 0.001$.

TABLE 2. Effect of dopa incubation on Na^+-K^+-ATPase activity in PCT

Condition of incubation	n	Na^+-K^+-ATPase activity, pmol $P_i \cdot mm^{-1} \cdot h^{-1}$
MHM	4	1.550 ± 74
MHM + 10^{-4} M dopa	4	993 ± 52*
MHM + 10^{-4} M dopa, animals pretreated with benserazide	5	1.638 ± 96

Values are means ± SE for n, number of observations. PCT, proximal convoluted tubule; MHM, modified Hank's medium.
* Significantly different from tubules incubated in MHM only and from tubules from benserazide-treated rats incubated with 10^4M dopa, $P < 0.02$.

Role of DA as an intrarenal natriuretic hormone

We have examined whether intrarenal DA production is of physiological importance for the modulation of PCT NaKATPase activity and for the control of sodium excretion by studying rats on normal salt (NS) diet and high salt (HS) diet (Bertorello et al. 1988). We used clearance studies to determine the acute effect of blocking dopadecarboxylase with benserazide on the natriuretic response to NS diet. The natriuretic response to NS diet was significantly attenuated by benserazide. The effect of NS diet on NaKATPase activity was studied both in PCT segments that have a high concentration of dopadecarboxylase and in thick ascending limbs (TAL) of Henle that contain no dopadecarboxylase. NaKATPase activity was depressed in PCT segments from HS rats but not in TAL segments from HS rats. The reduction in PCT NaKATPase activity in NS rats was reversed by benserazide treatment. These findings strongly suggest that during NS diet PCT NaKATPase activity is inhibited by locally produced DA. Furthermore, we determined the dose response effect of exogenous DA on NaKATPase activity in PCT segments from HS rats where endogenous production of DA had been blocked. We found that the inhibitory effect of DA on NaKATPase activity was enhanced by NS diet.

We have demonstrated that dopadecarboxylase activity in renal cortex is significantly increased during NS diet (Aperia et al. 1987b). This finding provides additional evidence for the concept that DA inhibition of NaKATPase is of physiological importance during HS diet. Taken together those results strongly suggest that during NS diet DA is an intrarenal natriuretic hormone which acts by inhibiting PCT NaKATPase activity.

G-protein modulation of PCT NaKATPase activity

The activation of cell surface receptors is coupled to the regulation of intracellular effectors by a family of guanine regulatory (G) proteins (Rodbell 1980; Gilman 1987). The stimulatory G (G_s) protein and the inhibitory G (G_i) protein are responsible for the control of adenylate cyclase activity. The role of other G-proteins is generally less well understood. The G-proteins can be characterized with regard to toxin sensitivity. We have examined the involvement of G-proteins in DA inhibition of NaKATPase activity in PCT segments. T he first step was to determine whether DA inhibition of NaKATPase in PCT segments is influenced by the presence of pertussis toxin and/or cholera toxin (Bertorello & Aperia 1987b, c).

In PCT segments from rats treated with pertussis toxin, NaKATPase activity was not significantly different from control. There was no significant inhibition of NaKATPase activity in DA incubated PCT segments from pertussis toxin treated rats. In rats pretreated with cholera toxin, NaKATPase activity was significantly higher than control. NaKATPase activity in DA incubated PCT segments from cholera toxin treated rats was significantly lower than in vehicle incubated PCT segments from cholera toxin treated rats.

The effect of DA_1 and DA_2 agonists on NaKATPase activity was also examined in the presence and the absence of pertussis toxin. NaKATPase activity was significantly lower in $DA_1 + DA_2$ incubated tubules than in vehicle incubated tubules. In PCT segments from rats treated with pertussis toxin, $DA_1 + DA_2$ caused no significant inhibition of NaKATPase.

GTP has a negative heterotropic effect on the binding of hormone to a complex consisting of hormone receptor G protein. This prompted us to study the inhibitory effect of DA on NaKATPase activity in the presence of different concentrations of GppNHp. GppNHp was found to interact with DA modulation of PCT Na-

KATPase activity. Permeabilized PCT segments were incubated with GppNHp from 5 to 80 /uM in the presence of DA 10^{-5} or DA^{-6}M. (DA 10^{-6}M gives a submaximal inhibition of NaKATPase). GppNHp competitively abolished DA inhibition of NaKATPase. K_i was approximately 24 /uM. The finding that GppNHp abolished DA inhibition of NaKATPase activity in a competitive way provides additional evidence to the concept that DA inhibition of NaKATPase is modulated via a G-protein.

SUMMARY

Studies of permeabilized rat renal proximal convoluted tubules have demonstrated that NaKATPase activity can be inhibited by locally produced DA. The effect is mediated by a pertussis toxin sensitive inhibitory G-protein.

DA inhibition of proximal tubule NaKATPase activity is of physiological importance. During high salt diet renal DA production is increased and the inhibitory effect of DA on NaKATPase is enhanced. DA inhibition of NaKATPase activity will contribute to the natriuresis during high salt diet.

REFERENCES

Alexander RW, Gill GR, Yamabe H, Lovenberg W, Keiser HR (1974). Effects of dietary sodium and of acute saline infusion on the interrelationship between dopamine excretion and adrenergic activity in man. J Clin Invest 54:194-200.

Aperia A, Bertorello A, Seri I (1987a). Dopamine causes inhibition of NaKATPase activity in proximal tubular segments. Am J Physiol 252:F39-F45.

Aperia A, Seri I, Brenner BM, Ballermann BJ (1987b). Regulation of aromatic L-amino acid decarboxylase (L-AADC) activity by changes in dietary sodium intake in the rat. Abstract. Washington: Am Soc Nephrol.

Ball SG, Lee MR (1977). The effects of carbidopa administration on urinary sodium excretion in man. Is dopamine an intrarenal natriuretic hormone? Br J Clin Pharmacol 4:115:119.

Beach RE, Schwab SJ, Brazy PC, Dennis VW (1987). Norepinephrine increases Na^+-K^+-ATPase and solute transport in rabbit proximal tubules. Am J Physiol 252:F215-F220.

Bertorello A, Aperia A (1987a). Effect of l-dopa, dopamine, dihydroxyphenyl acetic acid and homovanillic acid on Na^+-K^+ -

ATPase activity in rat proximal tubule segments. Acta Physiol Scand 130:571-574.

Bertorello A, Aperia A (1987b). NaKATPase activity in proximal convoluted tubule (PCT) segments is modulated by a G-protein which is Na-dependent and inhibitable with dopamine (DA). Abstract. Washington: Am Soc Nephrol.

Bertorello A, Aperia A 1987c. Pertussis toxin modulated dopamine inhibition of NaKATPase activity in rat proximal convoluted tubule segments. In A Maunsbach (ed): "Proc Fifth Int Conf NaKATPase," New York: Alan R. Liss.

Bertorello A, Aperia A (1988, in press). Both DA_1 and DA_2 receptor agonists are necessary to inhibit NaKATPase activity in proximal tubules from rat kidney. Rapid communication. Acta Physiol Scand.

Bertorello A, Hökfelt T, Goldstein M, Aperia A (1988, in press). Proximal tubule NaKATPase activity inhibited during high salt diet: Evidence for a dopamine mediated effect. Am J Physiol.

Clausen T, Kohn PG (1977). The effect of insulin on the transport of sodium and potassium in rat soleus muscle. J Physiol 265:19-42.

Desaiah D, Ho IK (1976). Effect of biogenic amines and GABA on ATPase activties in mouse tissue. Eur J Pharmacol 40:255-261.

Gilman AG (1987). G proteins: Transducers of receptor-generated signals. Ann Rev Biochem 56:615-649.

Goldstein M, Fuxe K, Hökfelt T (1972). Characterization and tissue localization of catecholamine synthetizing enzymes. Pharmacol Rev 24:293-309.

Hagege J, Richet (1985). Proximal tubule dopamine histofluorescence in renal slices incubated with dopa. Kidney Int 27:3-8.

Kometiani ZP, Tsakadze LG, Jariashvili TY (1984). Functional significance of the effects of neurotransmitters on the Na^+,K^+-ATPase system. J Neurochem 42:1246-1250.

Lytton J (1985). Insulin affects the sodium affinity of the rat adipocyte (Na^+,K^+)-ATPase. J Biol Chem 260:10075-1184.

Rodbell M (1980). The role of hormone receptors and GTP-regulatory proteins in membrane transduction. Nature 284:17-22.

Rodriguez de Lores Arnaiz G, Antonelli de Gómez de Lima M (1981). The effect of several neurotransmitter substances on nerve ending membrane ATPase. Acta Physiol Lat Am 31:43-47.

Sawas AH, Gilbert JC (1982). The effects of dopamine agonists and antagonists on Na^+,K^+-ATPase and Mg^{2+}-ATPase activities of synaptosomes. Biochem Pharmacol 31:1531-1533.

Seri I, Kone BC, Gullans SR, Aperia A, Brenner BM, Ballermann BJ (1987). Inhibition of Na-K-ATPase-dependent O_2 consumption by L-dopa (D) in rat renal cortical tubule cell suspension (Na_i). Abstract. Washington: Am Soc Nephrol.

Sowers JR, Crane PD, Beck FWJ, McClanahan M, King ME, Monanty PK (1984). Relationship between urinary dopamine production and natriuresis after acute intravascular volume expansion with sodium chloride in dogs. Endocrinology 115:2085-2090.

Walters JR, Bergström DA, Carlson JH, Chase TN, Braun AR (1987). D_1 dopamine receptor activation required for postsynaptic expression of D_2 agonist effects. Science 236:719-722.

The Ion Pumps: Structure, Function,
and Regulation, pages 435–450

REGULATION OF MUSCLE CONTRACTION AND RELAXATION IN HEART

Sidney Fleischer and Makoto Inui

Department of Molecular Biology, Vanderbilt University, Nashville, TN 37235

INTRODUCTION

In heart, as in skeletal muscle, contraction and relaxation are controlled by the free Ca^{2+} concentration in the myoplasm, which is regulated at least in part by the sarcoplasmic reticulum (SR), a specialized membrane system (Tada et al. 1978; Fleischer 1985). In skeletal muscle, the SR is an extensive system, whereas in cardiac muscle it is scarcer, and surface membranes (sarcolemma and transverse tubule) constitute a greater percentage of the membrane content (Fawcett et al. 1969). This is one of the reasons why SR from skeletal muscle has been easier to isolate and has been studied more extensively than that from heart, whereas the converse has been the case for plasmalemma.

In fast twitch skeletal muscle, depolarization at the transverse tubule induces Ca^{2+} release from the SR which triggers muscle contraction (Endo 1977). This process has been referred to as "depolarization induced calcium release". The energized reuptake of Ca^{2+} into the SR enables muscle to relax (Tada et al. 1978; Fleischer 1985; Endo 1977; Martonosi 1984; Inesi 1985). Thus, Ca^{2+} cycles between the SR compartment and the myoplasm during relaxation and contraction, respectively. A key difference in heart muscle is that a second pool of calcium is involved. Extracellular Ca^{2+} must first enter the heart cell via the slow inward Ca^{2+} channel following depolarization of the plasma membrane. The elevated Ca^{2+} then induces Ca^{2+} to be released from the SR compartment (Fabiato 1983). This latter process has been referred to as "Ca^{2+}-induced Ca^{2+} release".

The lowering of myoplasmic calcium concentration is then mediated both by energized uptake into the sarcoplasmic reticulum and by extrusion via the plasmalemma.

Thus, it would appear that important differences exist between heart and skeletal muscle with regard to the machinery for calcium uptake and release and its regulation. The approach of our laboratory to define these processes has involved the isolation and characterization of defined membrane fractions referable to membranous organelles from muscle (Fleischer 1985). Once the overall process has been described, the dissociation and reconstitution approach is used to isolate the molecular machinery involved in the calcium uptake and release process. We shall review our progress to study muscle contraction and relaxation in heart muscle, which parallels our ongoing program with fast twitch skeletal muscle (Fleischer 1985).

MODULATION OF CALCIUM UPTAKE BY SARCOPLASMIC RETICULUM

Contraction and relaxation of heart muscle are modulated by β-adrenergic action of catecholamines. Catecholamines and their intracellular messengers, cAMP and Ca^{2+}, are known to alter several aspects of excitation-contraction coupling of the myocardium (Tada et al. 1982). In cardiac SR, the Ca^{2+} pump, which transports Ca^{2+} from the cytoplasm into the lumen of the SR, can be modulated by cAMP and Ca^{2+}-calmodulin via protein kinases (Katz et al. 1978; Kirchberger et al. 1974). A number of findings indicate that the modulation of the Ca^{2+} pump parallels phosphorylation and dephosphorylation of phospholamban, a 27,000 dalton membrane protein (Tada et al. 1982). Phosphorylation of phospholamban is catalyzed by cAMP-dependent (Kirchberger et al. 1974) and Ca^{2+}-calmodulin-dependent protein kinases (Le Peuch et al. 1979). Ca^{2+}-phospholipid-dependent protein kinase (protein kinase C) from brain has also been found to phosphorylate phospholamban (Movsesian et al. 1984). Phosphorylation of phospholamban lowers the Km for Ca^{2+} and thereby to more effective Ca^{2+} uptake by the sarcoplasmic reticulum. Reconstitution studies were initiated to gain insight into the molecular mechanism of the modulation by protein kinase-catalyzed phosphorylation.

An important step in our progress was the isolation of a highly purified and stable preparation of heart sarcoplasmic reticulum (Chamberlain et al. 1983).

Membrane vesicles capable of energized Ca^{2+} pumping have been reconstituted from cardiac sarcoplasmic reticulum (SR) (Inui et al. 1986). Cardiac SR were solubilized with Triton X-100 in a detergent to protein weight ratio of 0.8, and membranous vesicles were reconstituted by removal of detergent with Bio-Beads SM-2 (a neutral porous styrenedivinylbenzene copolymer). The reconstituted vesicles exhibited ATP dependent oxalate-facilitated Ca^{2+} accumulation with rates and efficiency comparable to the best reconstituted skeletal muscle preparation (Ca^{2+}-loading rate = 1.65 ± 0.31 μmol mg^{-1} min^{-1}, and Ca^{2+} loading efficiency (Ca^{2+}/ATP) = 0.69 ± 0.09). Phospholamban in the reconstituted vesicles was phosphorylated with added catalytic subunit of cAMP-dependent protein kinase to almost the same extent as that in original vesicles. However, phosphorylation of phospholamban had no effect on the Ca^{2+} accumulation of the reconstituted vesicles. This is to be contrasted with a decrease in the half-maximal concentration of Ca^{2+} for Ca^{2+} accumulation (K_{Ca}) in the original vesicles from 1.35 ± 0.08 μM to 0.75 ± 0.12 μM by cAMP-dependent phosphorylation of phospholamban. On the other hand, K_{Ca} for the reconstituted vesicles was about 0.5 μM and was unchanged by phosphorylation, indicating that the Ca^{2+} pump in the reconstituted vesicles is already fully activated. These results suggest that in normal cardiac SR, phospholamban in the dephosphorylated state acts as a suppressor of the Ca^{2+} pump and that phosphorylation of phospholamban serves to reverse the suppression (Fig. 1).

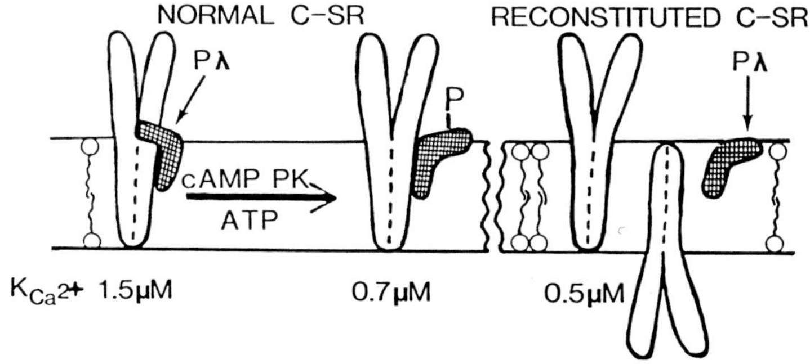

Fig. 1. Diagrammatic representation of the modulation of the calcium pump protein of cardiac sarcoplasmic reticulum (C-SR) by protein kinase phosphorylation of phospholamban (PΛ). -P refers to phosphorylation.

THE CALCIUM RELEASE PROCESS IN EXCITATION-CONTRACTION COUPLING

The definition of the molecular machinery involved in excitation-contraction coupling and the nature of the coupling process is one of the most important problems in cell biology and muscle physiology. Significant progress has been made for fast twitch skeletal muscle: 1) defined subfractions of terminal cisternae and longitudinal tubules of sarcoplasmic reticulum were isolated (Saito et al. 1984). The longitudinal tubules consist essentially of calcium pump membrane containing the calcium pump protein (~90% of the protein) (Chu et al. 1987) which is involved in energized Ca^{2+} uptake enabling muscle to relax. The terminal cisternae (junctional SR) consist of two types of membranes, the calcium pump membrane (~85% of the membrane) and the junctional face membrane with well defined feet structures (~15% of the membrane) (Saito et al. 1984). The latter are involved in junctional association with transverse tubules to form the triad junction. The compartment contains the Ca^{2+} binding protein (calsequestrin) which is involved in binding and storage of Ca^{2+} (MacLennan et al. 1983; Meissner et al. 1973); 2) heavy SR were incorporated by fusion, into bilayers, and channel gating, characteristic of Ca^{2+} release in SR, was observed (Smith et al. 1985); 3) ryanodine in nM concentration was found to modulate Ca^{2+} release which was localized to terminal cisternae of sarcoplasmic reticulum (Fleischer et al. 1985); 4) the ryanodine receptor was isolated and found to consist of a single high molecular weight polypeptide (Mr ~360 KD) (Inui et al. 1987a; Imagawa et al. 1987) and identified as the feet structures of the junctional face membrane (Inui et al. 1987a); and 5) the purified ryanodine receptor was incorporated into bilayers and found to have Ca^{2+} channel activity characteristic of the Ca^{2+} release in terminal cisternae (Hymel et al. 1988a; Lai et al. 1988). Thus, the ryanodine receptor has been isolated, and identified as the feet structures and as the Ca^{2+} release channel of SR. In molecular terms, the calcium release channel consists of an oligomer of the high molecular weight polypeptide (Mr ~360 KD).

The goals of our studies on excitation-contraction coupling in heart were to: 1) characterize the triad/dyad junction in molecular terms; 2) isolate and characterize the calcium release channels; and 3) define the calcium release mechanism and the nature of the modulation.

Table I Characteristics of Cardiac Subcellular SR Fractions

Fraction	Ca^{2+} loading rate -RR	Ca^{2+} loading rate +RR[a]	Ratio (+RR/-RR)	$[^3H]$Ryanodine binding[b] (pmol/mg)
	(μmol Ca^{2+}/mg.min)			
Microsomes	0.63 ± 0.02 (3)	0.91 ± 0.06 (3)	1.44	3.02 ± 0.41 (3)
Longitudinal SR	0.93 ± 0.08 (3)	1.01 ± 0.05 (3)	1.09	1.69 ± 0.53 (5)
Terminal cisternae of SR	0.16 ± 0.02 (4)	0.84 ± 0.16 (4)	4.75	8.78 ± 1.87 (5)

[a] +RR refers to 7 μM ruthenium red.
[b]Ryanodine binding was measured at 210 nM ryanodine in the presence of 1 M KCl and 25 μM $CaCl_2$.

Cardiac SR were subfractionated into junctional SR and longitudinal SR making use of calcium phosphate loading (Wang et al. 1987; Inui et al. 1988). The junctional SR are leaky to and hence do not load Ca^{2+}. Thus, they can be separated by sucrose gradient centrifugation from longitudinal SR which have efficient Ca^{2+} loading. The two types of SR in heart resemble those from skeletal muscle (Table I). Longitudinal SR have a high Ca loading rate which is not appreciably enhanced by ruthenium red. Junctional SR have a low Ca^{2+} loading rate which is enhanced approximately five-fold by ruthenium red. Ryanodine binding is enriched in the junctional SR and therefore appears to be localized to this domain of SR. Morphologically, the junctional SR are filled with electron opaque contents and have well defined feet structures (Fig. 2). The longitudinal SR consist essentially of only the calcium pump membrane and are devoid of compartmental contents. The polypeptide profile of cardiac SR is more complex than that of skeletal muscle (Fig. 3) (Chamberlain et al. 1983). The amount of calcium pump protein in cardiac is about 40% in contrast with 90% for skeletal muscle longitudinal SR. Junctional SR are enriched in the calcium binding protein (calsequestrin) and have a characteristic high molecular weight band (Mr 340 KD) (Wang et al. 1987). Thus, cardiac SR can be subfractionated into junctional and longitudinal SR which have similar characteristics to that of skeletal muscle (see also Jones et al. 1981; Seiler et al. 1984).

Ryanodine binding to junctional SR displays binding which can be resolved into tighter ($K_D \sim 8$ nM) and weaker (0.9 µM) binding (Wang et al. 1987; Inui et al. 1988b; Fig. 4). Ryanodine preincubated with junctional SR at low concentration blocks the action of ruthenium red to enhance calcium loading (Fig. 5A). This concentration dependence is

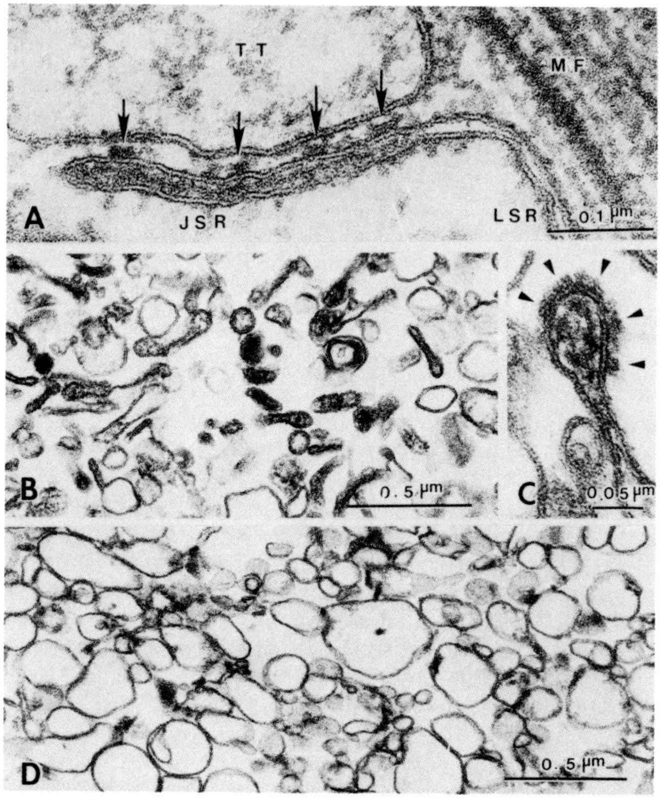

Fig. 2. Electron microscopy of cardiac membranes. A (x 168,000), thin section of cardiac tissue *in situ* showing junctional SR (JSR) associated with transverse tubules by way of the feet structures (arrows) and longitudinal SR (LSR). B Isolated junctional SR (x 38,700) and C (x 168,000). The feet structures are indicated by arrowheads. D, isolated longitudinal SR (x 38,700). MF, myofilaments (Inui et al. 1988).

hyperbolic as evidenced by a linear double reciprocal plot. The Ki is in the same range as the K_D for the high affinity ryanodine binding. Therefore, ryanodine at low concentration appears to lock the calcium release channels in the open state. Higher concentrations of ryanodine (to 10 μM)

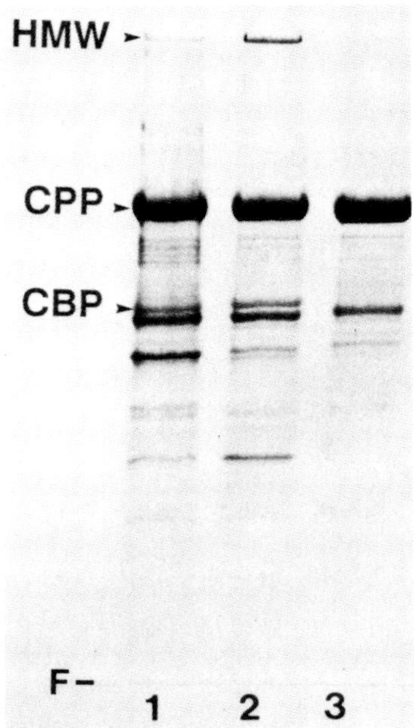

Fig. 3. SDS-polyacrylamide gel electrophoresis of cardiac membrane subfractions. Electrophoresis was carried out on a 5-15% polyacrylamide gradient gel in SDS and stained with Coomassie Blue. Lane 1, cardiac microsomes; lane 2, junctional SR; lane 3, longitudinal SR. High molecular weight protein (HMW); Ca^{2+} pump protein (CPP); calcium binding protein (CBP) (calsequestrin); F refers to the front (Inui et al. 1988).

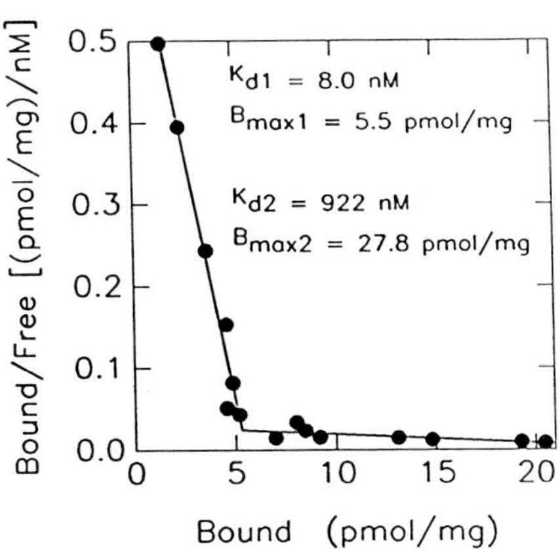

Fig. 4. Scatchard analysis of ryanodine binding to junctional SR from heart. [^{3}H]-ryanodine binding to junctional SR (0.1 mg/ml) was measured in 1 M KCl, 25 μM $CaCl_2$, 10 mM Hepes, pH 7.4. Ryanodine concentrations were varied from 3 nM to 2.4 μM. Total binding to junctional SR, after subtracting the radioactivity binding to 0.22 μM Millipore filters (type GSWP), was plotted. Two types of binding can be discerned (Inui et al. 1988).

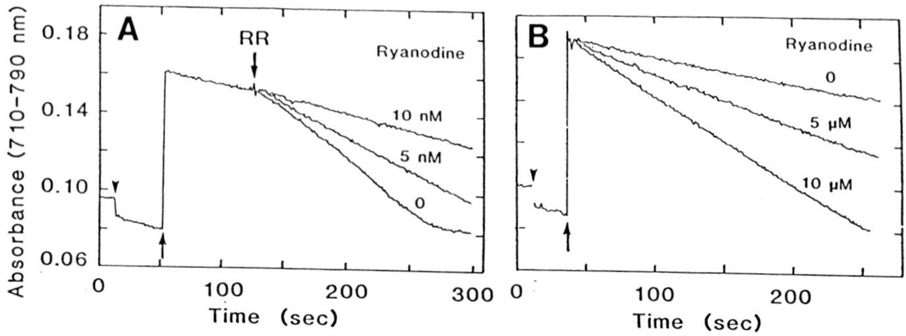

Fig. 5. The effect of low and higher ryanodine concentrations on the Ca^{2+} release channel of SR. A. Low ryanodine concentration locks the channel in the open state. The Ca^{2+} loading assay is shown. Junctional SR (3 mg/ml) were incubated at 37°C with various concentrations (1 to 10 nM) of ryanodine in 1 M KCl, 1 mM $MgCl_2$, 25 μM $CaCl_2$ and 10 mM Hepes, pH 7.4. After 30 min, an aliquot was added to 1 ml (final volume) of Ca^{2+} loading assay medium. Ca^{2+} loading in the presence of 100 mM potassium phosphate buffer, was determined at 37°C using antipyrylazo III. Na_2-ATP was added (arrowhead) to give a final concentration of 1 mM. Ca^{2+}-loading was initiated by adding $CaCl_2$ (upward arrow) to a final concentration of 50 μM. Ruthenium red (RR) (a final concentration of 7 μM) was added 70-80 sec after the start of Ca^{2+} loading. The apparent inhibition constant (Ki) for ryanodine is obtained from a double reciprocal plot. 5B. Effect of high concentration of ryanodine on Ca^{2+} loading in junctional SR. The junctional SR (3 mg/ml) were preincubated for 30 min at 37°C with various concentrations (0.5 to 10 μM) of ryanodine using the conditions as in Fig. 5A. Ca^{2+} loading was measured as described above except that ruthenium red was omitted. The apparent Km for stimulation for ryanodine was obtained from a double reciprocal plot (Inui et al. 1988).

enhance Ca^{2+} loading (Fig. 5B). The Km is ~0.8 μM. This higher concentration leads to additional binding of ryanodine which then closes the Ca^{2+} release channels. Thus, the pharmacological effects of ryanodine can be correlated with the binding. Low concentrations lock the channel in the open state, whereas higher concentrations close the channel (Table II) (Wang et al. 1987; Inui et al. 1988).

Table II Ryanodine Binding and Modulation of Function of Cardiac Junctional SR

Ryanodine Binding	Kd (nM)	Bmax (pmol/mg)	Ryanodine Action on Channel	Ki or Km (nM)
High affinity site	8	5.1	Locks channel open (Ki)	7
Low affinity site	1082	32.8	Closes channel (Km)	1,110

Ryanodine therefore appeared also to be a specific ligand for the Ca^{2+} release channel of SR in heart. It was then only a matter of time before the ryanodine receptor from heart was isolated. Similar methodology as for skeletal muscle (Inui et al. 1987a) could be applied to isolate the cardiac receptor (Inui et al. 1987b). The procedure makes use of solubilization with detergent (CHAPS) and three sequential types of calcium chromatography (Table III). The purification of the ryanodine receptor resulted in essentially a single band of Mr 340,000 (Fig. 6). The purified receptor retained the same ryanodine binding characteristics in that high affinity (K_D ~8 nM) and weaker binding (K_D ~0.8 µM) are observed similar to that of heart terminal cisternae (Fig. 7). The binding is distinct from that of skeletal muscle in that it is tighter binding (K_D ~10 nM vs 50 nM for

Table III Purification of Ryanodine Receptor from Heart Muscle SR

Fraction	Protein (mg)	Ryanodine binding[a] Specific activity (pmol/mg)	Ryanodine binding[a] Total activity (pmol)	Yield (%)	Purification (-fold)
1. Cardiac SR	100	8.64	864	100	1
2. Solubilized SR	74.2	9.35	694	80	1.1
3. Heparin column chromatography	1.80	181	326	38	20.9
4. p-Aminobenzamidine column chromatography	0.21	782	164	19	90.5
5. Gel permeation chromatography	0.084	983[b]	83	9.6	114

[a] Ryanodine binding was measured with 300 nM [^{3}H]ryanodine. The binding contains both high and low affinities.

[b] Bmax for high affinity binding is 455 pmol/mg protein.

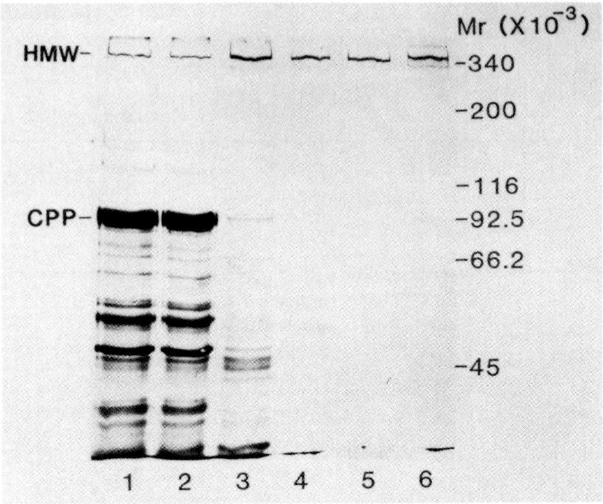

Fig. 6. Purification of the ryanodine receptor from cardiac microsomes as observed on SDS-PAGE. Lane 1, cardiac microsomes; lane 2, solubilized cardiac microsomes with CHAPS and soybean phospholipid; lane 3, pooled fractions from a heparin-agarose column; lane 4, pooled fractions from a p-aminobenzamidine-agarose column; lane 5, purified ryanodine receptor from cardiac SR (pooled fractions from a Fractogel TSK HW65F column); lane 6, purified ryanodine receptor from junctional SR of skeletal muscle. CPP, Ca^{2+} pump protein, HMW, high molecular weight protein. Note that the high molecular weight protein from heart migrates somewhat faster than that from skeletal muscle (Inui et al. 1987b).

skeletal muscle), and skeletal muscle exhibits only the high affinity binding in this concentration range (Inui et al. 1988b). The purified receptor from heart is morphologically identical to the feet structures as it is for skeletal muscle (Fig. 8). The purified ryanodine receptor from heart has recently been incorporated into bilayers and found to have channel behavior characteristics of Ca^{2+} release in cardiac SR (Hymel et al. 1988b and c). Hence, it can be concluded that the feet structures of cardiac SR are equivalent to the calcium release channel as had already been demonstrated for skeletal muscle.

A comparison of the characteristics of the purified ryanodine receptors from skeletal muscle and heart is given

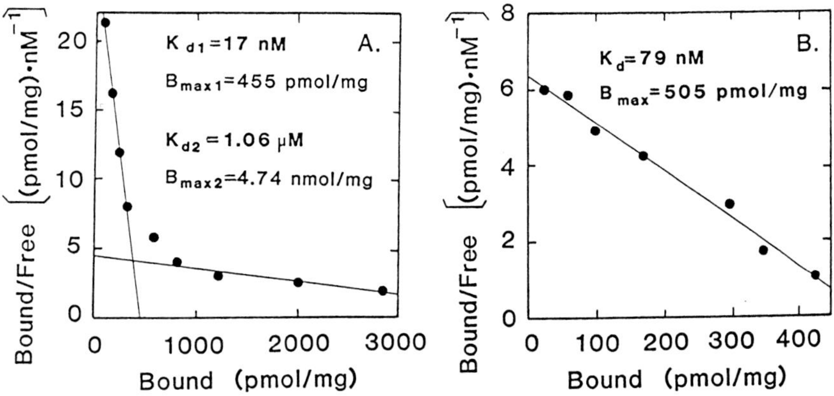

Fig. 7. Scatchard analysis of ryanodine binding to the purified receptor. The purified samples (15-25 μg of protein/ml) from cardiac (A) and skeletal muscle SR (B) were incubated at 24°C for 40 min with various concentrations (4-1500 nM) of [^{3}H]ryanodine under comparable conditions (Inui et al. 1987b).

in Table IV. The characteristics are similar but not identical.

SUMMARY

The calcium uptake and release machinery in heart SR have been characterized:

(1) The calcium pump membrane is involved in energized Ca^{2+} uptake enabling muscle to relax. The calcium pump protein (CPP) in heart SR is modulated by protein kinase phosphorylation of phospholamban lowering the $K_{Ca^{2+}}$. We conclude that in the membrane, phospholamban elevates $K_{Ca^{2+}}$ of calcium pump protein. Phosphorylation of phospholamban attenuates the influence of phospholamban. In the limit, the intrinsic $K_{Ca^{2+}}$ of calcium pump protein in heart and skeletal muscle are approximately the same.

(2) The junctional face membrane is involved in calcium release which triggers muscle contraction. Ryanodine is a specific modulator of the Ca^{2+} release channels of SR which are involved in excitation-contraction coupling. The

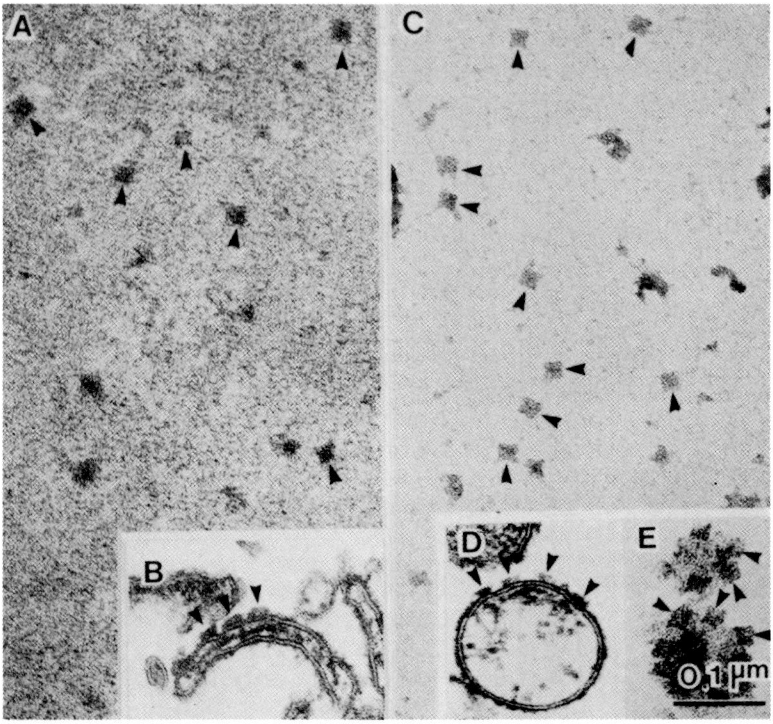

Fig. 8. Morphology of the purified ryanodine receptor from heart and skeletal muscle. The insets (B and D) show the terminal cisternae of SR from heart and skeletal muscle; the arrowheads point to the feet structures (Inui et al. 1987a and b).

ryanodine receptor has been isolated, found to be equivalent to the feet structures, and on reconstitution into bilayers, identified as the calcium release channel of SR.

The calcium release channel of SR is closed by ruthenium red and Mg^{2+} and opened by Ca^{2+} and ATP and low ryanodine concentration. The calcium release channel of SR is not effected by drugs such as nitrendipine, diltiazem and D-600 which modulate the slow inward Ca^{2+} channel of the plasmalemma/transverse tubule.

(3) The calcium release channels from heart and skeletal muscle SR are similar but not identical (Table IV).

Table IV Comparison of Ca^{2+} Release Channel from Heart vs Skeletal Muscle SR

Characteristic	Heart	Skeletal Muscle
Polypeptide Composition (M_R)	340K	360K
Function		
Terminal Cisternae		
Ryanodine binding (high affinity site)	K_D = 8.0 nM Bmax = 5.5 pmol/mg	K_D = 59 nM Bmax = 20 pmol/mg
Pharmacological action (ryanodine locks Ca^{2+} release channel in the open state)	K_I = 7.1 nM	K_I = 56 nM
Purified Receptor		
Ryanodine binding (high affinity site)	K_D = 17 nM Bmax = 455 pmol/mg	K_D = 79 nM Bmax = 505 pmol/mg
Ca^{2+}-activated Ca^{2+} channel activity	+	+
Morphology	Feet structures (square rectangles)	Feet structures (square rectangles)

Important differences distinguish the calcium release machinery in heart from that of skeletal muscle.

1. In heart there are two sources of calcium fluxes: a) extracellular Ca^{2+} enters via the plasmalemma slow inward calcium current; and b) "Ca^{2+} induced Ca^{2+} release" from SR. In skeletal muscle, SR is the single main source of calcium which enters via "Depolarization induced calcium release".

2. The calcium release channel from heart SR has a lower Mr ~340,000 vs 360,000 for skeletal muscle.

3. Ryanodine binding in cardiac SR is distinct from that in skeletal muscle (Fig. 7).

4. The isolated calcium release channel from heart SR is more sensitive to Ca^{2+} for calcium release (Hymel et al. 1988c).

Significant progress has been achieved in identifying the calcium release channel of SR in heart and skeletal muscle. The focus of excitation-contraction coupling now shifts to defining the precise nature of the coupling of excitation to contraction.

ACKNOWLEDGEMENTS

These studies were supported in part by grants from the National Institutes of Health HL 32711 and DK 14632 and by the Muscular Dystrophy Association. MI is Investigator of the American Heart Association, Tennessee Affiliate. We are grateful to our many coworkers and collaborators who participated in these studies (cf references). Special thanks go to Akitsugu Saito who carried out the electron microscopy and provided the illustrations.

REFERENCES

Chamberlain BK, Levitsky DO, Fleischer S (1983). Isolation and characterization of canine cardiac sarcoplasmic reticulum with improved Ca^{2+} transport properties. J Biol Chem 258:6602-6609.

Chu A, Saito A, Fleischer S (1987). Preparation and characterization of longitudinal tubules of sarcoplasmic reticulum from fast skeletal muscle. Arch Biochem Biophys 258: 13-23.

Endo M (1977). Calcium release from sarcoplasmic reticulum. Physiol Rev 57:71-108.

Fabiato A (1983). Calcium induced release of calcium from cardiac sarcoplasmic reticulum. Am J Physiol 245:C1-C14.

Fawcett DW, McNutt NS (1969). The ultrastructure of the cat myocardium. J Cell Biol 42:1-45.

Fleischer S (1985). Sarcoplasmic reticulum and other membranes in the regulation of muscle contraction and relaxation, a multilevel approach. In Fleischer S, Tonomura Y (eds): "Structure and Function of Sarcoplasmic Reticulum," Academic Press, pp 119-145.

Fleischer S, Ogunbunmi EM, Dixon MC, Fleer EAM (1985). Localization of Ca^{2+} release channels with ryanodine in terminal cisternae of sarcoplasmic reticulum of fast skeletal muscle. Proc Natl Acad Sci USA 82:7256-7259.

Hymel L, Inui M, Fleischer S, Schindler HG (1988a). Purified skeletal muscle ryanodine receptor forms Ca^{2+}-activated multisubunit Ca^{2+} channels in planar bilayers. Proc Natl Acad Sci USA 85:441-445.

Hymel L, Schindler H, Inui M, Nagasaki K, Fleischer S (1988b). The purified cardiac muscle ryanodine receptor from sarcoplasmic reticulum (SR) is a Ca^{2+}-activated oligomeric Ca^{2+} channel. Biophys J 53:470a.

Hymel L, Schindler H, Inui M, Fleischer S (1988c). Reconstitution of purified cardiac muscle calcium release channel (ryanodine receptor) into planar bilayers. Biochem Biophys Res Communs, in press.

Imagawa T, Smith JS, Coronado R, Campbell KP (1987). Purified ryanodine receptor from skeletal muscle sarcoplasmic reticulum is the Ca^{2+} permeable pore of the calcium release channel. J Biol Chem 262:16636-16643.

Inesi G (1985). Mechanism of calcium transport. Annu Rev Physiol 47:573-601.

Inui M, Chamberlain BK, Saito A, Fleischer S (1986). The nature of the modulation of Ca^{2+} transport as studied by reconstitution of cardiac sarcoplasmic reticulum. J Biol Chem 261:1794-1800.

Inui M, Saito A, Fleischer S (1987a). Purification of the ryanodine receptor and identity with feet structures of junctional terminal cisternae of sarcoplasmic reticulum from fast skeletal muscle. J Biol Chem 262:1740-1747.

Inui M, Saito A, Fleischer S (1987b). Isolation of the ryanodine receptor from cardiac sarcoplasmic reticulum and identity with the feet structures. J Biol Chem 262:15637-15642.

Inui M, Wang S, Saito A, Fleischer S (1988). Characterization of junctional and longitudinal sarcoplasmic reticulum from heart muscle. Submitted.

Jones LR, Cala SE (1981). Biochemical evidence for functional heterogeneity of cardiac sarcoplasmic reticulum vesicles. J Biol Chem 256:11809-11818.

Katz S, Remtulla MA (1978). Phosphodiesterase protein activator stimulates calcium transport in cardiac microsomal preparations enriched in sarcoplasmic reticulum. Biochem Biophys Res Commun 83:1373-1379.

Kirchberger MA, Tada M, Katz AM (1974). Adenosine-3',5' monophosphate-dependent protein kinases-catalyzed phosphorylation reaction and its relationship to calcium transport in cardiac sarcoplasmic reticulum. J Biol Chem 249:6166-6173.

Lai FA, Erickson HP, Rousseau E, Liu Q-Y, Meissner G (1988). Purification and reconstitution of calcium release channel from skeletal muscle. Nature 331:315-319.

Le Peuch CJ, Haiech J, Demaille JG (1979). Concerted regulation of cardiac sarcoplasmic reticulum calcium transport by cyclic adenosine monophosphate dependent and calcium-calmodulin-dependent phosphorylations. Biochemistry 18: 5150-5157.

MacLennan DH, Campbell K, Reithmeyer RAF (1983). Calsequestrin. In Cheung WY (ed): "Calcium and Cell Functions," New York: Academic Press, pp 151-173.

Martonosi A (1984). Mechanisms of Ca^{2+} release from sarcoplasmic reticulum of skeletal muscle. Physiol Rev 64: 1240-1320.

Meissner G, Conner GE, Fleischer S (1973). Isolation of sarcoplasmic reticulum by zonal centrifugation and purification of Ca^{++}-pump and Ca^{++}-binding proteins. Biochim Biophys Acta 298:246-269.

Movsesian MA, Nishikawa M, Adelstein RS (1984). Phosphorylation of phospholamban by calcium-activated, phospholipid-dependent protein kinase. J Biol Chem 259:8029-8032.

Saito A, Seiler S, Chu A, Fleischer S (1984). Preparation and morphology of SR terminal cisternae from rabbit skeletal muscle. J Cell Biol 99:875-885.

Seiler S, Wegener AD, Whang DD, Hathaway DR, Jones LR (1984). High molecular weight protein in cardiac and skeletal muscle junctional reticulum vesicles bind calmodulin, are phosphorylated and are degraded by Ca^{2+}-activated protease. J Biol Chem 259:8550-8557.

Smith JS, Coronado R, Meissner G (1985). Single-channel calcium and barium currents of large and small conductance from sarcoplasmic reticulum. Nature 316:446-449.

Tada M, Yamamoto T, Tonomura Y (1978). Molecular mechanism of active calcium transport by sarcoplasmic reticulum. Physiol Rev 58:1-79.

Tada M, Katz AM (1982). Phosphorylation of the sarcoplasmic reticulum and sarcolemma. Annu Rev Physiol 44:401-423.

Wang S, Inui M, Ogunbunmi E, Saito A, Fleischer S (1987). Localization of ryanodine receptor (RyR) in a subfraction of cardiac sarcoplasmic reticulum (CSR). Biophys J 51:354a.

Index